INTERNATIONAL UNION OF THEORETICAL AND APPLIED MECHANICS

CREEP IN STRUCTURES

COLLOQUIUM
HELD AT STANFORD UNIVERSITY, CALIFORNIA
JULY 11–15, 1960

EDITED BY

NICHOLAS J. HOFF

WITH 156 FIGURES

SPRINGER-VERLAG
BERLIN / GÖTTINGEN / HEIDELBERG
1962

ISBN 978-3-642-86016-4 ISBN 978-3-642-86014-0 (eBook)
DOI 10.1007/ 978-3-642-86014-0

Softcover reprint of the hardcover 1st edition 1962
Library of Congress Catalog Card Number: 62-16170

Foreword

The Colloquium on Creep in Structures was organized as part of the activities of the International Union on Theoretical and Applied Mechanics. It was supported financially by IUTAM, as well as by the National Science Foundation, Washington, D. C., U. S. A. The Scientific Committee charged with the work of organization consisted of the following persons:

Nicholas J. Hoff, Stanford University, Stanford, California, Chairman

Folke K. G. Odqvist, Royal Institute of Technology, Stockholm, Sweden, Vice-Chairman

R. Mazet, Office National d'Etudes et de Recherches Aéronautiques, Châtillon-sous-Bagneux, France

Y. N. Rabotnov, Academy of Sciences, Novosibirsk, U. S. S. R.

Shuji Taira, Kyoto University, Kyoto, Japan.

The details of the organization of the sessions were entrusted to the Local Committee consisting of the following persons:

Max Anliker	W. H. Horton
K. Binford	B. Lempriere
C. C. Chao	L. Nickel

W. G. Flügge

In order to insure good discussion and, in general, a broad exchange of ideas among the participants in the meeting, the number of persons invited to attend was kept to a minimum. Participants not living in the Stanford area were housed in Donner House of Stern Hall on the Stanford campus, and were served meals in the dining room of Donner House. The speakers and observers started to arrive on July 9 and the last ones left on July 18.

All the scientific sessions were held in the Engineering Building of Stanford University, where the Colloquium was officially opened at 9.00 A. M. on July 11. The program of the meeting was as follows:

Monday, July 11

9.00 A. M.: **Official Opening of Colloquium**

Chairman:

Donald L. Putt, Lt. Gen. (ret.) U. S. Air Force; Chairman Advisory Committee, Industrial Affiliates of Stanford University in Aeronautics and Astronautics

Speakers:

Nicholas J. Hoff, Chairman, Colloquium on Creep in Structures

Frederick E. Terman, Provost, Stanford University

JOSEPH M. PETTIT, Dean, School of Engineering, Stanford University
FOLKE K. G. ODQVIST, President, International Union on Theoretical and Applied Mechanics

9.30 A. M.: **Linear Viscoelasticity**

Chairman:
OLEG D. SHERBY, Stanford University

E. H. LEE and A. H. CORNELIUSSEN, Brown University, Providence, Rhode Island, U. S. A.: The influence of viscoelasticity on stress distribution
H. H. HILTON, University of Illinois, Urbana, Illinois, and E. L. WALSH, Boeing Aircraft Co., Seattle, Washington, U. S. A.: Torsional Response of Linear Viscoelastic Plates Subjected to Thermal Stresses

2.00 P. M.: **Stress Distribution**

Chairman:
HAROLD LIEBOWITZ, Office of Naval Research, U. S. Navy

S. A. PATEL and B. VENKATRAMAN, Polytechnic Institute of Brooklyn, Brooklyn, New York, U. S. A.: On the creep stress analysis of some structures
A. PHILLIPS, Yale University, New Haven, Connecticut, U. S. A.: The shear center in creep of beams of thinwalled open cross section

5.00 P. M.: **Reception**

in honor of speakers given by Dr. and Mrs. N. J. HOFF at 782 Esplanada Way, Stanford, California

Tuesday, July 12

9.00 A. M.: **Creep Laws**

Chairman:
JAMES J. MURRAY, Office of Ordnance Research

J. MARIN, Pennsylvania State University, University Park, Pennsylvania, U. S. A.: Mechanics of creep and combined stresses
S. TAIRA, Kyoto University, Kyoto, Japan: The lifetime of structures subjected to varying load and temperature
E. T. ONAT and T. T. WANG, Brown University, Providence, Rhode Island, U. S. A.: The effect of incremental loading on creep behavior of metals

Wednesday, July 13

9.00 A. M.: **Large Deformations**

Chairman:
JOHN A. JOHNSON, Lockheed Missiles and Space Division

F. K. G. ODQVIST, Royal Institute of Technology, Stockholm, Sweden: Applicability of the elastic analogue to creep problems of plates, membranes and beams
J. HULT, Royal Institute of Technology, Stockholm, Sweden: Oil canning problems in creep

12.30 P.M.: **Excursion by Chartered Bus**

Thursday, July 14

9.00 A. M.: **Disks and Plates**

Chairman:

J. N. GOODIER, Stanford University

J. F. BESSELING, Institute of Technology, Delft, Holland: Investigation of transient creep in tubes and disks under radially symmetric loading

A. M. WAHL, Westinghouse Electric Corporation, Pittsburgh, Pennsylvania, U. S. A.: A comparison of flow criteria applied to elevated temperature creep of rotating disks with consideration of the transient condition

T. H. LIN, University of California, Los Angeles, California, U. S. A.: Bending of a circular plate with non-linear and strain-hardening creep

2.00 P. M.: **Cylindrical Shells**

Chairman:

WILHELM FLÜGGE, Stanford University

H. PORITSKY, The General Electric Company, Schenectady, New York, U. S. A.: Creep in cylindrical shells

C. R. CALLADINE, The English Electric Company, Ltd., Whetstone, Leicester, England: The creep of a wrinkle

6.00 P. M.: Social Hour at Adobe Creek Lodge, followed by **Banquet** at 7.00 p. m.

Toastmaster:

NICHOLAS J. HOFF

Guest Speaker:

RONALD SMELT: "Satellite Environment"

Friday, July 15

9.00 A. M.: **Columns**

Chairman:

RALPH H. LONG, National Science Foundation

R. L. CARLSON and W. W. BREINDEL, Battelle Memorial Institute, Columbus, Ohio, U. S. A.: On the mechanics of column creep

B. M. LEMPRIERE, Stanford University, Stanford, California, U. S. A.: Comparison of ranges of applicability of predictions of creep buckling time

M. ŻYCZKOWSKI, Technical University of Kraków, Kraków, Poland: Geometrically non-linear creep buckling of bars

2.00 P. M.: **Assemblies-Torsional Buckling-Damping**

Chairman:

F. K. G. ODQVIST, Royal Institute of Technology, Stockholm, Sweden

R. MAZET, Office National d'Etudes et de Recherches Aéronautiques, Châtillon-sous-Bagneux (Seine), France: Sur un modèle apte a traduire le fluage sous charge constante des assemblages

A. H. CHILVER, Cambridge University, Cambridge, England: Some simple models for torsional creep buckling

N. J. HOFF, Stanford University, Stanford, California, U. S. A.: Damping of the vibrations of a coiled spring due to creep

4.45 P. M.: **Closure**

A list of the participants and their affiliations follows:

Professor J. F. Besseling	Institute of Technology, Delft, Holland
Professor R. J. Bollard	California Institute of Technology
Mr. E. B. Borek	Marquardt Corporation
Dr. J. M. Brown	Hughes Aircraft Company
Dr. C. R. Calladine	English Electric Company, Ltd.
Mr. R. L. Carlson	Battelle Memorial Institute
Dr. K. T. Chang	AiResearch Manufacturing Company, Division of Garrett Corporation
Dr. Robert A. Chase	United Technology Corporation
Dr. A. H. Chilver	Cambridge University, Cambridge, England
Monsieur H. de l'Estoile	Service Technique Aéronautique, French Air Force
Monsieur E. Fage	Service Technique Aéronautique, French Air Force
Dr. Robert P. Felgar	Space Technology Laboratories, Inc.
Professor Wilhelm Flügge	Stanford University
Professor Y. C. Fung	California Institute of Technology
Dr. Robert Gatts	General Electric Company
Professor Norman Goodier	Stanford University
Mr. Charles A. Hermach	Ames Research Center NASA
Professor H. H. Hilton	University of Illinois
Professor N. J. Hoff	Stanford University
Dr. Oscar Hoffman	Lockheed Missiles and Space Division
Dr. Jan Hult	Royal Institute of Technology, Stockholm, Sweden
Dr. William E. Jahsman	Stanford University
Mr. John A. Johnson	Lockheed Missiles and Space Division
Professor Makoto Kikukawa	Osaka University, Osaka, Japan
Mr. Bertram Klein	Convair, Division of General Dynamics Corporation
Professor Albert Kobayashi	Boeing Airplane Company, Aero Space Division
Professor E. H. Lee	Brown University
Mr. B. M. Lempriere	Stanford University
Dr. Harold Liebowitz	Office of Naval Research
Professor T. H. Lin	University of California at Los Angeles
Professor Hsu Lo	Purdue University
Dr. Ralph H. Long	National Science Foundation
Professor Joseph Marin	Pennsylvania State University
Mr. H. G. McComb	Langley Research Center, NASA
Mr. James J. Murray	Office of Ordnance Research, U. S. Army
Dr. J. Murzewski	Institute of Technology, Kraków, Poland
Dr. William Nachbar	Stanford University
Professor F. K. G. Odqvist	Royal Institute of Technology, Stockholm, Sweden
Professor E. T. Onat	Brown University
Professor S. A. Patel	Polytechnic Institute of Brooklyn
Professor Aris Phillips	Yale University
Professor Yves Pirroneau	University of Nantes, France
Dr. H. Poritsky	General Electric Company
Lt. Gen. D. L. Putt	Stanford University
Professor Cedric W. Richards	Stanford University

Dr. CARLO RIPARBELLI	General Atomic, Division of General Dynamics Corporation
Dr. E. E. SECHLER	California Institute of Technology
Professor OLEG SHERBY	Stanford University
Mr. RONALD SMELT	Stanford University
Dr. W. STUIVER	IBM Research Laboratory
Professor Y. SUEZAWA	Institute of Technology, Tokyo, Japan
Professor S. TAIRA	Kyoto University, Kyoto, Japan
Dr. KICHINOSUKE TANAKA	Kyoto University, Kyoto, Japan
Provost FREDERICK TERMAN	Stanford University
Dr. SITARAMARAO VALLURI	California Institute of Technology
Professor B. VENKATRAMAN	Polytechnic Institute of Brooklyn
Dr. A. M. WAHL	Westinghouse Electric Corporation
Professor HARRY WILLIAMS	Stanford University
Dr. JOHN ZICKEL	Aerojet-General Corporation
Dr. M. ŻYCZKOWSKI	Technical University of Kraków, Kraków, Poland

To provide relief from the scientific activities, a bus trip was organized for the afternoon of July 13. The 17 persons who took part in the excursion were shown some of the scenic beauties of California and some of the oldest buildings, dating back to the days of Spanish rule, in Monterey ıd in Carmel.

The banquet was held at Adobe Creek Lodge in a pleasant setting among the steep foothills of the Coast Range. It was attended by 58 persons, including many of the wives of the participants. In the after-dinner talk, Mr. RONALD SMELT, Chief Scientist of the Lockheed Missiles and Space Division and a lecturer in the Department of Aeronautical Engineering of Stanford University, gave an interesting report on efforts to investigate the environment in which satellites travel. Although most informative, the talk was given in a manner understandable to the ladies present; it was illustrated by an interesting sound film.

The sessions took place in accordance with the prearranged program, except that the paper written by Professor MAZET was delivered by Monsieur HUGUES DE L'ESTOILE of the Service Technique Aéronautique of the French Air Force, and that the titles of some of the papers were changed slightly.

The meeting closed in the afternoon of July 15. N. J. HOFF voiced the thanks of Stanford University for the work done by the participants, and FOLKE K. G. ODQVIST, President of the International Union for Theoretical and Applied Mechanics, expressed the appreciation of the participants for the hospitality shown them by Stanford.

Speakers and discussors enjoyed the possibility offered to them for the discussion of problems caused by creep in structures. It is hoped that readers of these proceedings will derive equal benefits from this volume.

Stanford University, California **N. J. Hoff**

Contents

Stress Distribution Analysis for Linear Viscoelastic Materials[1]

By

A. H. Corneliussen and E. H. Lee

Brown University, Providence, Rhode Island, USA

Abstract

The formulation of the differential or integro-differential equations governing the quasi-static stress analysis problem for linear viscoelastic bodies is considered, with particular reference to the initial conditions associated with sudden loading. If classical methods of solving the differential equations, such as the LAPLACE transform or integrating factor, are used, the need to evaluate the initial conditions after load application at $t = 0^+$ presents difficulties. These can be avoided by working with the lower limit $t = 0^-$, before load application, when the body is still undisturbed, and, in the case of direct integration of the equations, delta functions and their derivatives must then be incorporated into the analysis. Examples of both methods of approach are contrasted. The stress distribution in a spinning hollow circular cylinder with annihilating inner cavity is evaluated as an example which does not fall within the scope of the LAPLACE transform method of analysis. The procedures are justified through the application of the corresponding integral operators.

1. Introduction

In analysing the stress distributions in viscoelastic bodies, a system of partial differential equations or integro-differential equations containing derivatives or integrals with respect to both the space variables and the time must be solved [*1, 2*][2]. It is common to consider applied surface tractions in the form of the HEAVISIDE step function in the time variable:

$$H(t) = 0, \; t < 0$$
$$= 1, \; t > 0 \tag{1.1}$$

---

[1] Sponsored under Department of Defense Contract DA-19-020-ORD-4750, between the Office of Ordnance Research, US Army, and Brown University.

[2] Numbers in square brackets refer to the bibliography at the end of the paper.

this being the mathematical idealisation of a rapidly applied load. Many such problems can be adequately treated on the basis of quasi-static theory, in which the inertia forces associated with the deformation of the body can be neglected in comparison with the applied forces. In such an analysis the effect of the surface tractions, applied suddenly at $t = 0$, is felt instantaneously throughout the body. Thus, even though the body was unstressed for $t < 0$, the dependent variables, the stress and strain components, may have non-zero initial values after the instantaneous distribution of the loads. These values will be associated with the time $t = 0^+$, the situation just before load application being associated with $t = 0^-$. Initial time derivatives of first and higher orders will in general also be non-zero at $t = 0^+$. Thus the solution of the system of differential or integro-differential equations must be sought for $t > 0$, with non-zero initial values at $t = 0^+$. Moreover, these values are not known a priori.

For an isotropic linear viscoelastic body, the stress-strain relations contain two pairs of linear operators which are analogous to the two elastic constants required to specify linear isotropic elasticity. For convenience, separating the shear and dilatational effects, the viscoelastic laws take the form:

$$P\, s_{ij} = Q\, e_{ij}, \tag{1.2}$$

$$P'\, \sigma_{ii} = Q'\, \varepsilon_{ii}, \tag{1.3}$$

where P, Q, P', Q' are linear differential or integral operators; σ_{ij}, ε_{ij} the stress and strain tensors respectively, and s_{ij} and e_{ij} the corresponding deviators which represent shear effects. The summation convention for repeated suffixes is utilized, so that (1.3) is the relation between the average hydrostatic tension, $\sigma_{ii}/3$, and the dilatation, ε_{ii}. If a relation of type (1.2) or (1.3) between a stress component σ and a strain component ε corresponds to a viscoelastic model with a finite number of springs and dashpots, it can be represented by differential operators of finite order, for example:

$$\sum_{r=0}^{p} p_r \frac{\partial^r \sigma}{\partial t^r} = \sum_{r=0}^{q} q_r \frac{\partial^r \varepsilon}{\partial t^r}. \tag{1.4}$$

General linear viscoelasticity can be represented in terms of the creep compliance $J(t)$, by the integral relation:

$$\varepsilon(t) = \int_{-\infty}^{t} J(t-\tau) \frac{\partial \sigma}{\partial \tau} d\tau \tag{1.5}$$

and if the material is undisturbed for $t < 0$, this can be written in the form:

$$\varepsilon(t) = \int_{0}^{t} J(t-\tau) \frac{\partial \sigma}{\partial \tau} d\tau. \tag{1.6}$$

Equivalent relations in terms of the relaxation modulus $G(t)$ are:

$$\sigma(t) = \int_{-\infty}^{t} G(t-\tau)\frac{\partial \varepsilon}{\partial \tau}\,d\tau \tag{1.7}$$

and

$$\sigma(t) = \int_{0}^{t} G(t-\tau)\frac{\partial \varepsilon}{\partial \tau}\,d\tau. \tag{1.8}$$

In the special case of the differential law (1.4), $J(t)$ and $G(t)$ take the form of the sum of exponential terms.

For quasi-static stress analysis, we need, in addition to the stress-strain relations, the equations of equilibrium:

$$\frac{\partial \sigma_{ij}}{\partial x_j} + f_i = 0, \tag{1.9}$$

where $f_i(x_j, t)$ is the body force distribution, and x_j are CARTESIAN coordinates (x_1, x_2, x_3). We also need the expressions for the strain components in terms of the displacement components u_i. For infinitesimal strain theory, to which this discussion is confined, this relation takes the form:

$$\varepsilon_{ij} = \frac{1}{2}\left(\frac{\partial u_i}{\partial x_j} + \frac{\partial u_j}{\partial x_i}\right) \tag{1.10}$$

(1.2) and (1.3), with the operator pairs P, Q and P', Q' each taking one of the forms (1.4) to (1.8), combined with (1.9) and (1.10) comprise the system of differential or integro-differential equations which determine stress distributions when combined with boundary and initial conditions. If the surface tractions $T_i(x_j, t)$ are prescribed over portions of the surface S_1, the stress there must satisfy:

$$T_i = \sigma_{ij}\, n_j, \tag{1.11}$$

where n_j are the direction cosines of the outward normal. The displacement components $u_i(x_j, t)$ may be prescribed over the rest of the surface S_2, or combinations of displacement and traction components may be prescribed as in elasticity theory.

A restricted class of stress analysis problems for linear viscoelastic bodies can be treated by applying the LAPLACE transform ([*3*, *4*]) to the system of equations and boundary conditions. The dependent variables then change to the transformed variables denoted by a bar over the corresponding symbol; for example, the transformed stress is given by:

$$\bar{\sigma}_{ij}(x_k, s) = \int_0^{\infty} \sigma_{ij}(x_k, t)\, e^{-st}\, dt, \tag{1.12}$$

where s is the transform parameter. This transformation of the whole system can be utilized if at all material points the type of condition demanded by the system of relations given above does not change during the loading process. For example, a boundary point must not change from surface type S_1 to S_2, that is from a point of prescribed traction to one of prescribed displacement, for then neither the transform of the traction $\bar{T}_i(x_j, s)$ nor of the displacement $\bar{u}_i(x_j, s)$ can be obtained from an integral of the form (1.12), using only prescribed data. Moreover to permit the transform operation the body shape must not change during the loading process through, for example, penetration by an extending crack, or annihilation of material by ablation, for then governing conditions would change at fixed material points or material points at which stress and strain were defined in the early stages of the process would caese to exist, thus preventing the formulation of the LAPLACE transformations. When the system of equations and boundary conditions is amenable to transformation, after application of the transformation, it takes the form of the relations between transforms given below, with the corresponding equations barred (see, for example, [3, 4]). The assumption that the material is undisturbed for $t < 0$, and that zero initial conditions obtain for all variables, has been made.

$$\bar{P}\,\bar{s}_{ij} = \bar{Q}\,\bar{e}_{ij}, \tag{$\overline{1.2}$}$$

$$\bar{P}'\,\bar{\sigma}_{ii} = \bar{Q}'\,\bar{\varepsilon}_{ii}, \tag{$\overline{1.3}$}$$

$$\sum_0^p p_r\, s^r\, \bar{\sigma} = \sum_0^q q_r s^r\, \bar{\varepsilon}, \tag{$\overline{1.4}$}$$

$$\bar{\varepsilon} = \bar{J}\, s\, \bar{\sigma}, \tag{$\overline{1.6}$}$$

$$\bar{\sigma} = \bar{G}\, s\, \bar{\varepsilon}, \tag{$\overline{1.8}$}$$

$$\frac{\partial \bar{\sigma}_{ij}}{\partial x_j} + \bar{f}_i = 0, \tag{$\overline{1.9}$}$$

$$\bar{\varepsilon}_{ij} = \frac{1}{2}\left(\frac{\partial \bar{u}_i}{\partial x_j} + \frac{\partial \bar{u}_j}{\partial x_i}\right), \tag{$\overline{1.10}$}$$

$$\bar{T}_i = \bar{\sigma}_{ij}\, n_j. \tag{$\overline{1.11}$}$$

An appropriate selection from these transformed equations corresponding to a viscoelastic stress analysis problem comprises a linear elastic problem, with the elastic constants, the boundary conditions and the body force appearing as functions of the transform parameter s. Solution of this so-called "associated elastic problem" for the transformed stress distribution, $\sigma_{ij}(x_k, s)$, and inversion to give $\sigma_{ij}(x_k, t)$ determines the variation of the stress distribution for the original viscoelastic stress analysis problem.

For the classical LAPLACE transform ([3, 4]), the limits of the integral in (1.12) are $0^+, \infty$. Thus if $\sigma_{ij}(x_k, t)$ exhibits time variation in the form of a step function $H(t)$ as defined in (1.1), which will correspond to step function traction variation through (1.11), the initial value at $t = 0^+$ is not zero, and additional terms will be needed in the transformed relation $\overline{(1.4)}$. For example the transform of $\frac{\partial \sigma}{\partial t}$ is given in terms of the transform of σ by the relation [(3, 4]):

$$\overline{\left(\frac{\partial \sigma}{\partial t}\right)} = s\,\bar{\sigma} - \sigma(0^+) \tag{1.13}$$

and $\sigma(0^+)$ will be non-zero in the case of step function type behaviour. As developed, for example, in [3], (1.13) applies for a continuous function $\sigma(t)$ with a piece-wise continuous derivative. However, it is possible to interpret the theory for discontinuous functions, and by taking the limits of the integral in (1.12) to be 0^-, ∞, the zero initial conditions at $t = 0^-$ can be used, and the discontinuity between $t = 0^-$ and $t = 0^+$ taken care of in a unified manner with other discontinuities that may occur for $t > 0$. This approach to application of the LAPLACE transform is presented in Chapter XI of [4].

As an example, consider the HEAVISIDE step function $H(t)$, (1.1). Its transform is:

$$\bar{H}(s) = \frac{1}{s}. \tag{1.14}$$

Application of (1.13) gives:

$$\overline{\left(\frac{\partial H}{\partial t}\right)} = 1 - 1 = 0 \tag{1.15}$$

since $H(0^+) = 1$. Inversion of (1.15) gives:

$$\frac{\partial H}{\partial t} = 0, \quad t > 0 \tag{1.16}$$

which of course agrees with the usual derivative of $H(t) = 1$, $t > 0$.

If the lower limit $t = 0^-$ is used in the definition of the LAPLACE transform, no change occurs in (1.14). The last term of (1.13) is modified, and since $H(0^-) = 0$, the transform of $\frac{\partial H}{\partial t}$ becomes:

$$\overline{\left(\frac{\partial H}{\partial t}\right)} = 1. \tag{1.17}$$

From the definition of the LAPLACE transform (1.12) with the lower limit 0^-, (1.17) leads to:

$$\frac{\partial H}{\partial t} = \delta(t), \tag{1.18}$$

where $\delta(t)$ is the DIRAC delta function:

$$\delta(t) = 0, \quad t \neq 0, \quad \int_{0^-}^{t} \delta(\tau)\,d\tau = 1, \quad t > 0. \tag{1.19}$$

Thus the modified definition of the LAPLACE transform determines a "generalized" derivative which includes the δ function to represent the derivative of the discontinuity in the step function at $t = 0$. The vertical step yields the infinite spike of the δ function, and the integral of the generalized derivative from 0^- to 0^+ gives the discontinuity in the step function. For $t > 0$, (1.16) remains unchanged, and the classical derivative is obtained.

Use of this modified form of the Laplace transform introduces considerable advantage for viscoelastic stress analysis problems for material behaviour represented by differential operators (1.4), since, for bodies undisturbed for $t < 0$, initial conditions are all zero, with the corresponding simplification in the analysis. This formal method of dealing with the jump conditions associated with suddenly applied tractions can be justified by means of the classical approach by showing, with the use of the equivalent integral operator, that it leads to the solution obtained with non-zero initial conditions at $t = 0^+$. This is considered in Section 4 of this paper. Alternatively, use of the two sided Laplace transform as presented in [5] leads to a similar generalized interpretation of the problem, if all dependent variables are taken to be zero for $t < 0$.

A study of viscoelastic stress analysis solutions presented in the literature, shows that this broader approach to the interpretation of transform analysis has been tacitly assumed without explanation. For example, the solution for a step function point load on the surface of a semi-infinite body [1] is based on zero initial conditions for stress, in spite of the fact that at $t = 0^+$ the stresses are non-zero. The solution, however, is correct because of the broader interpretation of transform analysis discussed above. It should perhaps be emphasised that the generalization associated with the utilization of step functions and delta functions is not confined to the conditions at $t = 0$, but embraces the analysis of solutions with discontinuities for any values of t. This of course is clear since the particular choice of the time origin can have no significance for material behaviour.

It was mentioned above that the Laplace transform approach can be utilized only for a restricted class of problems. In order to benefit from the convenience of zero initial conditions in the analysis for materials represented by differential operators in cases when the Laplace transform method does not apply, integration must be carried out from $t = 0^-$ with the introduction of step functions and delta functions, and their derivatives to permit problems involving sudden loading to be treated. A discussion of the use of such symbolic functions is given in [6], Chapter 3, and a convenient method of utilizing them in viscoelastic stress analysis problems is presented below. The use of delta functions in determining the response of viscoelastic materials to homogeneous stress has been discussed by Gross and Güttinger [7], but it is suggested that the integrating factor method for integration detailed in this paper provides a more unified method of treatment.

2. The Analysis of Homogeneous Stress and Strain

Let us first consider an almost trivial example, which, however, illustrates the various approaches discussed in the Introduction. Since we are concerned with linear viscoelasticity, superposition of solutions applies, and we can separate out individual corresponding components of stress and strain, such as shear, from the combined stress system. Thus we may have a relation of the type (1.4) for individual components of stress and strain, such as shear, or simple tension and the associated extension. A simple particular case is the MAXWELL material represented by the differential relation:

$$\dot{\varepsilon} = \frac{\dot{\sigma}}{G} + \frac{\sigma}{\nu}. \tag{2.1}$$

The dot represents differentiation with respect to time, and G and ν are viscoelastic material contants. Thus in this particular case of (1.4), the operators on both sides are of first order. The corresponding viscoelastic model is shown in Fig. 1, with spring constant G, and dashpot viscosity ν.

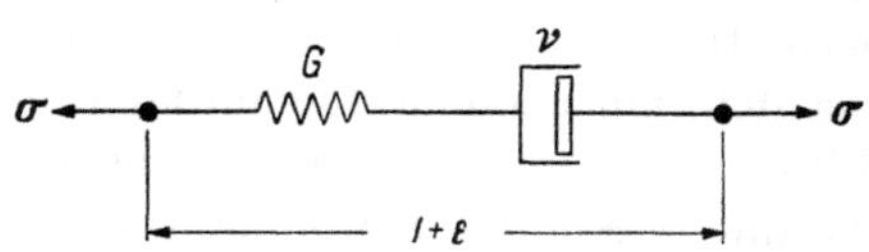

Fig. 1. The MAXWELL material, spring constant G, dashpot viscosity ν

In a relaxation test, a strain ε_0 is applied suddenly at $t = 0$, and maintained constant thereafter. The resulting variation of stress is given by $\sigma(t) = \varepsilon_0 G(t)$, where $G(t)$ is the relaxation modulus. Below, $\sigma(t)$ is obtained by various methods.

Using classical methods for the solution of the simple differential equation, $\dot{\varepsilon}$ is zero for $t > 0$, since ε has the constant value ε_0, so that (2.1) reduces to a homogeneous equation for σ, with the solution

$$\sigma = c\, e^{-t/\tau}, \tag{2.2}$$

where c is an arbitrary constant, and $\tau = \nu/G$. A glance at the viscoelastic model, Fig. 1, indicates that since only the spring stretches during the sudden initial straining, $\sigma(0^+) = G\,\varepsilon_0$, which determines the constant c to yield the result:

$$\sigma = G\,\varepsilon_0\, e^{-t/\tau}. \tag{2.3}$$

For the classical LAPLACE transform approach, we again need $\sigma(0^+) = G\,\varepsilon_0$, and with $\varepsilon(0^+) = \varepsilon_0$, transformation of (2.1) using relations of the type (1.13) yields:

$$s\,\varepsilon_0/s - \varepsilon_0 = (s\,\bar{\sigma} - G\,\varepsilon_0)/G + \bar{\sigma}/\nu, \tag{2.4}$$

$$\bar{\sigma} = \frac{G\,\varepsilon_0}{s + \frac{G}{\nu}} \tag{2.5}$$

and inversion again gives (2.3).

Using the generalized LAPLACE transform approach, $\varepsilon(0^-) = \sigma(0^-) = 0$, and the transform of (2.1) gives directly:

$$s\,\varepsilon_0/s = \frac{s\,\bar{\sigma}}{G} + \frac{\bar{\sigma}}{\nu} \tag{2.6}$$

which again leads to (2.5) and (2.3), written in the form:

$$\sigma = G\,\varepsilon_0\,e^{-t/\tau}\,H(t) \tag{2.3'}$$

in order to emphasize the inclusion of $t = 0^-$ in the range of t covered by the solution.

The advantage of the generalized LAPLACE transform approach is not merely that the analysis is less involved because non-zero initial conditions do not occur, but that it eliminates the need for the separate determination of the initial conditions at $t = 0^+$. In the case considered, the initial condition is obtained readily by means of intuition concerning the instantaneous deformation of the model in Fig. 1, but for higher order operators, for which initial conditions of the unknown function and its derivatives to an order one less than that of the operator are needed, the determination of the initial conditions at $t = 0^+$ may present a difficult problem. It is interesting to note that the terms arising from the initial conditions of ε and σ in (2.4) cancel, and in Section 4 it will be shown that this is true in general, thus establishing the justification for using initial conditions at $t = 0^-$.

Although the generalized LAPLACE transform method provides a convenient means of stress analysis, it was pointed out in the Introduction that it can be applied only for a restricted class of problems, and a more general approach is needed which retains the ease of treating the initial loading situation. A consistent use of step functions, their derivatives: delta functions, and higher derivatives provides this tool.

For the relaxation test of a MAXWELL material considered above

$$\varepsilon(t) = \varepsilon_0\,H(t). \tag{2.7}$$

Substitution in (2.1) gives:

$$\frac{\dot{\sigma}}{G} + \frac{\sigma}{\nu} = \varepsilon_0\,\delta(t). \tag{2.8}$$

The theory of linear first order equations determines the integrating factor $e^{Gt/\nu}$, and application of this leads to

$$\sigma\,e^{G\tau/\nu}\Big|_{0^-}^{t} = G\,\varepsilon_0 \int_{0^-}^{t} e^{G\tau/\nu}\,\delta(\tau)\,d\tau \tag{2.9}$$

with zero the initial value of σ at $t = 0^-$. Thus:

$$\sigma\,e^{Gt/\nu} = G\,\varepsilon_0\,H(t) \tag{2.10}$$

and $\sigma(t)$ is again given by (2.3'). As presented in [6], Chapter 3, derivatives of δ functions can be interpreted through successive integrations.

by parts. This permits such problems to be treated in a similar way for higher order differential operator relations of the type (1.4). Such solutions merely demand repeated application of the integrating factor procedure.

To provide a more significant example, we consider the material represented by the four element model shown in Fig. 2, and the corresponding differential operator relation:

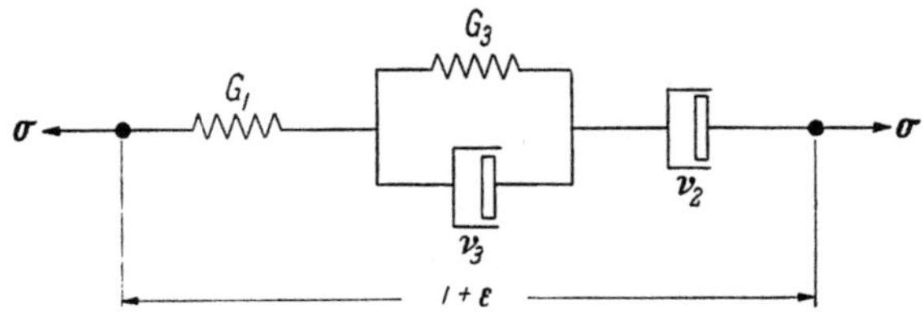

Fig. 2. The four element viscoelastic model

$$\left[\frac{1}{G_1} D^2 + \left(\frac{G_3}{G_1 \nu_3} + \frac{1}{\nu_2} + \frac{1}{\nu_3}\right) D + \frac{G_3}{\nu_2 \nu_3}\right] \sigma = \left[D^2 + \frac{G_3}{\nu_3} D\right] \varepsilon, \quad (2.11)$$

where D is the derivative operator $\partial/\partial t$. This is the simplest viscoelastic material which exhibits the three major types of viscoelastic response to stress: instantaneous elasticity, delayed elasticity, and viscous flow. To determine the variation of strain in a creep test in which the specimen is subjected to the suddenly applied stress σ_0, this condition is substituted into (2.11):

$$\sigma = \sigma_0 H(t) \quad (2.12)$$

giving:

$$\left[D^2 + \frac{G_3}{\nu_3} D\right] \varepsilon = \sigma_0 \left[\frac{1}{G_1} \delta'(t) + \left(\frac{G_3}{G_1 \nu_3} + \frac{1}{\nu_2} + \frac{1}{\nu_3}\right) \delta(t) + \frac{G_3}{\nu_2 \nu_3} H(t)\right]. \quad (2.13)$$

One integration from 0^- to t gives:

$$\left(D + \frac{G_3}{\nu_3}\right) \varepsilon = \sigma_0 \left[\frac{1}{G_1} \delta(t) + \left(\frac{G_3}{G_1 \nu_3} + \frac{1}{\nu_2} + \frac{1}{\nu_3}\right) H(t) + \frac{G_3}{\nu_2 \nu_3} t\, H(t)\right] \quad (2.14)$$

and application of the integrating factor $e^{G_3 t/\nu_3}$ gives

$$\begin{aligned} \varepsilon\, e^{G_3 t/\nu_3} = \sigma_0 \Big[\frac{1}{G_1} H(t) &+ \left(\frac{G_3}{G_1 \nu_3} + \frac{1}{\nu_2} + \frac{1}{\nu_3}\right) \frac{\nu_3}{G_3} (e^{G_3 t/\nu_3} - 1) H(t) \\ &+ \frac{1}{\nu_2} \left(t\, e^{G_3 t/\nu_3} - \frac{\nu_3}{G_3} \{e^{G_3 t/\nu_3} - 1\}\right) H(t)\Big] \end{aligned} \quad (2.15)$$

which simplifies to:

$$\varepsilon = \sigma_0 H(t) \left[\frac{1}{G_1} + \frac{1}{G_3} (1 - e^{-G_3 t/\nu_3}) + \frac{1}{\nu_2} t\right]. \quad (2.16)$$

While this result can be simply deduced by adding the contributions of the three series units in Fig. 2, the formal method using delta functions applies for a general operator pair of type (1.4) without the need to express it in the form of a series model as depicted in Fig. 2.

For a relaxation test

$$\varepsilon = \varepsilon_0 H(t) \tag{2.17}$$

and (2.11) becomes

$$\left[\frac{1}{G_1} D^2 + \left(\frac{G_3}{G_1 \nu_3} + \frac{1}{\nu_2} + \frac{1}{\nu_3}\right) D + \frac{G_3}{\nu_2 \nu_3}\right] \sigma = \varepsilon_0 \left[\delta'(t) + \frac{G_3}{\nu_3} \delta(t)\right]. \tag{2.18}$$

The left hand operator is factorized to give:

$$(D + K_1)(D + K_2)\sigma = G_1 \varepsilon_0 \left[\delta'(t) + \frac{G_3}{\nu_3} \delta(t)\right]. \tag{2.19}$$

Applying the first integrating factor $e^{K_1 t}$, and integrating from 0^- to t gives:

$$e^{K_1 t}(D + K_2)\sigma = G_1 \varepsilon_0 \left[\delta(t) - K_1 H(t) + \frac{G_3}{\nu_3} H(t)\right]. \tag{2.20}$$

The final integrating factor is $e^{(K_2 - K_1) t}$, and on applying this and integrating

$$e^{K_2 t}\sigma = G_1 \varepsilon_0 \left[H(t) - \frac{K_1 - \frac{G_3}{\nu_3}}{K_2 - K_1}\left(e^{(K_2 - K_1)t} - 1\right) H(t)\right] \tag{2.21}$$

leading to $\sigma(t)$ in the form:

$$\sigma = \frac{G_1 \varepsilon_0 H(t)}{K_2 - K_1}\left[\left(K_2 - \frac{G_3}{\nu_3}\right) e^{-K_2 t} - \left(K_1 - \frac{G_3}{\nu_3}\right) e^{-K_1 t}\right]. \tag{2.22}$$

Since $K_1 < G_3/\nu_3 < K_2$, the stress variation in the relaxation test is given as the sum of exponentials with positive coefficients, which agrees with the obvious solution based on the equivalent viscoelastic model of two Maxwell units in parallel as shown in Fig. 3, since in a relaxation test each Maxwell unit will supply a contribution to the stress independently of the other. The delta function approach gives the solution directly in terms of the differential operator representation of type (1.4). Repeated application of the integrating factor procedure will yield the stress or strain response for higher order differential operators when the strain or stress variation is prescribed respectively.

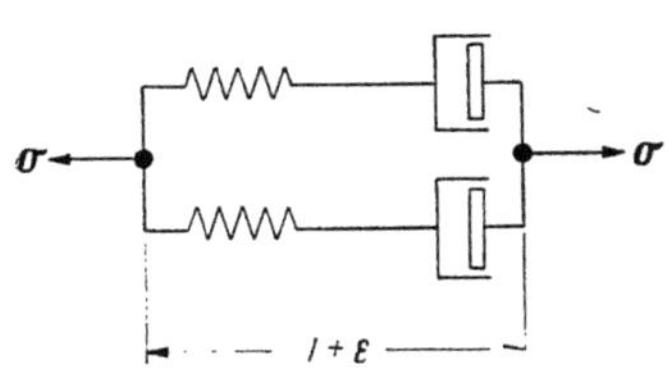

Fig. 3. A viscoelastic model equivalent to the four element model of Fig. 2.

3. The Analysis of Stress Distributions

It was shown above that in the analysis of homogeneous stress in viscoelastic bodies considerable difficulties can be anticipated if initial conditions at $t = 0^+$ have to be evaluated. The use of the Laplace transform with lower limit $t = 0^-$, or of delta function analysis with integration from $t = 0^-$ permits the substitution of zero initial conditions, thereby eliminating the difficulty. The problem of initial conditions at $t = 0^+$ can be much more involved in stress distribution problems since initial time derivatives of the stresses at $t = 0^+$ will depend not only on the derivatives of applied tractions or displacements and the material characteristics as in the homogeneous case, but may also depend on geometry changes of the system considered. For example, in the problem discussed below of a hollow cylinder with an annihilating inner radius, stress derivatives at $t = 0^+$ will in general depend on the initial motion of the surface of the inner cavity. In many cases the initial value of the stress following sudden loading can be deduced from an elastic analysis, since it is intuitively evident that the instantaneous elastic response of the material will be the governing factor in determining the initial stress distribution. But in most cases it would be virtually impossible to deduce stress derivatives on the basis of such physical intuition.

For the restricted class of problems amenable to solution by means of the LAPLACE transform, it was pointed out in the Introduction that examples of the use of the transform with lower limit $t = 0^-$ exist in the literature, although this aspect is not specifically mentioned there. Zero initial conditions are assumed in these analyses, even though this requirement is clearly violated for $t = 0^+$, where the initial conditions are considered to apply according to the usual theory. In addition to the example referred to in [*1*], the stress analysis for a hollow cylinder subjected to internal pressure [*8*] also can only be interpreted correctly on the basis of a transform defined for the lower limit $t = 0^-$. In view of these examples, and the added difficulty introduced by moving boundaries, an example of the latter type only will be presented in this section.

A long hollow circular cylinder of viscoelastic material is considered with constant external radius R_2, and an annihilating inner radius $R_1(t)$. The annihilation of the inner material is due to ablation, and the resulting displacement of the boundary is of course quite distinct from the particle displacement which is considered to be sufficiently small to produce no appreciable change in the geometry of the body in accordance with the small dipslacement theory adopted. At $t = 0$, constant internal pressure p_0 is suddenly applied and maintained, and the cylinder is considered to be subjected to a suddenly applied constant angular velocity ω about the axis. The distortional effects due to the shear stresses which would be needed to produce the angular acceleration during the short spin-up time are neglected, in view of the short period during which they are acting, but because of the linearity of the system the associated strains could be computed separately and superposed on the solution given below. When steady rotation is occurring, inertia forces due to the mass-density of the material produce an equivalent centrifugal body force. Plane strain with no axial displacement is assumed for the long cylinder, and the stress analysis presented below applies for a material which is elastic in dilatation and exhibits MAXWELL behaviour in shear. The ablation of the inner cavity prevents the application of the LAPLACE transform method of solution, and direct integration using delta functions is therefore adopted.

The stress-strain relations are:

$$\left(\frac{D}{G}+\frac{1}{\nu}\right)s_{ij}=D\,e_{ij}, \tag{3.1}$$

$$\sigma_{ii}/3=K\,\varepsilon_{ii}, \tag{3.2}$$

where K is the bulk modulus. Using cylindrical polar coordinates (r, θ, z), and the corresponding stresses, which are principal stresses by symmetry, (3.1) and (3.2) determine the following stress-strain relations in terms of the single displacement component, u, the radial displacement.

$$A\,\varepsilon_\theta=A\,\frac{u}{r}=B\,\sigma_\theta-C(\sigma_r+\sigma_z), \tag{3.3}$$

$$A\,\varepsilon_r=A\,\frac{\partial u}{\partial r}=B\,\sigma_r-C(\sigma_\theta+\sigma_z), \tag{3.4}$$

$$A\,\varepsilon_z=0=B\,\sigma_z-C(\sigma_r+\sigma_\theta), \tag{3.5}$$

where A, B, and C are appropriate combinations of the operators in (3.1) and (3.2) deduced by decomposing the strain field into dilatational and shear components.

$$A=D, \tag{3.6}$$

$$B=\frac{2}{3}\left(\frac{D}{G}+\frac{1}{\nu}\right)+\frac{D}{9K}, \tag{3.7}$$

$$C=\frac{1}{3}\left(\frac{D}{G}+\frac{1}{\nu}\right)-\frac{D}{9K}. \tag{3.8}$$

By differentiating (3.3) with respect to r, u can be eliminated between (3.3) and (3.4) to give:

$$(B+C)(\sigma_r-\sigma_\theta)=r\frac{\partial}{\partial r}[B\,\sigma_\theta-C(\sigma_r+\sigma_z)]. \tag{3.9}$$

The equation of motion for the cylinder of density ϱ set in rotation with constant angular velocity ω at $t=0$ is

$$r\frac{\partial\sigma_r}{\partial r}+\sigma_r-\sigma_\theta+\varrho\,\omega^2 r^2 H(t)=0. \tag{3.10}$$

Elimination of σ_z between (3.5), (3.9) and (3.10) yields the relation

$$(B^2-C^2)\frac{\partial}{\partial r}(\sigma_r+\sigma_\theta)+B(B+C)\,\varrho\,\omega^2 r\,H(t)=0. \tag{3.11}$$

Since $(B+C)$ is a factor of both terms, and integration is considered from $t=0^-$ at which all initial conditions are zero, (3.11) implies:

$$(B-C)\frac{\partial}{\partial r}(\sigma_r+\sigma_\theta)+B\,\varrho\,\omega^2 r\,H(t)=0. \tag{3.12}$$

Inserting the expressions for $(B - C)$ and B from (3.7) and (3.8), and evaluating $B[H(t)]$, (3.12) takes a form amenable to integration with an integrating factor:

$$\left(D + \frac{\frac{1}{\nu}}{\frac{1}{G} + \frac{2}{3K}}\right)\frac{\partial}{\partial r}(\sigma_r + \sigma_\theta) + \frac{\varrho\,\omega^2 r}{\frac{1}{G} + \frac{2}{3K}}\left\{\left(\frac{2}{G} + \frac{1}{3K}\right)\delta(r) + \frac{2}{\nu}H(t)\right\} = 0. \tag{3.13}$$

Integration with respect to t from 0^- then determines:

$$\frac{\partial}{\partial r}(\sigma_r + \sigma_\theta) = -\varrho\,\omega^2 r\,H(t)\left\{2 - \frac{\frac{1}{K}}{\frac{1}{G} + \frac{2}{3K}}\,e^{-\frac{\frac{1}{\nu}}{\frac{1}{G}+\frac{2}{3K}}t}\right\}. \tag{3.14}$$

Integration with respect to r determines $\sigma_r + \sigma_\theta$ in terms of an arbitrary function of t:

$$\sigma_r + \sigma_\theta = -\varrho\,\omega^2 r^2 H(t)/2\left\{2 - \frac{\nu}{K}\alpha\,e^{-\alpha t}\right\} + f_1(t), \tag{3.15}$$

where for convenience the parameter α has been introduced:

$$\alpha = \frac{\frac{1}{\nu}}{\frac{1}{G} + \frac{2}{3K}}. \tag{3.16}$$

σ_θ can now be eliminated between (3.10) and (3.15) giving:

$$r\frac{\partial\sigma_r}{\partial r} + 2\,\sigma_r = -\varrho\,\omega^2 r^2 H(t)/2\left\{4 - \frac{\nu}{K}\alpha\,e^{-\alpha t}\right\} + f_1(t). \tag{3.17}$$

This can be integrated directly using the integrating factor r, to give:

$$r^2\sigma_r = -\frac{\varrho\,\omega^2 r^4 H(t)}{8}\left\{4 - \frac{\nu}{K}\alpha\,e^{-\alpha t}\right\} + \frac{r^2}{2}f_1(t) + f_2(t) \tag{3.18}$$

$f_1(t)$ and $f_2(t)$ must now be chosen to satisfy the boundary conditions of prescribed internal pressure p_0 on the surface of the cavity $r = R_1(t)$, and zero pressure on the outside, $r = R_2$.

$$\sigma_r(R_1[t], t) = -p_0 H(t), \quad \sigma_r(R_2, t) = 0 \tag{3.19}$$

(3.18) finally reduces to:

$$\sigma_r = -\left\{\frac{\varrho\,\omega^2}{8}\left[4 - \frac{\nu}{K}\alpha\,e^{-\alpha t}\right]\left[\{R_1(t)\}^2 - r^2\right] + \frac{\{R_1(t)\}^2 p_0}{R_2^2 - [R_1(t)]^2}\right\}\left(\frac{R_2^2}{r^2} - 1\right)H(t). \tag{3.20}$$

σ_θ can be obtained from (3.15), σ_z by integration of (3.5), and the radial displacement u from (3.3). In obtaining this solution, the integrals with respect to time were greatly simplified by the zero initial conditions for $t = 0^-$. Moreover, in general, the separate determination of the initial conditions at $t = 0^+$ would pose a difficult problem. They are, however, determined incidentally by the solution given. Higher order operators would call for repeated time integrals, but no difficulties in principle would be encountered in applications involving materials represented by higher order differential operators.

It is clear that the expression for σ_r, (3.20), consists of the sum of two terms, the first representing the contribution due to the body force associated with rotation, and the second the contribution due to the prescribed internal pressure. The first term embodies a factor which contributes an exponential time variation due to the viscoelasticity of the material. Such an effect would usually be expected for a viscoelastic body subjected to constant load. The second term, which represents the contribution of the internal pressure p_0, is identical to the LAMÉ solution for an elastic body with the current internal cavity radius $R_1(t)$. That there is no influence of the viscoelasticity of the material in this term is a result of the independence of the elastic solution for σ_r of the elastic constants. This represents a particular example of a theorem due to ALFREY [9] concerning situations in which the stress distribution for a viscoelastic body is identical with that in the corresponding elastic problem. The present example is noteworthy since the elastic solution for σ_r due to internal pressure is not modified by viscoelasticity even though the geometry of the system is changing with time. This property is exhibited by the stress components σ_r and σ_θ only. σ_z will be influenced by the viscoelasticity of the material.

4. Justification of the Analytical Procedures

The methods of solution developed in this paper must be considered to be only formal, with the need to check each solution obtained, unless they are brought within the scope of an established theory, such as the theory of distributions, or unless they are justified by methods of classical analysis. The latter procedure is adopted here, and the justification for the methods utilized is presented in this section. The aspect of the solutions which violates the constrictions of classical analysis is the treatment of the discontinuous behaviour between $t = 0^-$ and $t = 0^+$. A similar difficulty could of course arise at any value of the time t, and the present study will be generally applicable merely by shifting the origin of time. The influence of the jump in the initial conditions between $t = 0^-$ and $t = 0^+$ can be removed by basing the analysis on the equivalent integral operator representation of viscoelastic behaviour (1.6), from which initial conditions can be removed as shown below.

Considering first problems for which a solution may be obtained with the aid of the LAPLACE transform, we wish to show that the elimination of the non-zero initial conditions by replacing the lower limit $t = 0^+$ of the LAPLACE transform integral by $t = 0^-$ does not change the solution. For this purpose it suffices to show that the transformed stress-strain laws are the same in both cases. Without loss of generality we consider a single stress-strain law of the type (1.4), restricted to physically sensible materials for which the initial strain response to unit stress applied at $t = 0$ is finite. This implies $p \leq q$, and we may write for convenience

$$P(D) = \sum_{j=0}^{q} p_j D^j, \; p_{p+1}, \ldots, p_q = 0, \qquad Q(D) = \sum_{j=0}^{q} q_j D^j, \tag{4.1}$$

where D is the differential operator $\partial/\partial t$. Taking the classical LAPLACE transform of (1.4), a number of terms in σ and ε and their derivatives evaluated at $t = 0^+$ are obtained in addition to $P(s)\,\bar{\sigma}$ and $Q(s)\,\bar{\varepsilon}$, since, for example

$$\overline{D^j \varepsilon} = s^j \bar{\varepsilon} - s^{j-1} \varepsilon(0^+) \cdots - D^{j-1} \varepsilon(0^+).$$

In Section 2 the relaxation behaviour of a MAXWELL material was determined by using the classical LAPLACE transform (2.4) and also, (2.6), using a LAPLACE transform based on the lower limit of integration, $t = 0^-$. As was mentioned, these methods give the same result, since in the former, the initial conditions at $t = 0^+$ on one side of the equation cancel with those on the other side [see (2.4)]. Thus, at least for a MAXWELL material under relaxation,

$$P(s)\,\bar{\sigma} = Q(s)\,\bar{\varepsilon}. \tag{4.2}$$

If we can prove that in general the classical LAPLACE transform of (1.4) also gives (4.2), the use of the transform with lower limit $t = 0^-$ will be justified.

Since classical methods of analysis are utilized in this section, $t = 0$ will have the significance $t = 0^+$, and the integral relation (1.5), for example, will be written in the form:

$$\varepsilon(t) = \sigma(0)\, J(t) + \int_0^t J(t-\tau) \frac{\partial \sigma}{\partial \tau}(\tau)\, d\tau. \tag{4.3}$$

The equations in the first section were intended to be interpreted in the spirit of symbolic functions, so that the $\partial\sigma/\partial\tau$ included a delta function associated with the initial value of σ. This gives the $\sigma(0)\,J(t)$ term in (4.3) and $\partial\sigma/\partial\tau$ represents only the classical derivative assumed to exist for $t > 0$.

Integrating (4.3) by parts:

$$\varepsilon(t) = \sigma(t)\,J(0) + \int_0^t \sigma(\tau)\,\frac{\partial J}{\partial t}(t-\tau)\,d\tau. \tag{4.4}$$

Operating on (4.4) with $Q(D)$:

$$[P(D) - J(0)\,Q(D)]\,\sigma(t) = Q(D)\int_0^t \sigma(\tau)\,\frac{\partial J}{\partial t}(t-\tau)\,d\tau.$$

If now the Laplace transform of the above is taken, making use of (4.1), it follows after considerable manipulation that:

$$\bar{\sigma}\sum_{k=0}^{q}[p_k - s\,\bar{J}\,q_k]\,s^k$$
$$= \sum_{k=1}^{q}\sum_{i=0}^{k-1} s^{k-1-i}\,D^i[\sigma(0)]\,[p^k - \sum_{l=k}^{q} D^{l-k}[J(0)]\,q_l]. \tag{4.5}$$

Operating on (4.3) with $Q(D)$, using (4.1) and the fact that $Q(D)\,J(t) = p_0$, we have at $t = 0$:

$$\sum_{k=1}^{q} p_k\,D^k[\sigma(0)] = \sum_{k=1}^{q} q_j\left[D^k\int_0^t J(t-\tau)\,\frac{\partial\sigma}{\partial\tau}(\tau)\,d\tau\right]_{t=0}.$$

Performing the integration:

$$\sum_{k=1}^{q} p_k\,D^k[\sigma(0)] = \sum_{k=1}^{q}\sum_{l=1}^{k} q_k\,D^{k-l}[J(0)]\,D^l[\sigma(0)].$$

Since σ is arbitrary, we may equate terms in $D^l(\sigma)$ to obtain:

$$p_k = \sum_{l=k}^{q} D^{l-k}[J(0)]\,q_l,\ j = l, \ldots, q. \tag{4.6}$$

Using (4.6) and (4.1), it follows from (4.5) that

$$P(s) = s\,\bar{J}\,Q(s).$$

However, the Laplace transform of (4.4) gives:

$$\bar{\varepsilon} = s\,\bar{J}\,\bar{\sigma}$$

so that

$$P(s)\,\bar{\sigma} = Q(s)\,\bar{\varepsilon}. \tag{4.7}$$

Thus the terms in σ and ε at $t = 0^+$ cancel from both sides when the classical Laplace transform is applied to (1.4), and the formal use of the lower limit $t = 0^-$ is justified.

We now turn to the related question for problems in which the Laplace transform is not applicable. The use of delta functions in Section 3 determines the initial conditions for the stresses, strains and their derivatives at $t = 0^+$ which would be needed for integration of the differential equations by classical methods. Again we can make use of the integral representation (1.6) to determine these initial values, and show that they are identical with those deduced on the basis of the formal delta function theory. The situation is much more complicated than for the homogeneous case, since the geometrical form of the body and its initial variation is involved, so that instead of being able to prove a generally applicable result such as (4.7) above, we check the particular solution of the rotating cylinder treated in Section 3, with the implication that the same approach carries through in other cases.

For the introduction of the integral operator (1.6) into the stress analysis of Section 3, it is convenient to substitute formally for the combined operator $P/Q = \varkappa$, the representation:

$$\varkappa\, g(t) = J(0)\, g(t) - \int_0^t g(\tau) \frac{\partial J}{\partial \tau}(t-\tau)\, d\tau . \tag{4.8}$$

This procedure reproduces the equations for the determination of stress distributions which would have been obtained if the theory had been developed ab initio on the basis of the integral operator (1.6). Again, since we are now working in the spirit of classical analysis, the term $J(0)\, g(t)$ is needed to incorporate the initial value of g.

When using the integral operator (1.6), it is convenient to integrate (3.11) with respect to r, instead of delaying this integration until those with respect to time have been completed [see (3.15)]. This gives:

$$(B+C)\,[(B-C)(\sigma_r+\sigma_\theta) + B\,\varrho\,\omega^2 r^2/2] = f(t), \tag{4.9}$$

where $f(t)$ is an arbitrary function. Substituting for $\varkappa$ (4.8) and using (3.10) for $t>0$ to eliminate σ_θ, the following relation results:

$$\begin{aligned} r\, f(t) = {} & J(0)\left\{J(0)+\frac{2}{3K}\right\}\frac{\partial}{\partial r}(r^2\sigma_r) - 2\left\{J(0)+\frac{1}{3K}\right\}\int_0^t \frac{\partial J}{\partial \tau}(t-\tau)\frac{\partial}{\partial r}(r^2\sigma_r)\, d\tau \\ & + \int_0^t \frac{\partial J}{\partial \tau'}(t-\tau')\int_0^{\tau'}\frac{\partial J}{\partial t}(\tau'-\tau)\frac{\partial}{\partial r}(r^2\sigma_r)\, d\tau\, d\tau' \\ & + \varrho\,\omega^2 r^3\left[J(t)\left\{2\,J(0)+\frac{5}{6K}\right\} - 2\int_0^t J(\tau)\frac{\partial J}{\partial \tau}(t-\tau)\, d\tau\right]. \end{aligned} \tag{4.10}$$

Integrating with respect to r, this becomes:

$$\begin{aligned} h(t) + \frac{k(t)}{r^2} = {} & J(0)\left\{J(0)+\frac{2}{3K}\right\}\sigma_r - 2\left\{J(0)+\frac{1}{3K}\right\}\int_0^t \frac{\partial J}{\partial \tau}(t-\tau)\,\sigma_r\, d\tau \\ & + \int_0^t \frac{\partial J}{\partial \tau'}(t-\tau')\int_0^{\tau'}\frac{\partial J}{\partial \tau}(\tau'-\tau)\,\sigma_r(r,\tau)\, d\tau\, d\tau' \\ & + \frac{\varrho\,\omega^2 r^2}{4}\left[J(t)\left\{2\,J(0)+\frac{5}{6K}\right\} - 2\int_0^t J(\tau)\frac{\partial J}{\partial \tau}(t-\tau)\, d\tau\right], \end{aligned} \tag{4.11}$$

where $h(t)$ and $k(t)$ are arbitrary functions. $h(t)$ and $k(t)$ can now be eliminated by applying the boundary conditions. In the discussion of the solution obtained in Section 3 it was pointed out that the stresses comprised the sum of two parts, one due to the body force, and one due to the surface tractions. This, of course, is to be expected because of the linearity of the problem and the consequent applicability of the superposition principle. For brevity we will apply the present consideration only to the component due to body forces, and so introduce homogeneous boundary conditions,

$$\sigma_r = 0, \; r = R_1(t), \; R_2. \tag{4.12}$$

Eliminating h and k from (4.12), (4.11) becomes:

$$\begin{aligned}
J(0)\left\{J(0)+\frac{2}{3K}\right\}\sigma_r &= \frac{\varrho\,\omega^2}{4}\left[R_1^2+R_2^2-\left(\frac{R_1 R_2}{r}\right)^2-r^2\right] \\
&\times\left[J(t)\left\{2\,J(0)+\frac{5}{6K}\right\}-2\int_0^t J(\tau)\frac{\partial J}{\partial \tau}(t-\tau)\,d\tau\right] \\
&+2\left\{J(0)+\frac{1}{3K}\right\}\int_0^t \frac{\partial J}{\partial \tau}(t-\tau)\,\sigma_r(r,\tau)\,d\tau \\
&-\int_0^t \frac{\partial J}{\partial \tau'}(t-\tau')\int_0^{\tau'} \frac{\partial J}{\partial \tau}(\tau'-\tau)\,\sigma_r(r,\tau)\,d\tau\,d\tau' \\
&+2\,\frac{R_1^2}{R_2^2-R_1^2}\left[1-\left(\frac{R_2}{r}\right)^2\right] \\
&\times\left[J(0)+\frac{1}{3K}\right]\int_0^t \frac{\partial J}{\partial \tau}(t-\tau)\,\sigma_r(R_1(t),\tau)\,d\tau \\
&-\frac{R_1^2}{R_2^2-R_1^2}\left[1-\left(\frac{R_2}{r}\right)^2\right]\int_0^t \frac{\partial J}{\partial \tau'}(t-\tau')\int_0^{\tau'} \frac{\partial J}{\partial \tau}(\tau'-\tau) \\
&\times\sigma_r(R_1(\tau),\tau)\,d\tau\,d\tau'.
\end{aligned} \tag{4.13}$$

Evaluating this at $t = 0$ gives:

$$\sigma_r(r,0) = \frac{\varrho\,\omega^2}{8}\,\frac{4\,J(0)+\dfrac{5}{3K}}{J(0)+\dfrac{2}{3K}}\left[R_1^2+R_2^2-\left(\frac{R_1 R_2}{r}\right)^2-r^2\right] \tag{4.14}$$

which agrees with (3.20) evaluated at $t = 0$ when use is made of the relation $J(0) = 1/G$ for the instantaneous compliance in shear. This result shows that the initial stress depends only on the initial geometry of the problem and the instantaneous elastic compliance, $J(0)$, and is not influenced by other viscoelastic characteristics. This confirms the prediction based on physical intuition mentioned in Section 3.

Eq. (4.13) can be repeatedly differentiated to obtain initial derivatives of σ_r of any desired order. Combining these with the equilibrium Eq. (3.10) for $t > 0$, and derivatives of it, initial values of σ_θ and its derivatives can also be obtained. The initial value of σ_z follows from (3.5) expressed in terms of the integral operator (4.8):

$$\begin{aligned}\left[\frac{1}{3K} + 2J(0)\right]\sigma_z = {} & \left[J(0) - \frac{1}{3K}\right](\sigma_r + \sigma_\theta) \\ & - \int_0^t (\sigma_r + \sigma_\theta)(\tau)\frac{\partial J}{\partial \tau}(t-\tau)\,d\tau \\ & + 2\int_0^t \sigma_z(\tau)\frac{\partial J}{\partial \tau}(t-\tau)\,d\tau.\end{aligned} \tag{4.15}$$

Initial derivatives of σ_z are obtained by differentiation of (4.15). Thus all derivatives required for the integration of the equations of stress distribution can be evaluated whatever the order of the differential operators considered.

As an example, a single differentiation of (4.13) yields:

$$\begin{aligned}J(0)\left\{J(0) + \frac{2}{3K}\right\}\dot{\sigma}_r(r,0) = {} & \frac{\varrho\,\omega^2}{8}\left[1 - \left(\frac{R_2}{r}\right)^2\right]\frac{d(R_1)^2}{dt}(0)\,J(0)\left\{4J(0) + \frac{5}{3K}\right\} \\ & + \frac{\varrho\,\omega^2}{8}\left[R_1^2 + R_2^2 - \left(\frac{R_1 R_2}{r}\right)^2 - r^2\right]\frac{dJ}{dt}(0)\left\{8J(0) + \frac{5}{3K}\right\} \\ & - 2\left[J(0) + \frac{1}{3K}\right]\frac{dJ}{dt}(0)\,\sigma_r(r,0).\end{aligned} \tag{4.16}$$

This result is in conformity with (3.20), writing $\frac{dJ}{dt}(0) = \frac{1}{\nu}$ for MAXWELL behaviour in shear, and it shows, as would be expected, that the initial derivative of σ_r depends on the initial velocity of annihilation of the cavity, $\frac{dR_1}{dt}(0)$, and on $\frac{dJ}{dt}(0)$, a viscoelastic characteristic other than the instantaneous compliance.

We have seen that, using the integral operator representation, initial conditions can be determined at $t = 0^+$ to permit integration of the differential equations of viscoelastic stress analysis by classical methods, and that in the particular problem studied the delta function approach enabled these initial conditions to be determined along with the solution for $t > 0$. The involved dependence of these initial conditions on characteristics of the problem illustrates the value of integrating the equations from $t = 0^-$, with zero initial conditions, and so avoiding the separate evaluation of the discontinuities at $t = 0$.

References

[1] Lee, E. H.: Quart. Appl. Math. **13**, 183 (1955).
[2] Lee, E. H.: Proc. 1st Symp. Naval Struct. Mech. 456 (1960).
[3] Churchill, R. V.: Operational Mathematics. McGraw Hill, 1958.
[4] Carslaw, H. S., and J. C. Jaeger: Operational Methods in Applied Mathematics, 2nd Edition, Oxford University Press, 1953.
[5] van der Pol, B., and H. Bremmer: Operational Calculus based on the Two-sided Laplace Integral. Cambridge University Press, 1950.
[6] Friedman, B.: Principles and Techniques of Applied Mathematics, Wiley, 1956.
[7] Gross, B., and W. Güttinger: Appl. Sci. Res. B **6**, 189.
[8] Lee, E. H., J. R. Radok and W. B. Woodward: Transaction of the Society of Rheology **3**, 41 (1959).
[9] Alfrey, T.: Quart. Appl. Math. **2**, 113 (1944).

Torsional Response of Linear Viscoelastic Plates Subjected to thermal Stresses

By

Harry H. Hilton

University of Illinois Urbana, Illinois, U. S. A.

and

Edward L. Walsh

Boeing Aircraft Company Seattle, Washington, U. S. A.

Abstract

The govering differential equations are derived for a flat plate obeying the general linear viscoelastic stress-strain relations subjected to external torques and temperature gradients. The equations describe the induced thermal stresses and the angle of twist due to the combined effects of the applied torsional moment, induced thermal stresses, and creep. Particular solutions of the general differential equations are presented for a MAXWELL body plate under steady state and time dependent temperature gradients. It is shown that chordwise temperature distributions increase at all times the twist deformations beyond those induced by torsion alone. In addition, the angle of twist of the MAXWELL body plate is greater, even though the thermal stresses relax, than the deformation of a corresponding elastic plate.

1. Introduction

Prior to World War II aircraft structures were generally rigid enough so that the effects of structural deformations on aerodynamic loads were relatively minor. As the speed of modern aircraft increased, the lifting surface became thinner and the consequences of flexibility had to be taken into account [*1*]. However, at high supersonic speeds, in addition to flexibility, aerodynamic heating is also encountered. It is well known that elevated temperatures and temperature gradients induce material inelasticity and nonhomogeneity, thermal stresses, loss of strength and stiffness, creep and thermal buckling etc. An excellent discussion of these problems is given by HOFF [*2*].

Another problem associated with these high temperature phenomena that has received attention in recent years is the reduction in effective stiffness due to the presence of thermal stresses. GOODIER [3] has investigated the effect of axial stresses on elastic torsion. His analysis shows that axial compressive stresses in the presence of torsion appreciably increase the twist deformation of the section. Several years later BUDIANSKY and MEYERS [4] investigated the effects of thermal stresses on the torsional stiffness of thin elastic wings. A chordwise temperature gradient, due to aerodynamic heating, induced detrimental axial compressive thermal stresses in the region of the leading and trailing edges and, for the configurations considered, led to an appreciable reduction, and in some cases to complete loss, of the effective elastic torsional stiffness. Later SINGER and HOFF [5] have shown that the use of large deflection theory results in appreciably smaller reduction in stiffness and in finite deflections.

Unfortunately, at elevated temperatures materials do not behave elastically, and aeroelasticity must give way to aeroinelasticity. There are numerous analytical expressions used to represent inelastic material response to loads. Probably the most popular is viscoelasticity [6] since it adequately describes high temperature material behavior [7] and is mathematically easy to handle.

The actual inelastic behavior of most materials at very large stresses is nonlinear and linear viscoelastic laws do not account for such phenomena. In particular, in instability problems (creep buckling and divergence) the case of linear viscoelastic stress-strain laws has been proved unable to predict the physically observable finite critical times. However, prior to the onset of instability linear viscoelasticity can describe stress-strain behavior, since there are an unlimited number of coefficients available in the linear stress-strain laws to match experimental creep behavior. ALLEN *et al* [7] have demonstrated that some linear viscoelastic stress-strain relationships, such as those of the MAXWELL-KELVIN body, can be used to represent actual creep properties under constant temperatures. It has been further shown [8] that under constant stress and temperature conditions a MAXWELL-KELVIN linear viscoelastic body is analogous to an experimentally determined nonlinear creep law.

Thermal stresses in nonhomogeneous viscoelastic media were investigated in Reference [9]. Analyses and examples were presented for thick, and thin walled cylinders and flat plates and an elastic-viscoelastic analogy for thermal stresses was derived. In the cases considered, the stresses were, at some time, greater than the stresses calculated for elastic models under identical conditions.

Solutions to some aeroinelastic problems are presented in References [10] and [11]. In the first paper torsional creep divergence of rigid lifting surfaces mounted on viscoelastic supports was treated. Nonlinear viscoelastic relations had to be used to obtain a finite divergence time, although linear viscoelastic relations were found to be useful for deflection calculations prior to divergence. Secondly, the analysis of divergence of linear viscoelastic lifing surfaces, although more complex mathematically, also does not yield finite divergence times but gives information on the time-deflection history prior to divergence.

The present investigation of the effects of thermal stresses and creep on the torsional response of flat plates is not a true aeroinelastic problem since aerodynamic forces were not considered acting on the plate. This was done in order to be able to use linear viscoelastic stress-strain relations and at the same time avoid the difficulties associated with the lack of finite divergence times. In this manner, it is possible to obtain a fundamental understanding of the interaction of thermal stresses, creep and the torsional response of plates. Consequently, in the present analysis constant torque has been used, rather than the aerodynamic torsion which is a function of the angle of twist, in order to simplify the mathematical formulation sufficiently to allow general analytic solutions instead of having to resort to approximations by some numerical methods. It is believed that the experience and information gained from this idealized study will prove beneficial in planning subsequent investigations.

List of Symbols

a_{mn}	constants defined by Eqs. (26)—(39)
a_k, b_k	viscoelastic material properties
c	non-dimensional half-chord distance
c_1, c_2	constants
E_{ij}	strain deviator on the ith face in the j direction
E_i	exponential integral
F_x, F_y	forces in the x and y directions
G	shear modulus
J_e	effective polar moment of inertia
k_1,k_2	as defined by Eqs. (47) and (48)
$L_n^\alpha(z)$	Laguerre polynomial
M_a	torsional moment about center of twist due to aerodynamic loads
M_0	moment about the origin
M_T	total applied torsional moment about center of twist
M_σ	torsional moment about center of twist due to thermal stresses
P, Q	differential operators
r	radial distance from center of twist
S_{ij}	stress deviator on the ith face in the j direction
t	time
t^*	t/τ, non-dimensional time
$T(x, t)$	temperature
u_i	displacement in the ith direction
x, y, z	non-dimensional coordinates

α	coefficient of thermal expansion
δ_{ij}	KRONECKER delta
ε	mean strain defined by Eq. (16)
ε_{ij}	strain on the ith face in the j direction
$\theta(y, t)$	angle of twist
λ	constant
σ	mean stress as defined by Eq. (15)
σ_{ij}	stress on the ith face in the j direction
τ	relaxation time of a Maxwell viscoelastic body
φ	stress function
∇^2	LAPLACIAN operator in two-dimensional space

Superscripts

e	elastic function
s	spacial variation of the function
v	viscoelastic function

2. General Analysis

(a) Thermal Stresses in a Viscoelastic Plate

Consider a thin, rectangular plate with coordinate system as shown in Fig. 1. The plate is considered rigidly supported at $y = 0$, and all measurements have been non-dimensionalized by dividing them by the half span. The temperature distribution has been assumed a separable function of time and space to allow solution of the governing differential equation by separation of variables. As a first approximation the material is considered homogeneous with constant viscoelastic material properties. The real, and more complex problem of variation of material properties with temperature has been forsaken in order to present general analytic solutions.

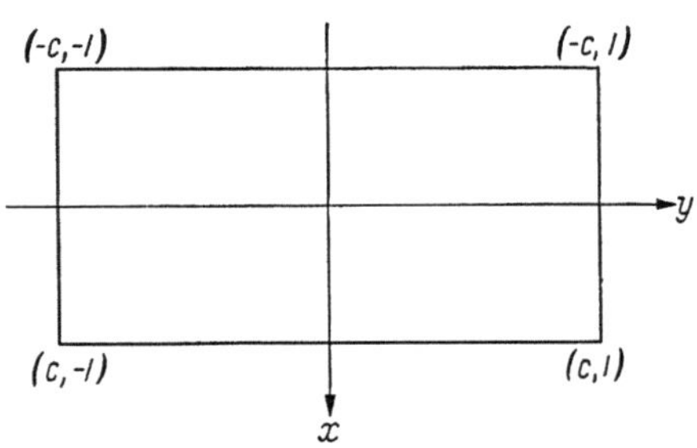

Fig. 1. Coordinate system of the plate

Since the plate is thin, the usual plane stress assumptions can be made

$$\sigma_{zz} = \sigma_{xz} = \sigma_{yz} = 0 \tag{1}$$

and

$$\left.\begin{aligned} &\partial\sigma_{ij}/\partial z = \partial\varepsilon_{ij}/\partial z = 0 \\ &(i, j = x, y, z). \end{aligned}\right\} \tag{2}$$

From Eq. (2), the strain definitions reduce to

$$\begin{aligned} \varepsilon_{xx} &= u_{x,x}, \\ \varepsilon_{yy} &= u_{y,y}, \\ 2\,\varepsilon_{xy} &= u_{x,y} + u_{y,x} \end{aligned} \tag{3}$$

and the compatibility conditions become

$$\varepsilon_{xx,yy} + \varepsilon_{yy,xx} - 2\,\varepsilon_{xy,xy} = 0, \tag{4}$$

where a subscript after a comma indicates differentiation with respect to that coordinate.

In the absence of body forces and moments the equilibrium equations reduce to

$$\begin{aligned} \sigma_{xx,x} + \sigma_{xy,y} &= 0, \\ \sigma_{xy,x} + \sigma_{yy,y} &= 0. \end{aligned} \tag{5}$$

Let there be a stress function $\varphi = \varphi(x, y, z, t)$ defined by

$$\begin{aligned} \sigma_{xx} &= \varphi_{,yy}, \\ \sigma_{yy} &= \varphi_{,xx}, \\ \sigma_{xy} &= -\varphi_{,xy}. \end{aligned} \tag{6}$$

This function will satisfy the equilibrium equations (5) identically.

Since the stresses are separable functions of time and space [9]. The stress function must also be considered a separable function of time and space

$$\varphi = g(t)\,\varphi^s(x, y, z). \tag{7}$$

Repeated differentiation with respect to space coordinates will not affect the time function, therefore, for this type of stress function, the stresses must be

$$\sigma_{ij}(x, y, z, t) = g(t)\,\sigma_{ij}^s(x, y, z). \tag{8}$$

Let $g(t)$ be dimensionless so that σ^s has the dimensions of stress.

The general linear, incompressible, viscoelastic stress-strain relations are given by

$$P\{s_{ij}\} = 2\,Q\{E_{ij}\}, \tag{9}$$

$$\varepsilon = \alpha\,T, \tag{10}$$

where P and Q are differential operators defined as

$$P = \sum_{k=0}^{r_1} a_k \frac{\partial^k}{\partial t^k}, \tag{11}$$

$$Q = \sum_{k=0}^{r_2} b_k \frac{\partial^k}{\partial t^k} \tag{12}$$

and the stress and strain deviators are

$$S_{ij} = \sigma_{ij} - \delta_{ij}\,\sigma, \tag{13}$$

$$E_{ij} = \varepsilon_{ij} - \delta_{ij}\,\varepsilon, \tag{14}$$

where

$$\sigma = \sigma_{ii}/3, \tag{15}$$

$$\varepsilon = \varepsilon_{ii}/3. \tag{16}$$

A repeated index, i or j, indicates summation according to tensor convention, and δ_{ij} is the KRONECKER delta.

The temperature distribution

$$T = T_0 [1 + (x/c)^2] \, H(t) \tag{17}$$

a separable function of time and space, has been assumed to vary parabolically in the chordwise direction only. This variation is mathematically easy to work with, and results in a distribution not unlike that expected on a supersonic lifting surface [4]. The constant T_0 will carry the dimensions of temperature, while the time function will be dimensionless.

By combining Eqs. (6), (12), (13), (14), (15), (16) and substituting into Eq. (9), three equations are obtained

$$P\{2\, \varphi_{,yy} - \varphi_{,xx}\} = 6\, Q\{\varepsilon_{xx} - \alpha\, T\}, \tag{18}$$

$$P\{2\, \varphi_{,xx} - \varphi_{,yy}\} = 6\, Q\{\varepsilon_{yy} - \alpha\, T\} \tag{19}$$

$$- P\{\varphi_{,xy}\} = 2\, Q\{\varepsilon_{xy}\}. \tag{20}$$

Operating on Eq. (4) with the operator Q, yields

$$Q\{\varepsilon_{xx,yy} + \varepsilon_{yy,xx} - 2\, \varepsilon_{xy,xy}\} = 0. \tag{21}$$

By differentiating Eq. (18) twice with respect to y, Eq. (19) twice with respect to x, and Eq. (20) with respect to x and y, then combining these results with Eq. (17) and Eq. (21), the governing differential equation of the problem is obtained

$$P\{\nabla^2 \nabla^2 \varphi\} + (6\, \alpha\, T_0/c^2)\, Q\{H\} = 0. \tag{22}$$

Substitution of Eq. (7) in Eq. (22) permits separation of the time and space dependent functions

$$\frac{P\{g(t)\}}{Q\{H(t)\}} = \frac{-6\, \alpha\, T_0}{c^2 \nabla^2 \nabla^2 \varphi^s} = \frac{1}{\lambda}, \tag{23}$$

where λ is, as yet, some undefined constant. A relation similar to Eq. (23) was derived in References 9 for the three dimensional case, but does not reduce to the two dimensional case because of the assumptions made in the plane stress solution.

The governing differential equation for the time function $g(t)$ may now be restated as a total differential equation in time

$$\lambda\, P\{g(t)\} = Q\{H(t)\} \tag{24}$$

and a partial differential equation in space only,

$$\nabla^2 \nabla^2 \varphi^s = -\, 6\, \alpha\, T_0\, \lambda/c^2. \tag{25}$$

Eq. (25) may be solved by a method suggested by NEOU [12], where the spacial variation of the stresses is assumed to be representable by power series. Then the spacial variation of the stress function must also be written in series form

$$\varphi^s = \sum_{m=0}^{\infty} \sum_{n=0}^{\infty} a_{mn} x^m y^n . \tag{26}$$

By substituting the series, Eq. (26), into Eq. (25) and properly shifting the series indices, one obtains

$$\begin{aligned} &\sum_{m=0}^{\infty} \sum_{n=4}^{\infty} m(m-1)(m-2)(m-3)\, a_{m\,n-4}\, x^{m-4} y^{n-4} \\ &\quad + 2 \sum_{m=2}^{\infty} \sum_{n=2}^{\infty} (m-2)(m-3)(n-2)(n-3)\, a_{m-2\,n-2}\, x^{m-4} y^{n-4} \\ &\quad + \sum_{m=4}^{\infty} \sum_{n=0}^{\infty} n(n-1)(n-2)(n-3)\, a_{m-4\,n}\, x^{m-4} y^{n-4} = -6 \lambda \alpha T_0/\lambda . \end{aligned} \tag{27}$$

The boundary conditions of the plate are given by

$$\sigma_{xx} = 0 \quad x = \pm c \text{ for all } y, \tag{28}$$

$$\sigma_{xy} = 0 \quad x = \pm c \text{ for all } y. \tag{29}$$

These boundary conditions can be satisfied if

$$a_{mn} = 0 \quad n \geq 3. \tag{30}$$

Since the stresses are polynomials, it follows that

$$a_{01} = a_{00} = a_{10} = 0. \tag{31}$$

Then from Eq. (27) the recurrence relations for some of the remaining coefficients are obtained by equating coefficients of terms of like powers of x and y.

For powers corresponding to $m = n = 4$

$$24\, a_{40} + 8\, a_{22} = -6 \lambda \alpha T_0/c^2 . \tag{32}$$

For powers corresponding to $m \geq 4$, $n \geq 4$ with the exception of the above case

$$\begin{aligned} &m(m-1)(m-2)(m-3)\, a_{m\,n-4} \\ &\quad + 2(m-2)(m-3)(n-2)(n-3)\, a_{m-2\,n-2} \\ &\quad + n(n-1)(n-2)(n-3)\, a_{m-4\,n} = 0 . \end{aligned} \tag{33}$$

With the aid of Eqs. (28), (29), (32), (33) the function φ^s becomes

$$\varphi^s = a_{20} x^2 + a_{30} x^3 + a_{31} (x_y^3 y - 3 c^2 x y) - \lambda \alpha T_0 x^4/4 c^2 \tag{34}$$

and the functions σ_{ij}^s are obtained by the use of Eq. (16)

$$\sigma_{xx}^s = 0, \tag{35}$$

$$\sigma_{xy}^s = 3\, a_{31} (c^2 - x^2), \tag{36}$$

$$\sigma_{yy}^s = 2\, a_{20} + 6\, a_{30}\, x + 6\, a_{31}\, x\, y - 3\, \lambda\, \alpha\, T_0\, x^2/c^2. \tag{37}$$

The remaining coefficients are evaluated by summing the forces and moments on the plate

$$\begin{aligned} \sum F_x &= \int_{-c}^{c} \sigma_{xy}\big|_{y=0}\, dx = 0, \\ \sum F_y &= \int_{-c}^{c} \sigma_{yy}\big|_{y=0}\, dx = 0, \\ \sum M_0 &= \int_{-c}^{c} \sigma_{yy}\big|_{y=0}\, x\, dx = 0. \end{aligned} \tag{38}$$

Evaluation of these integrals yields

$$\begin{aligned} a_{30} &= 0, \\ a_{31} &= 0, \\ a_{20} &= \lambda\, \alpha\, T_0/2. \end{aligned} \tag{39}$$

The final solution of Eq. (25) therefore is

$$\varphi^s = (\lambda\, \alpha\, T_0/2)\, x^2\, [1 - (1/2)\, (x/c)^2] \tag{40}$$

which results in

$$\sigma_{xx} = \sigma_{zz} = \sigma_{xy} = \sigma_{yz} = \sigma_{xz} = 0, \tag{41}$$

$$\sigma_{yy}^s = \lambda\, \alpha\, T_0 [1 - 3 (x/c)^2]. \tag{42}$$

The solution of Eq. (24) defining the time function $g(t)$ will depend on the particular definitions of the differential operators.

(b) Viscoelastic Torsion with Thermal Stresses

The general governing differential equation defining viscoelastic torsion, neglecting chordwise bending, has been derived in Reference [*11*] in order to evaluate torsional divergence time. This same equation may be used to describe the effects of creep on torsional deflection, neglecting chordwise bending, experienced when a viscoelastic body is subjected to thermal stresses

$$\frac{\partial}{\partial y} \left[J_e\, Q \left\{ \frac{\partial \theta}{\partial y} \right\} \right] = - P\{M_T\} = - P\{M_a + M_\sigma\}. \tag{43}$$

The applied torsional moment (M_T) must be the sum of the applied external torques (M_a) plus the torque due to the thermal stresses (M_σ), and is considered positive when it acts in the direction of positive twist.

Since the thermal stress σ_{yy} acts along lines originally parallel to the spanwise coordinate, any torsional deformation of the plate will result in a corresponding change in the direction of action of the thermal stress (Fig. 2). The moment about the axis of twist, due to the thermal stress σ_{yy}, will be

$$M_\sigma = \iint_A \sigma_{yy}\, r \sin(r\,\partial\vartheta/\partial y)\, dx\, dz, \tag{44}$$

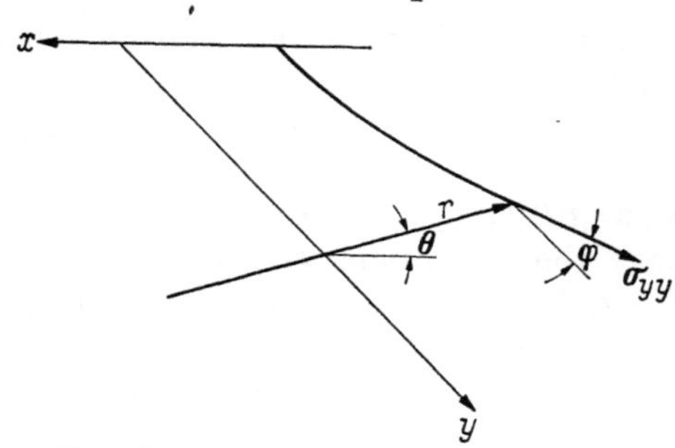

Fig. 2. Free body diagramm of the termal stresses

where r is the distance from the center of twist, along the displaced chordwise direction of the plate. For small angles of rotation one can make the usual approximation

$$\sin(r\,\partial\vartheta/\partial y) \cong \tan(r\,\partial\vartheta/\partial y) \sim r\,\partial\vartheta/\partial y. \tag{45}$$

Then, since θ and $g(t)$ are independent of the coordinates x and z, Eq. (44) may be written as

$$M_\sigma = (\partial\theta/\partial y)\, g(t) \iint_A \sigma^s_{yy}\, r^2\, dx\, dz. \tag{46}$$

Now, letting

$$k_1 = \Big[\iint_A \sigma^s_{yy}\, r^2\, dx\, dz\Big] / G\, J_e \tag{47}$$

and

$$k_2 = M_a / G\, J_e. \tag{48}$$

Eq. (43) may be rewritten as

$$Q\left\{\frac{\partial^2\theta}{\partial y^2}\right\} + G\,P\left\{k_1\, g(t)\,\frac{\partial\theta}{\partial y} + k_2\right\} = 0, \tag{49}$$

where $g(t)$ is defined by Eq. (24). The solution of these equations will naturally be dependent on the particular chosen viscoelastic stress-strain relation, which defines the differential operators P and Q.

3. Particular Solutions of the General Equations

(a) Elastic Case — Steady State Temperature Distribution

Since the elastic case forms the initial conditions of the viscoelastic problems it will be discussed first.

The particular definitions of the differential operators for the elastic body are

$$P = 1 \quad Q = G. \tag{50}$$

Eq. (24) then becomes

$$g(t) = G\,H(t)/\lambda. \tag{51}$$

For a steady state temperature distribution $H(t) = 1$ and Eq. (51) results in

$$g(t) = G/\lambda. \tag{52}$$

Since $g(t)$ and $H(t)$ were defined as nondimensional quantities, λ must have the dimensions of the shear modulus, and for the elastic case $\lambda = G$. Through use of this substitution and of the operators of Eq. (50), Eq. (49) reduces to

$$\frac{\partial^2 \theta^e}{\partial y^2} + k_1 \frac{\partial \theta}{\partial y} = -k_2. \tag{53}$$

The boundary conditions of the problem

$$\theta^e(0, t) = 0 \tag{54}$$

and

$$\partial \theta^e(1, t)/\partial y = 0 \tag{55}$$

require that the twist be zero at the rigid support, and that the shear be zero at the free end of the twisted plate. The solution of Eq. (53), subject to the above boundary conditions, is

$$\theta^e = (k_2/k_1^2)\, e^{k_1}(1 - e^{-k_1 y}) - k_2\, y/k_1 \tag{56}$$

and the existing thermal stress is defined by Eqs. (8) and (42) as

$$\sigma^e_{yy} = G\,\alpha\,T_0[1 - 3(x/c)^2]. \tag{57}$$

(b) Elastic Case — Time Dependent Temperature Distribution

As an example of a typical time dependent temperature distribution, $H(t) = e^{c_1 t}$ was chosen, where C_1 is any real number with dimensions of $1/t$, because of the advantages when used in Eq. (24). A great majority of viscoelastic bodies have definitions of the P and Q operators such that the complementary solutions of Eq. (24) give $g(z)$ in exponential form. This particular selection of $H(t)$ also allows, by choice of the value of c_1, good control over the time variation of the temperature. Using Eq. (51) and the chosen time variation of the temperature,

$$g(t) = G\,e^{c_1 t}/\lambda. \tag{58}$$

Since this is still the elastic case, the substitution $\lambda = G$ is mandatory, and with the operators defined by Eq. (50), Eq. (49) becomes

$$\frac{\partial^2 \theta^e}{\partial y^2} + k_1\, e^{c_1 t} \frac{\partial \theta^e}{\partial y} = -k_2. \tag{59}$$

The boundary conditions are given by Eqs. (54) and (55), resulting in the solution to the above differential equation of

$$\theta^e = (k_2/k_1^2)\, e^{(k_1 e^{c_1 t} - 2c_1 t)} \left[1 - e^{-k_1 y e^{c_1 t}}\right] - (k_2\, y/k_1)\, e^{-c_1 t}. \tag{60}$$

The existing thermal stress is defined by Eqs. (8) and (42) as

$$\sigma^c_{yy} = G\,\alpha\, T_0\, e^{c_1 t}\, [1 - 3\,(x/c)^2]. \tag{61}$$

At time $t = 0$, the solutions given by Eqs. (60) and (61) reduce to the steady state temperature distribution elastic solutions given by Eqs. (56) and (57). Furthermore, when $c_1 = 0$ all of the equations presented for the time dependent temperature distribution reduce to those given for the steady state temperature distribution. This serves as a check on the behavior of the differential equations as well as supplying the initial condition for the viscoelastic body considered.

(c) Maxwell Body Case — Steady State Temperature Distribution

The MAXWELL body was chosen to illustrate creep type behavior since the MAXWELL body is representative of a linear creep rate. It is mathematically simpler than other bodies of equal interest, and experimental data [7] are available with which to calculate the material constants. The particular definitions of the differential operators for this case are

$$P = (1/\tau)\,[1 + \partial/\partial t^*], \tag{62}$$

$$Q = (G/\tau)\, \partial/\partial t^*,$$

where the substitution $t^* = t/\tau$ has been made. If these differential operators are used Eq. (24) becomes

$$(dg/dt^*) + g = (G/\lambda)\,(dH/dt^*). \tag{63}$$

Letting $H(t^*) = 1$ and requiring $g(0) = G/\lambda$ in order to satisfy the initial condition of the MAXWELL body, which is an elastic deformation, Eq. (63) has the solution

$$g = (G/\lambda)\, e^{-t^*}. \tag{64}$$

Since the terms g and σ^s_{yy}, Eq. (42), always appear as linear products, the value of λ is unimportant in determining stresses and deformations, and therefore, the substitution $\lambda = G$ will be made as a matter of convenience to simplify the writing of the following equations.

With the operators defined by Eq. (62), Eq. (49) becomes

$$\frac{\partial^3 \theta^v}{\partial y^2\, \partial t^*} + k_1\, e^{-t^*}\, \frac{\partial^2 \theta^v}{\partial y\, \partial t^*} = -\, k_2 \tag{65}$$

subject to the boundary conditions requiring, (1) the twist to be zero at the rigid support and, (2) the shear to be zero at the free end, for all time.

A solution may be assumed in the form

$$\theta^v = \sum_{n=0}^{\infty} a_n(t^*)\, y^n. \tag{66}$$

Substitution of Eq. (66) into the differential equation (65) and proper shifting of the series indices will yield the recurrence relations for the a_n terms as

$$\dot{a}_0,\ \dot{a}_1 \text{ arbitrary}$$

$n \geq 2$:

$$\dot{a}_n = (-1)^{n-1}\, k_1^{n-2}\, e^{-(n-2)t^*}\, [k_2 + k_1\, \dot{a}_1\, e^{-t^*}]\, (n!)^{-1}, \tag{67}$$

where the dot above the term indicates differentiation with respect to t^*. The initial condition is the elastic solution for the constant temperature distribution and the $a_n(0)$ terms are obtained by expanding the elastic solution, Eq. (56), in an infinite series and equating the coefficients of each power term in y. This process results in

$$a_1(0) = -\,(k_2/k_1)\,(e^{k_1} + 1)$$

$n \geq 2$:

$$a_n(0) = [(-1)^n\, k_2\, k_1^{n-2}\, e^{k_1}]/n. \tag{68}$$

The boundary conditions together with Eq. (66) give

$$\theta^v(0, t^*) = \sum_{n=0}^{\infty} a_n \cdot (0)^n = 0 \tag{69}$$

l. c. Which requires that $a_0 = 0$, and

$$Q\left\{\frac{\partial \theta^v(1, t^*)}{\partial y}\right\} = G \sum_{n=0}^{\infty} n \cdot a_n = 0. \tag{70}$$

Substituting Eqs. (67) into Eq. (70) and solving for a_1 one obtains

$$a_1 = -\,(k_2/k_1)\, e^{t^*}\left[1 - e^{k_1 e^{-t^*}}\right]. \tag{71}$$

Use of Eqs. (67) and Eq. (71) will yield the remaining derivatives of the coefficients as

$$\dot{a}_2 = -\,(k_2/2)\, e^{k_1 e^{-t^*}} \tag{72}$$

$n \geq 3$:

$$\dot{a}_n = \left\{(-1)^{n-1}\, k_2\, k_1^{n-2}\, e^{[-(n-2)t^* + k_1 e^{-t^*}]}\right\}/n!. \tag{73}$$

Integration of Eqs. (71) and (73) subject to the initial conditions (68) gives

$$\begin{aligned} a_1 = {} & (k_2/k_1)\left[e^{t^*} e^{k_1 e^{-t^*}} - e^{t^*} - e^{k_1} + 1\right] \\ & + k_2\,[-E\,i(k_1\, e^{-t^*}) + E\,i(k_1)] + a_1(0), \end{aligned} \tag{74}$$

$$a_2 = (k_2/2)\,[E\,i(k_1\, e^{-t^*}) - E\,i(k_1)] + a_2(0) \tag{75}$$

$n \geq 3$:

$$Q_n = \frac{(-1)^n k_2}{n(n-1)(n-2)} \left[e^{k_1 e^{-t^*}} \sum_{m=0}^{n-3} \frac{(-1)^m k_1^{n-m-3} e^{-(n-m-3)t^*}}{(n-m-3)!} - e^{k_1} \sum_{m=0}^{n-3} \frac{(-1)^m k_1^{n-m-3}}{(n-m-3)!} \right] + a_n(0). \tag{76}$$

In order to establish convergence of the series for θ^v, the convergence of the first time derivative of θ^v will be proven, since in this case it is easier to work with, and assures convergence of the series. The ratio test, where Eq. (73) has been used to determine the ratio of the time derivatives of the coefficients, gives

$$\lim_{n\to\infty} \left| \frac{\dot{a}_n y^n}{\dot{a}_{n-1} y^{n-1}} \right| = \lim_{n\to\infty} \left| \frac{\dot{a}_n}{\dot{a}_{n-1}} \right| |y| = \lim_{n\to\infty} \left| \frac{k_1 e^{-t^*}}{n} \right| |y|. \tag{77}$$

Using the maximum value of y we obtain

$$\lim_{n\to\infty} \left| \frac{k_1 e^{-t^*}}{n} \right| \to 0 \tag{78}$$

for all positive t^*. Therefore the series converges uniformly and absolutely for all useful values of t^* and y.

The thermal stress, defined by Eqs. (8) and (42), is

$$\sigma_{yy} = e^{-t^*} G \alpha T_0 [1 - 3(x/c)^2]. \tag{79}$$

(d) Maxwell Body Case — Time Dependent Temperature Distribution

If $H(t^*) = e^{c_2 t^*}$ to obtain the same form as used in the elastic case and if $g(0) = G/\lambda$ as before, Eq. (63) leads to the solution

$$g = [G/\lambda(c_2 + 1)] [c_2 e^{c_2 t^*} + e^{-t^*}], \tag{80}$$

where the substitution $t^* = t/\tau$ has again been used. The special case of $c_2 = -1$ has not been considered since computations show that $|c_2| \ll 1$ in order to confine the temperatures to realistic values. In the case of aluminum [13] the values of τ range from $6.2 \cdot 10^{-5}$ sec at 200°C to $1.3 \cdot 10^{-11}$ sec at 550°C. Since the maximum value of $H(t^*)$ must be less than three for aluminum or twelve for steel, in order to keep the temperatures below the melting point of the metal, one can readily see that $|c_2| \ll 1$ for any useful time range considered.

If the operators defined by Eq. (62) due used, Eq. (49) becomes

$$\frac{\partial^3 \theta^v}{\partial y^2 \partial t^*} + k_1 g \frac{\partial^2 \theta^v}{\partial y \partial t^*} + k_1 (\dot{g} + g) \frac{\partial \theta}{\partial y} = -k_2. \tag{81}$$

The solution of this differential equation with $g(t^*)$ as defined by Eq. (80), by a method similar to that used in the previous sections, leads to cumbersome high order integral equations for the coefficients of the series.

Therefore, in order to be able to solve Eq. (81) without recourse to numerical methods, the function $g(t^*)$ and its first time derivative were approximated by step functions. A similar procedure has been used in Reference [9]. The resulting differential equation is then solved in each time interval used in defining the step functions. For any interval

$$t_q^* \leq t^* < t_{q+1}^* \quad q = 0, 1, 2, \ldots \tag{82}$$

let the value of $t^* = t_q$ be used to calculate the step function of $g(t^*)$ and its time derivative where

$$\bar{t} = [t_q^* + t_{q+1}^*]/2. \tag{83}$$

Then in any time interval the step functions will be defined as

$$g(t^*) = g(t_q) = A_q \tag{84}$$

and

$$\dot{g}(t^*) = \dot{g}(t_q) = C_q. \tag{85}$$

In order to simplify the writing of the differential equation and its solution, let

$$B_q = 1 + C_q/A_q. \tag{86}$$

The resulting solutions $\theta_q(y, t^*)$ will then piecewise approximate the function $\theta^v(y, t^*)$. In order that the step functions θ_g represent a continuous function θ^v, one must require that the end conditions of each interval represent the initial conditions of each succeeding interval

$$\begin{aligned}
&\lim_{t^* \to t_{q+1}^*} \theta_q(y, t^*, A_q, B_q) = \theta_{q+1}(y, t_{q+1}^*, A_{q+1}, B_{q+1}) \\
&\lim_{t^* \to t_{q+1}^*} \frac{\partial \theta_q}{\partial y}(y, t^*, A_q, B_q) = \frac{\partial \theta_{q+1}}{\partial y}(y, t_{q+1}^*, A_{q+1}, B_{q+1}) \\
&\lim_{t^* \to t_{q+1}^*} \frac{\partial^2 \theta_q}{\partial y^2}(y, t^*, A_q, B_q) = \frac{\partial^2 \theta_{q+1}}{\partial y^2}(y, t_{q+1}^*, A_{q+1}\, B_{q+1}).
\end{aligned} \tag{87}$$

The Laplace transform is defined as

$$L\left\{\frac{\partial \theta_q}{\partial y}(y, t^*)\right\} = \frac{\partial \bar{\theta}_q}{\partial y}(y, p). \tag{88}$$

Application of the Laplace transform to Eq. (81) yields for each interval

$$\begin{aligned}
p\,\frac{\partial^2 \theta_q}{\partial y^2} - \frac{\partial^2 \theta_q}{\partial y^2}(y, 0) + k_1 A_q (p + B_q) \frac{\partial \theta_q}{\partial y} \\
- k_1 A_q \frac{\partial \theta_q}{\partial y}(y, 0) = -k_2/p.
\end{aligned} \tag{89}$$

In each interval the time variable is changed to

$$\tilde{t} = t^* - t_q^* \quad (t_q^* \leq t^* < t_{q+1}^*) \tag{90}$$

so that each interval begins with $\tilde{t} = 0$ and the Laplace transform may then be properly evaluated. The initial condition of the first interval is the equivalent elastic solution at $t^* = 0$ and the initial conditions of each succeeding interval are given by Eq. (87). The equation describing the equivalent elastic solution is obtained from Eq. (59) as

$$\frac{\partial^2 \theta^e}{\partial y^2} + k_1 A_0 \frac{\partial \theta}{\partial y} = - k_2 . \tag{91}$$

Substitution of Eq. (91) into Eq. (89) gives for the first interval

$$\frac{\partial^2 \bar{\theta}_0}{\partial y^2} + \frac{k_1 A_0}{p} (p + B_0) \frac{\partial \bar{\theta}_0}{\partial y} = - \frac{k_2}{p} (p + 1) . \tag{92}$$

The boundary condition requiring the shear at the free end of the plate to be zero is

$$\frac{\partial^2 \theta}{\partial y \, \partial \tilde{t}} (1, \tilde{t}) = 0 \tag{93}$$

and the LAPLACE transform of this boundary condition yields

$$p \frac{\partial \bar{\theta}_q}{\partial y} (1, p) = \frac{\partial \theta_q}{\partial y} (1, 0) . \tag{94}$$

For the first time interval the right side of Eq. (94) is the elastic condition and is equal to zero due to Eq. (70). The solution of Eq. (92) subject to the boundary condition, Eq. (94) with the right hand member equal to zero, is

$$\frac{\partial \bar{\theta}_0}{\partial y} = \frac{k_2}{k_1 A_0} \frac{p + 1}{p (p + B_0)} \left[- 1 + e^{\frac{k_1 A_0}{p} (p + B_0)(1 - y)} \right] . \tag{95}$$

The solution of Eq. (89) for all other intervals except the first, subject to the boundary condition of Eq. (94), is

$$\frac{\partial \bar{\theta}_q}{\partial y} = \frac{k_2}{k_1 A_q p (p + B_q)} \left[e^{\frac{k_1 A_q}{p} (p + B_q)(1 - y)} - 1 \right] \tag{96}$$

$$- \frac{K_q}{k_1 A_q (p + B_q)} \left[e^{\frac{k_1 A_q}{p} (p + B_q)(1 - y)} - 1 \right] + \frac{1}{p} \frac{\partial \theta_q (1, 0)}{\partial y} e^{\frac{k_1 A_q}{p} (p + A_q)(1 - y)}$$

$$(q > 0) ,$$

where

$$K_q = \lim_{t^* \to t_q^*} \left[\frac{\partial^2 \theta_{q-1}}{\partial y} (y, t^*) + k_1 A_q \frac{\partial \theta_{q-1}}{\partial y} (g, t^*) \right] \tag{97}$$

the value of $\frac{\partial \theta_q}{\partial y} (1, 0)$ must be obtained from the final value of θ_{q-1}. Eq. (95) is expanded in an infinite series and the inverse LAPLACE transform used to obtain the unknown function

$$\frac{\partial \theta_0}{\partial y} = k_2 \sum_{n=0}^{\infty} \frac{(k_1 A_0)^n (1 - y)^{n+1}}{(n + 1)!} \left[L_n (- B_0 t^*) + \frac{t^*}{n + 1} L_n^1 (- B_0 t^*) \right.$$

$$(0 \leq t^* < t_1^*) . \tag{98}$$

Similarly Eq. (96) is expanded and inverted to obtain

$$\frac{\partial\theta_q}{\partial y}=\sum_{n=0}^{\infty}\frac{(k_1 A_q)^n (1-y)^{n+1}}{(n+1)!}\Big\{k_2 L_n[B_q(t_q^*-t^*)] - \frac{K_q(t^*-t_q^*)}{n+1} L_n^2[B_q(t_q^*-t^*)]+\frac{n+1}{1-y}\frac{\partial\theta_q(1,t_q^*)}{\partial y} L_n[B_q(t_q^*-t^*)]\Big\} \tag{99}$$

$$q>0$$
$$t_q^*\leq t^*<t_{q+1}^*$$

From Eq. (98), for any value of t^* in that interval including in the limit the end point one obtains

$$\frac{\partial\theta_0}{\partial y}(1,t^*)=0 \qquad (0\leq t^*<t_1^*). \tag{100}$$

Then from Eq. (99) by induction, starting with $\partial\theta_1/\partial y$, one finds that for any value of t^* in that interval including in the limit the end point

$$\frac{\partial\theta_q}{\partial y}(1,t^*)=0 \qquad q>0,\quad t_q^*\leq t^*<t_{q+1}^* \tag{101}$$

provided that K_q is finite which will be shown subsequently.

The consequence of Eq. (101) is to reduce Eq. (99) to

$$\frac{\partial\theta_q}{\partial y}=\sum_{n=0}^{\infty}\frac{(k_1 A_q)^n (1-y)^{n+1}}{(n+1)!}\Big\{k_2 L_n[B_q(t_q^*-t^*)] - \frac{K_q(t^*-t_q^*)}{n+1} L_n^1[B_q(t_q^*-t^*)]\Big\} \qquad (t_q^*\leq t^*<t_{q+1}^*). \tag{102}$$

If Eq. (98) and Eq. (97) are used, the value of K_q for the second interval is

$$K_1=k_2\sum_{n=0}^{\infty}\frac{(k_1 A_0)^n (1-y)^n}{n!}\Big[\frac{k_1 A_1(1-y)}{n+1}-1\Big]\Big[L_n(-B_0 t_1^*)+\frac{t_1^*}{n+1}L_n^1(-B_0 t_1^*)\Big]. \tag{103}$$

Then Eq. (102), for the second time interval, results in

$$\begin{aligned}\frac{\partial\theta_1}{\partial y}&=k_2\sum_{n=0}^{\infty}\frac{(k_1 A_1)^n (1-y)^{n+1}}{(n+1)!}L_n[B_1(t_1^*-t^*)]\\&-k_2\sum_{n=0}^{\infty}\sum_{m=0}^{\infty}\frac{(k_1 A_1)^{m+1}(k_1 A_0)^{n-m}(t^*-t_1^*)}{(m+1)(m+1)!(n-m+1)!}L_m^1[B_1(t_1^*-t)]\,(1-y)^{n+1}\\&\times\Big\{L_{n-m}(-B_0 t_1^*)+\frac{t_1^*}{n-m+1}L_{n-m}^1(-B_0 t_1^*)\\&-k_1 A_0\Big[L_{n-m+1}(-B_0 t_1)+\frac{t_1}{n-m+2}L_{n-m+1}(-B_0 t_1^*)\Big]\Big\}\\&+k_2\sum_{n=0}^{\infty}\frac{(k_1 A_1)^{n+1}(1+t_1^*)(t^*-t_1^*)(1-y)^{n+1}}{(n+1)(n+1)!}L_n^1[B_1(t_1^*-t)].\end{aligned} \tag{104}$$

Subsequent use of Eqs. (97), (102) and (104) will lead to $\partial\theta_q/\partial y$ for the third time interval. The process is then repeated to obtain the desired number of $\partial\theta_q/\partial y$ terms.

The other boundary condition which requires zero twist at the rigid support, or

$$\theta(0, t^*) = 0, \tag{105}$$

when combined with direct integration of Eqs. (98) and (104) results in θ_q for the first two intervals.

$$\theta_0 = k_2 \sum_{n=0}^{\infty} \frac{(k_1 A_0)^n [1-(1-y)^{n+2}]}{(n+2)} \left[L_n(-B_0 t^*) + \frac{t^*}{n+1} L_n^1(-B_0 t^*) \right. \quad 0 \leq t^* < t_1^* \tag{106}$$

and

$$\theta_1 = k_2 \sum_{n=0}^{\infty} \frac{(k_1 A_1)^n [1-(1-y)^{n+2}]}{(n+2)!} L_n[B_1(t_1^* - t^*)]$$

$$+ k_2 \sum_{n=0}^{\infty} \sum_{m=0}^{n} \frac{(k_1 A_1)^{m+1} (k_1 A_0)^{n-m} (t^* - t_1^*) [1-(1-y)^{n+2}]}{(n+2)(m+1)(m+1)(n-m+1)!} L_m^1 [B_1(t_1^* - t^*)]$$

$$\times \left\{k_1 A_0 \left[L_{n-m+1}(-B_0 t_1^*) + \frac{t_1^*}{n-m+2} L_{n-m+1}(-B_0 t_1^*)\right.\right.$$

$$\left.\left. - L_{n-m}(-B_0 t_1^*) - \frac{t_1^*}{n-m+1} L_{n-m}^1(-B_0 t_1^*)\right]\right\}$$

$$+ k_2 \sum_{n=0}^{\infty} \frac{(k_1 A_1)^{n+1} (1+t_1^*)(t^* - t_1^*) [1-(1-y)^{n+2}]}{(n+1)(n+2)!} L_n[B_1(t_1^* - t^*)]$$

$$(t \leq t^* < t_2^*). \tag{107}$$

Direct integration of all succeeding $\partial\theta_q/\partial y$ and application of the boundary condition Eq. (105) will result in θ_q for each time interval desired. The existing thermal stress is obtained from Eqs. (8) and (42) as

$$\sigma_{yy} = G \alpha T_0 [1 - 3(x/c)^2] [c_2 e^{c_2 t^*} + e^{-t^*}]/(c_2 + 1). \tag{108}$$

The numerical results will be discussed in the subsequent section.

4. Discussion and Conclusions

In this analysis the general equations describing the angle of twist ofr a thin, flat, linear viscoelastic plate subjected to torque and thermal stresses have been derived. Particular solutions for the elastic case and MAXWELL body case using time dependent and independent temperature distributions have been presented. Numerical examples of the particular solutions of thermal stresses and angle of twist are presented in graphical form in Figs. (3), (4), (5), and (6) for the temperature variations considered and are in terms of non-dimensional quantities independent of the parametric value of the applied torque. Fig. (4) indicates that, for the

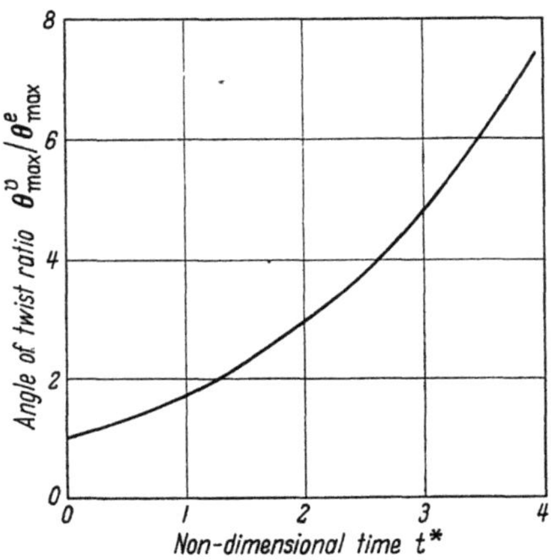

Fig. 3.

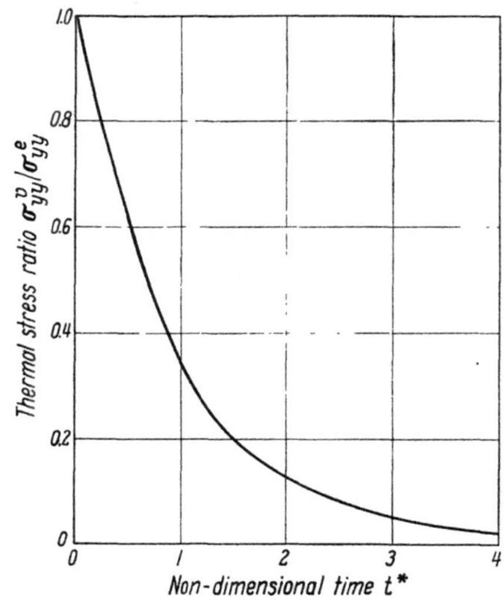

Fig. 4.

time independent temperature distribution considered, the thermal stresses decrease with time in the MAXWELL body plate due to relaxation. Therefore, the thermal stresses have a decreasing effect with time

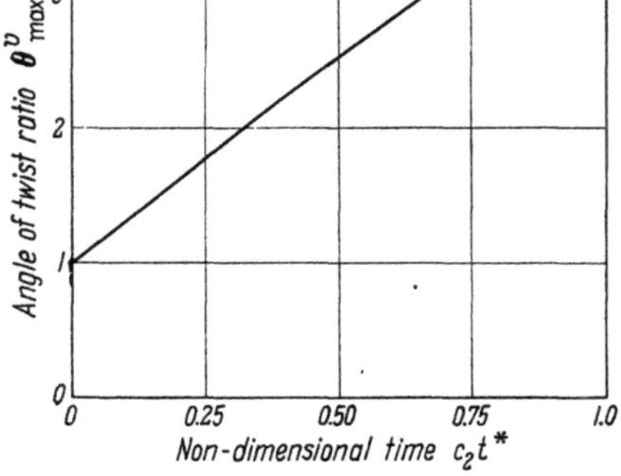

Fig. 5.

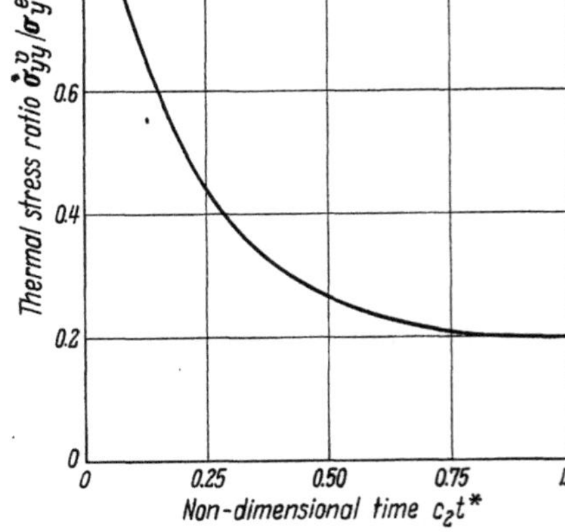

Fig. 6.

on the twist deformation Fig. (3). Figs. (5) and (6) indicate the same trend for the time dependent temperature variation. In Figs. (7) and (8) comparisons between angle of twist of the MAXWELL body plate due to applied torque and thermal stresses, and due to applied torque alone,

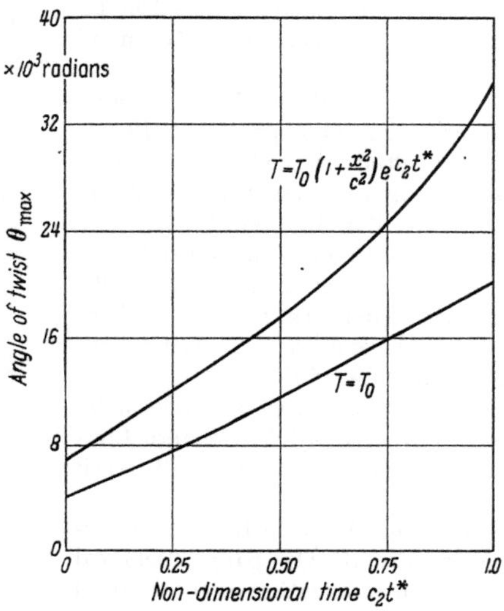

Fig. 7.

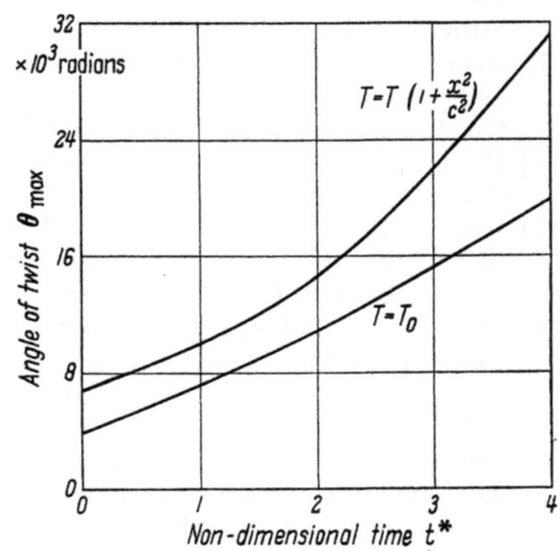

Fig. 8.

are presented for the two particular temperature variations considered. It is seen that in all cases the angle of twist increases in time due to the creep effects.

The consequence of increasing the temperature is to increase the creep rate. Increase in the temperature gradient will increase the torsion due to thermal stress and result in greater torsional deformations. The importance of the creep phenomenon may be seen from the constant temperature curves of Figs. (7) and (8). The thermal stress in this case is zero but a significant increase with time of the torsional deformation occurs. For the temperature gradients considered, creep predominates over the effects of the thermal stress. For a given combination of thermal stress and applied torque, the twist will increase in time, which will adversely affect creep divergence by shortening the critical time, and which will change torsional vibration frequencies.

References

[1] BISPLINGHOFF, R. L., H. ASHLEY and R. L. HALFMAN: Aeroelasticity, Cambridge: Addison-Wesley Publishing Co., Inc. Mass., pp. 7—13, (1955).

[2] HOFF, N. J.: High Temperature Effects in Aircraft Structures, Applied Mechanics Reviews, 8, No. 11, 453—456 (1955).

[3] GOODIER, J. N.: Elastic Torsion in the Presence of Initial Axial Stress, Journal of Applied Mechanics, **17**, No. 4, 383 (1950).
[4] BUDIANSKY, B., and J. MEYERS: Influence of Aerodynamic Heating on the Effective Torsional Stiffness of Thin Wings, Journal of Aero. Sciences, **23**, No. 12, 1081—1092, 1108 (1956).
[5] SINGER, J., and N. J. HOFF: Effect of the Change in Thermal Stresses Due to Large Deflections on the Torsional Rigidity of Wings, Journal of the Aeronautical Sciences, **24**, No. 4, 310—311 (1957).
[6] FREUDENTHAL, A. M.: The Inelastic Behavior of Engineering Materials and Structures, New York: John Wiley and Sons, Inc., 1950, pp. 305—319.
[7] ALLEN, H. F., M. A. BRULL and W. J. WILKIE: A Study of the Stress-Analysis and Structural-Testing Procedures Applicable to Aircraft Structures at Elevated Temperatures, WADC Technical Report 54—499, October 1954.
[8] HILTON, H. H.: On the Representation of Nonlinear Creep by a Linear Viscoelastic Model, Journal of the Aero/Space Sciences, **26**, No. 5, 311—312 (1959).
[9] HILTON, H. H., H. A. HASSAN and H. G. RUSSEL: Analytical Studies of Thermal Stresses in Media Possessing Temperature Dependent Viscoelastic Properties. WADC Technical Report 53—322, September 1953.
[10] HILTON, H. H.: Pitching Instability of Rigid Lifting Surfaces on Viscoelastic Supports in Subsonic or Supersonis Potential Flow, Third Midwestern Conference on Solid Mechanics, University of Michigan Press, pp. 1—19, April 1957.
[11] HILTON, H. H.: The Divergence of Supersonic Linear Viscoelastic Lifting Surfaces Including Chordwise Bending, Hughes Aircraft Company Technical Report ,S. F. 3.3—4, July, 1959. Also accepted for publication in the Journal of Aero/Space Sciences.
[12] NEOU, C. Y.: Direct Method of Determining Any Polynomial Stress Functions, Journal of Applied Mechanics, **24**, No. 3, 387—389 (1957).
[13] ZENER, C. M.: Elasticity and Anelasticity of Metals, First Edition, University of Chicago Press, pp. 147—158, Chicago 1948.
[14] HILTON, H. H.: Thermal stresses in thick-walled cylinders exhibiting temperature-dependent viscoelastic properties of the KELVIN type. Proc. of the Second US National Congress of Applied Mechanics, pp. 547—553, 1954.

Discussion

W. FLÜGGE, Stanford University: The authors solved their problem using the method published by NEOU [12]. Unfortunately, the basic idea of this method is in error. The fallacy can be shown in the following manner: When (26) is inserted into (28), (29) it follows that

$$\Sigma\,\Sigma\, a_{mn}\, m(m-1)\, c^{m-2}\, y^{n} \equiv 0$$

and

$$\Sigma\,\Sigma\, a_{mn}\, m\, n\, c^{m-1}\, y^{n-1} \equiv 0$$

for all values of y. Evidently, in each of the series the coefficient of every single power of y must vanish, hence

$$\sum_m m(m-1)\, c^{m-2}\, a_{mn} = 0$$

and

$$\sum_m m\, c^{m-1}\, a_{mn} = 0$$

for every n.

NEOU and the authors believe also that every coefficient of a power of c should be zero in these series, and their arguments are based on this erroneous conclusion.

E. H. LEE, Brown University: In the presentation, Professor HILTON mentioned that the solution given was for an idealized problem which incorporated features of particular interest within the framework of a simple analytical formulation. There are two idealizations which seem to me to be particularly significant, and I would appreciate the authors' comments on them. Firstly: the inclusion of thermal expansion, Eq. (10), with the assumption of incompressible deformation due to stress. Although for viscoelastic materials it is common for the dilatation due to stress to be small compared with shear strains, it seems to me that it is likely to be of the same order as thermal expansion, and so have an appreciable influence on thermal stresses.

The second point is concerned with the neglect of the influence of temperature on the viscoelastic operators. In general, such operators are highly temperature sensitive. Some work on stress analysis for viscoelastic materials with varying temperature on the basis of the temperature-time shift hypothesis has been carried out at Brown University[1], and the influence of temperature on viscoelastic behavior has been shown to have an appreciable effect on the stress variation in particular cases. Moreover, the inclusion of the temperature influence on viscoelastic properties emphasizes the importance of a viscoelastic relation which covers a wide frequency band, since moderate temperature changes can correspond to wide frequency variations.

N. J. HOFF, Stanford University: It is perfectly true that the large number of parameters available with a high-order linear viscoelastic model permits a very accurate reproduction of the strain-time curve obtained in a constant-stress creep test of any structural metal. This was shown convincingly by the senior author in an earlier paper [*8*]. However, this fact does not establish a complete analogy between the behavior of the metal and that of the linear model because changes in stress or temperature have non-linear effects in the metal and linear effects in the model.

Authors' closing remarks: The authors are grateful to Professor FLÜGGE for his comments. They agree with him that the general solution can be obtained if the coefficients of c are not allowed to vanish identically. The present solution satisfies the differential equation and all the boundary conditions except $\sigma^s_{yy}(x, 1) = 0$. The objections to the solution can be removed if the word "only" is deleted from the sentence preceding Eq. (30). The solution for σ^s_{yy} [Eq. (42)] agrees with the solution given by TIMOSHENKO and GOODIER in their "Theory of Elasticity", p. 400, for $\lambda = G = E/3$, which is the incompressible elastic case. The authors feel that the accuracy of their solution is adequate as long as they are interested only in the trends in the effects of creep on torsion.

The authors also wish to thank Professor LEE for his kind discussion, and are in general agreement that material compressibility and temperature dependence of material properties in the viscoelastic operators play an important role in the problem considered.

---

[1] MORLAND, L. W., and E. H. LEE: Stress analysis for linear viscoelastic materials with temperature variation. Technical Report No. 1, Contract NOrd 18594, Div. of Appl. Math., Brown University, Sept. 1959, (to appear in Trans. Soc. Rheology, 4, 1960).

MUKI, R., and E. STERNBERG: On transient thermal stresses in viscoelastic materials with temperature dependent properties. Technical Report No. 18, Contract 562(20), Div. of Engineering, Brown University, July, 1960.

However, the authors wished to study the effects of creep on the torsional response and compare the viscoelastic deflections to the corresponding elastic ones, i. e., incompressible and homogeneous. The inclusion of material compressibility would not affect the operators in Eq. (49), but would, as it was shown in Reference [*9*], alter the thermal stresses. The relatively simple solution of Eqs. (22) and (43). based on a separation of the variables, would no longer be possible and the angle of twist would be affected by the M_σ term.

The senior author has also studied the influence of material non-homogeneity due to temperature dependence of the viscoelastic operators [*9, 14*] and has found this to be a significant phenomenon. The inclusion of this effect would again lead to more complex solutions than the relatively simple ones based on the separation of the variables. It would affect both the thermal stresses and the angle of twist.

In a less idealized problem, such as a wing for instance, where the thermal stress field is more complex and the present relatively simple solutions can no longer be obtained, material compressibility and nonhomogeneity should certainly be included in the formulation of the problem. These two phenomena would influence the actual numerical values, but it is doubtful that they would seriously affect the observed trends in the growth with time of the angle of twist.

On the Creep-Stress Analysis of some Structures[1]

By

Sharad A. Patel and B. Venkatraman

Polytechnic Institute of Brooklyn, Brooklyn, New York, USA

Summary

The present paper considers the analysis of some structures in which the deformations are caused exclusively by creep. The paper begins with a discussion of a uniaxial creep law and its generalizations for the analyses of problems in a triaxial state of stress. Moment-curvature relations are then developed from these laws for problems in bending, and an elementary device for the inclusion of compressibility effects in creep is briefly discussed. With these as basis, the particular problems considered are, the bemding of thin-walled beams, the bending of annular plates, and the stress distribution in a compressible sphere. In analysing these problems, use has been made of the elastic analogue. Further, in some cases, the similarity of creep to plasticity has also been used. In all cases, closed form solutions have been presented.

1. Introduction

With the advent of supersonic and hypersonic flight, structural analysts are confronted with problems which differ from more classical ones. For example, exposure of structural components to high temperatures is more often encountered. The mechanical behavior of a component under such conditions differs radically from those with which the structural analyst is familiar. For a realistic design then, it is imperative that new analytical techniques be devised for incorporating the effects of new operating conditions such as creep.

---

[1] The results reported in this paper were obtained in the course of research under Contract No. AF (49) (638) — 302, sponsored by the Air Force Office of Scientific Research and under Contract No. Nonr 839(23), sponsored by the Office of Naval Research of the US Government. The authors are grateful to these organizations for permission to publish the results.

With the foregoing in view, the present paper considers the analysis of some structures subject to creep. The attempt has been made to keep the mathematics relatively simple in order to make the results more readily usable.

The assumption has been made that structures considered are at a uniform temperature and hence, no thermal stresses exist. Further, compared to the creep deformations elastic deformations are considered negligibly small. Under these assumptions, the creep problems of the structures are reduced to analogous nonlinear elastic problems. The nonlinear elastic analyses are then carried out for particular structures.

For clarity, Sec. 2 discusses the creep law and its generalized formulations in some detail. The section begins with a discussion of the elastic analogue of creep laws in one and three dimensions. These laws are then used to develop moment-curvature relations for the analyses of creep-bending problems. A brief discussion of the TRESCA criterion and the associated flow rule as a basis for creep analysis is also included. Finally. an elementary device for the inclusion of compressibility of the material is briefly outlined.

With the formulations of Sec. 2 as bases, the rest of the paper presents the creep analyses of some structures. In particular, Sec. 3 considers the bending of thin-walled beams. The bending of annular plates is analysed in Sec. 4. Finally, the creep-stress analysis of a compressible hollow sphere in Sec. 5 concludes the paper.

In all cases closed form solutions are given. It is hoped that the techniques developed can be extended to the analyses of other types of structural problems.

2. Basic Discussion

Most theoretical analyses of structures for creep stresses and deformations are developed on the bases of stress-strain-time relations empirically formulated from uniaxial creep test data. For the creep analysis of structures in which the stresses are essentially uniaxial, HOFF [1] has proposed the use of the relation

$$\varepsilon = (\sigma/\lambda)^n \, t^{1/q}, \tag{2.1}$$

where ε is the strain, σ is the stress, t represents time, and λ, n, and q are material constants deduced from tensile creep tests at constant temperatures. These constants assume different values for the primary and secondary phases of creep. Further, if n is assumed an odd integer, Eq. (2.1) applies equally in tension and compression.

When structures subject to creep in two or three dimensional states of stresses are analysed, it is necessary that a description of their mechanical behavior under combined stresses be known. Since such empirical formulations from creep test data under combined stresses are extremely difficult, it is usual to base the analyses on a generalized formulation of the uniaxial law. Alternatively, a suitable basis may be devised from theories which have proved successful for other types of materials. To this end, theories of plasticity prove a fruitful source.

In seeking to generalize the uniaxial law, a reasonable assumption from experimental evidence is made that creep flow is incompressible. Now, PRAGER [2] formulated a general stress-strain law for an incompressible isotropic elastic material. Based on this, HOFF [3] proposed the following relatively simple law for the creep analysis of structures subject to a triaxial state of stress:

$$\varepsilon_{ij} = C\, J_2^m\, s_{ij}\, t^{1/q}, \tag{2.2}$$

where C and m are material parameters related to the constants λ and n of uniaxial law (2.1), ε_{ij} is the strain tensor, $J_2 = \frac{1}{2} s_{ij} s_{ij}$ is the second invariant of the stress deviation tensor

$$s_{ij} = \sigma_{ij} - \frac{1}{3}\sigma_{kk}\,\delta_{ij}, \tag{2.3}$$

and δ_{ij} is the KRONECKER delta. The relations of C and m to λ and n are obtained by the stipulation that Eq. (2.2) reduce to Eq. (2.1) for the uniaxial case. Thus,

$$2m = n - 1$$

$$2C = 3^{\frac{n+1}{2}}/\lambda^n. \tag{2.4}$$

For the creep analysis of structures governed by Eq. (2.1) or its generalization Eq. (2.2), HOFF [3] proposed a further simplification. This is based on the observation that, if elastic deformations are considered negligibly small compared to creep deformations, a large class of problems governed by either Eq. (2.1) or Eq. (2.2) does not explicitly contain the displacements in the equations involved. Therefore, the mathematical problem is formally identical with one in which the stresses remain unchanged, while the velocities are everywhere replaced by displacements. According to HOFF [3], it follows that the creep laws (2.1) and (2.2) may be replaced, respectively, by the elastic laws

$$\varepsilon = (\sigma/\lambda)^n, \tag{2.5}$$

$$\varepsilon_{ij} = C\, J_2^m\, s_{ij}, \tag{2.6}$$

with any velocity boundary conditions being replaced by corresponding conditions on displacement. The constants λ, n, C, and m retain the same values as in Eqs. (2.1) and (2.2). If the nonlinear elastic problems based on Eqs. (2.5) and (2.6) can be solved, then by a mere reversal of the above substitutions the creep solution is obtained. Thus, while the stress solutions for the two problems are identical, the elastic and creep strains are related by

$$(\varepsilon_{ij})_{cr} = (\varepsilon_{ij})_{el}\, t^{1/q}. \tag{2.7}$$

It follows that the velocities are obtained by the relation

$$u^q = \frac{d}{dt}\,(u_{cr}^q), \tag{2.8}$$

where u and u_{cr} are the elastic and creep displacements, respectively.

While the generalized flow law Eq. (2.2) or its elastic analogue Eq. (2.6) does provide a relatively simple basis for the creep analysis of structures subjected to combined stresses, there are problems where even the use of this law leads to intractable mathematical equations. In such instances, an attempt is made to obtain a solution by the adaptation of theories which have proved successful in plasticity. Such an approach was used by WAHL and his co-authors [*4*] in the creep analysis of rotating disks.

Now, the yield condition for a perfectly plastic material is the critical combination of the stress components which initiates plastic flow. Since this stipulation of the critical combination of stresses is not sufficient to characterize the mechanical behavior of the material, it is supplemented by a flow rule for the continuation of plastic flow. Thus, the yield condition is represented geometrically in a stress space whose coordinates are the stress variables. The plasticity flow law then states that the strain rate vector must be in the direction normal to the yield surface. For a perfectly plastic material the length of this vector remains undetermined, while for a strain hardening material an additional relation between the strain rate vector and the stress intensity must be provided.

Since the MISES and TRESCA criteria are the two most commonly used ones in plasticity, it is reasonable to adapt these to the analysis of creep problems. It is well known that in generalized plane stress problems of rotational symmetry, while the MISES condition is represented by an ellipse in a stress space whose coordinates are the principal stresses, the TRESCA condition is represented by a hexagon. Since the ellipse has a uniquely defined normal at every point on the circumference, the MISES flow states that the strain rate vector must be in the direction of this uniquely defined normal. On the other hand, for the TRESCA hexagon, while the strain rate vector must be in the direction of the uniquely defined normal at each point on the sides of the hexagon, at the corners where the normals are not unique, the strain rate vector may take on any direction between the limiting normals.

In using this approach in the formulation of creep flow laws, it is first observed that creep is assumed to be present at any non-zero stress intensity. Although, effectively, the yield surface is now reduced to a point, various regimes in the stress space retain their meaning, since these regimes are distinguished by different forms for the flow laws. It can be verified [*5*] that the use of the nonlinear elastic law (2.6) is formally identical to the use of the Mises criterion and the associated flow rule. The use of the Tresca criterion and the associated flow rule will be briefly discussed, with particular reference to circular plates, later in the section.

When it comes to the creep-bending analysis of beams and plates, the stress-strain laws (2.5) and (2.6) may be extended to include bending. In other words, with the use of these laws, basic moment-curvature relations can be derived for the anlysis.

For the case of a beam, the usual simplified assumptions of the elementary beam theory are made. In particular, it is assumed that shearing and extensional deformations are negligibly small compared to bending deformations, and that sections plane and normal to the longitudinal axis remain plane and normal to the deformed longitudinal axis. Thus, if y' denotes the distance from the neutral plane, the relation between the strain ε and the curvature $\varkappa$ is

$$\varepsilon = \varkappa\, y' . \tag{2.9}$$

This equation may be combined with the unidimensional stress-strain relation (2.5) to yield the bending moment about the neutral axis x'

$$M_{x'} = \int_A \sigma\, y'\, dA = \lambda\, \varkappa^{1/n} \int_A y'^{1+1/n}\, dA . \tag{2.10}$$

Therefore, the moment-curvature relation assumes the form

$$\varkappa = \left(\frac{M_{x'}}{\mu_{x'}}\right)^n , \tag{2.11a}$$

where

$$\mu_{x'} = \lambda \int_A y'^{1+1/n}\, dA . \tag{2.11b}$$

When it comes to the bending of thin plates, problems with stresses in more than one direction arise. It is therefore necessary to develop moment-curvature relations similar to Eq. (2.11). The development of these relations may be based on general stress-strain relations such as Eq. (2.6). The moment-curvature relation is now derived on the basis of this equation. It is assumed that the small-deflection theory of plates is valid.

The bending of thin plates out of their planes is a problem in generalized plane stress. Therefore, the normal stresses and shear stresses in the direction of the coordinate perpendicular to the plane of the plate are identically zero. Now, from the stress-strain law (2.6) and the definition $J_2 = \frac{1}{2} s_{ij} s_{ij}$ it can be readily seen that

$$J_2 = (\varepsilon_{ij}\, \varepsilon_{ij}/2C)^{\frac{1}{2m+1}}. \tag{2.12}$$

It follows then that

$$s_{ij} = \frac{\varepsilon_{ij}}{C^{\frac{1}{2m+1}} (\varepsilon_{ij}\, \varepsilon_{ij}/2)^{\frac{m}{2m+1}}}. \tag{2.13}$$

If it is assumed that the third coordinate is perpendicular to the plate, then the stress terms with the number 3 in their subscripts vanish identically. Let the distance from the neutral plane of the plate in the direction of this coordinate be denoted by z. Then, using the definition of s_{ij} in Eq. (2.3) it can be shown that Eq. (2.13) can be transformed to read

$$\sigma_{ij} = \frac{\varepsilon_{ij} + \varepsilon_{kk}\, \delta_{ij}}{C^{\frac{1}{2m+1}} \left(\frac{\varepsilon_{ij}\, \varepsilon_{ij} + \varepsilon_{ii}\, \varepsilon_{jj}}{2}\right)^{\frac{m}{2m+1}}},$$

$$= \frac{(\varkappa_{ij} + \varkappa_{kk}\, \delta_{ij})\, z^{\frac{1}{2m+1}}}{C^{\frac{1}{2m+1}} \left(\frac{\varkappa_{ij}\, \varkappa_{ij} + \varkappa_{ii}\, \varkappa_{jj}}{2}\right)^{\frac{m}{2m+1}}}, \quad i, j = 1, 2, \tag{2.14}$$

where the strain-curvature relation $\varepsilon_{ij} = z\, \varkappa_{ij}$ has been introduced. Eq. (2.14) is now used to obtain the moment-curvature relation as follows.

The moment tensor M_{ij} per unit length is given by the definition

$$M_{ij} = \int_{-h/2}^{+h/2} z\, \sigma_{ij}\, dz, \tag{2.15}$$

where h is the uniform thickness of the plate. Eqs. (2.14) and (2.15) are now combined and simplified to yield

$$M_{ij} = \frac{(\varkappa_{ij} + \varkappa_{kk}\, \delta_{ij})}{C^{\frac{1}{2m+1}} \left(\frac{\varkappa_{ij}\, \varkappa_{ij} + \varkappa_{ii}\, \varkappa_{jj}}{2}\right)^{\frac{m}{2m+1}}} \int_{-h/2}^{h/2} z^{\frac{2m+2}{2m+1}}\, dz,$$

$$= K \frac{\varkappa_{ij} + \varkappa_{kk}\, \delta_{ij}}{\left(\frac{\varkappa_{ij}\, \varkappa_{ij} + \varkappa_{ii}\, \varkappa_{jj}}{2}\right)^{\frac{m}{2m+1}}}, \quad i, j = 1, 2, \tag{2.16}$$

where

$$K = \frac{(4\,m + 2)}{(4\,m + 3)} \frac{(h/2)^{\frac{4m+3}{2m+1}}}{C^{\frac{1}{2m+1}}}. \tag{2.17}$$

With the use of Eq. (2.16) it can be shown that the invariant

$$(\varkappa_{ij}\,\varkappa_{ij} + \varkappa_{ii}\,\varkappa_{jj})/2 = [(3\,M_{ij}\,M_{ij} - M_{ii}\,M_{jj})/6\,K^2]^{2m+1}, \quad (2.18)$$

so that the curvature

$$\varkappa_{ij} = \left[\frac{(3\,M_{ij}\,M_{ij} - M_{ii}\,M_{jj})}{2\,\mu^2}\right]^m \frac{(3\,M_{ij} - M_{kk}\,\delta_{ij})}{\mu}, \quad i, j = 1, 2, \quad (2.19)$$

where

$$\mu = K\sqrt{3}\,(\sqrt{3})^{\frac{1}{2m+1}}. \quad (2.20)$$

Having developed the moment-curvature law (2.19) from the generalized law (2.6) as one basis for the bending analysis of plates, attention is now turned to an alternative formulation by the use of the Tresca criterion and the associated flow rule. Such a formulation for the analysis of circular plates was discussed in detail in Ref. [*6*]. For the purpose of clarity, a brief review is given in the following paragraph.

In the use of the Tresca criterion in generalized plane stress problems of rotational symmetry, the stress space is distinguished by twelve regimes. Based on this, moment-curvature rate relations (flow rule) corresponding to each regime (Fig. 6) were tabulated in Ref. [*6*] (see p. 166). For instance, the curvature rate corresponding to BOC is

$$\dot{\varkappa}_\vartheta = F(M_\vartheta)\,G(t), \quad (2.21)$$

where $F(M_\vartheta)$ is a function of M_ϑ only and $G(t)$ is a function of time. If it is assumed that

$$F(M_\vartheta) = (M_\vartheta/\mu)^{2m+1} \text{ and } G(t) = (1/q)\,t^{(1-q)/q}, \quad (2.22)$$

then a simple integration of Eq. (2.21) shows that

$$\varkappa_\vartheta = (M_\vartheta/\mu)^{2m+1}\,t^{1/q}. \quad (2.23)$$

It is readily seen that

$$\varkappa_\vartheta = (M_\vartheta/\mu)^n, \quad (2.24)$$

is the elastic analogue of Eq. (2.23). Therefore, in the analysis of the creep problem by the elastic analogue, all the moment-curvature rate relations listed in the table of Ref. [*6*] can be used as moment-curvature relations.

The generalized stress-strain and moment-curvature relations discussed thus far may be used in the analysis of structures whose material is incompressible. For such materials, the deformations are produced only by the deviations of stresses [Eq. (2.3)] from the hydrostatic state of stress [*7*]. On the other hand, for compressible materials even hydrostatic pressures will produce deformations. Therefore, if it is desired to include the effects of compressibility, the definition of the deviations of stresses given in Eq. (2.3) requires modification. One way of incorporating the effect of compressibility is as follows.

Let the stress deviations s_{ij} be defined as

$$s_{ij} = \sigma_{ij} - \alpha\, \sigma_{kk}\, \delta_{ij}, \tag{2.25}$$

where α, a compressibility constant, has replaced the factor 1/3 in Eq. (2.3). If the strains ε_{ij} are now required to depend upon this definition of s_{ij} in Eq. (2.25), the stress-strain law (2.6) will remain formally valid for elastic compressible materials. While the value of $\alpha = 1/3$ corresponds to incompressible materials, the fact that a hydrostatic pressure cannot increase the volume of the material shows that $\alpha \leq 1/3$. Finally, if the analysis is restricted to materials which do not expand laterally under longitudinal tension, α cannot be negative so that, $0 \leq \alpha \leq 1/3$.

Based on the formulations discussed in this section, particular problems outlined in the Introduction will be analysed in the remainder of this paper.

3. Bending of Thin-Walled Beams

(a) Thin-walled Open Sections

When thin-walled open section beams are subjected to bending, the bending and shear stress distributions and the corresponding deformations vary with the type of cross section considered. Since there are a number of variables involved in this problem, a relatively general analysis of an arbitrary section proves mathematically difficult. It is therefore usual to consider specific types of cross sections that are in current use in practical structures. With this in view, the investigation here is restricted to beams having channel sections. It is assumed that the material of the beam is governed by the stress-strain law (2.5), and the beam theory assumptions discussed in Sec. 2 apply for this cross section.

The stress analysis of a channel section beam may be conveniently begun by a consideration of the section under the action of a bending moment. As shown in Fig. 1, the section is idealized by its median line. The moment vector M acts in the plane of the cross section and at a given angle β to the axis of symmetry x. Since the material of the beam is assumed to follow a nonlinear stress-strain law, the neutral axis of the cross section does not pass through the centroid of the section as opposed to the case of the classical beam theory. The first step in the analysis is therefore to determine the location of this neutral axis.

Let the neutral axis make an angle α_0 with the axis of symmetry x. The point of intersection of these two axes is then chosen to be the origin 0 of both coordinate systems x, y, and x', y'. The sign convention is such that an angular coordinate is assumed positive in the counter-clockwise direction from the axis of symmetry x. It is apparent that the x', y' coordinate system may be obtained by rotating the x, y coordinate system through an angle α_0. The determination of α_0 and x_0 (Fig. 1) as functions of the given quantities β and M locates the neutral axis x'.

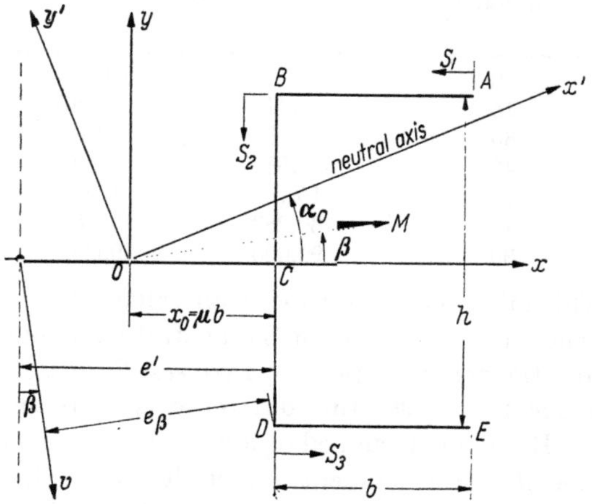

Fig. 1. Channel section

Since the beam is subjected to a pure bending moment, the stress distribution σ must give rise to a resultant moment M (or its components $M_{x'}$ and $M_{y'}$) and have a zero resultant force. These equilibrium conditions may be expressed with the use of Eqs. (2.9) and (2.10) in the form

$$M_{x'} = \int_A \sigma\, y'\, dA = \lambda\, \varkappa^{1/n} \int_A y'^{1+1/n}\, dA, \tag{3.1}$$

$$-M_{y'} = \int_A \sigma\, x'\, dA = \lambda\, \varkappa^{1/n} \int_A x'\, y'^{1/n}\, dA, \tag{3.2}$$

$$\sigma = M_{x'}\, y'^{1/n} \Big/ \int_A y'^{1+1/n}\, dA, \tag{3.3}$$

$$\int_A \sigma\, dA = \lambda\, \varkappa^{1/n} \int_A y'^{1/n}\, dA = 0, \text{ i.e., } \int_A y'^{1/n}\, dA = 0. \tag{3.4}$$

Fig. 1 shows that $-M_{y'}/M_{x'} = \tan(\alpha_0 - \beta)$. Consequently, Eqs. (3.1) and (3.2) may be combined to yield the relation

$$\tan(\alpha_0 - \beta) = \frac{\int^A x'\, y'^{1/n}\, dA}{\int_A y'^{1+1/n}\, dA} = -\frac{M_{y'}}{M_{x'}}. \tag{3.5}$$

The coupled nonlinear Eqs. (3.4) and (3.5) may be used to determine α_0 and x_0 and thus, the neutral axis x'. It can be shown that for $n = 1$ the neutral axis will always pass through the centroid of the section ($x_0 = 0$) and the solution for α_0 is reduced to that of the linear case. For other values of n, the solutions are somewhat involved and hence, only the value $n = 3$ for a particular channel section ($h/2\,b = 1$) has been considered. The variation of β with α_0 is given in Table 1 for $n = 1$ and $n = 3$.

Table 1. *Neutral Axis and Shear Center Variation* ($h/2\,b = 1$)

α_0, degrees	β degrees		e'/b	
	$n=1$	$n=3$	$n=1$	$n=3$
0	0	0	0.375	0.350
15	2.417	0.967	0.375	0.349
30	5.133	2.167	0.375	0.348
45	8.883	4.067	0.375	0.342
60	15.133	12.083	0.375	0.286
75	30.283	30.750	0.375	0.211
90	90.000	90.000	0.375	—

This table shows that for small values of β, there is a considerable difference in the location of the neutral axis obtained by the linear ($n = 1$) and the nonlinear ($n = 3$) analyses. For values of β approximately in the range 20° to 90° this difference is quite small.

Having determined α_0 and x_0, i. e. the neutral axis for a given direction β of the moment vector M, Eq. (3.3) and the relations between coordinates x, y, and x', y', may be used to obtain the bending stress distribution

$$\sigma = \frac{M\,[y \cos \alpha_0 - x \sin \alpha_0]^{1/n} \cos (\alpha_0 - \beta)}{\int\limits_A (y \cos \alpha_0 - x \sin \alpha_0)^{1+1/n}\, dA}\,. \tag{3.6}$$

In the preceding analysis of the channel section, the moment vector M was assumed to make an arbitrary angle β with the axis of symmetry. The resulting Eqs. (3.4) and (3.5) in α_0 and x_0 were difficult to solve for general values of n. However, for the particular cases when the vector M is parallel to the axis of symmetry ($M = M_x$, $M_y = 0$, $\beta = 0°$) or perpendicular to the axis of symmetry ($M = M_y$, $M_x = 0$, $\beta = 90°$), the analyses are considerably simpler. For example, when the moment vector is applied perpendicular to the axis of symmetry i. e., $\beta = 90°$, it can be shown that Eq. (3.5) is satisfied for $\alpha_0 = 90°$ and Eq. (3.4) assumes the relatively simple form

$$(x_0/b)^{1+1/n} - (1 + x_0/b)^{1+1/n} - (1 + 1/n)\,(h/2\,b)\,(x_0/b)^{1/n} = 0\,. \tag{3.7}$$

This equation determines x_0/b for given values of $(h/2\,b)$ and n. For the particular value of $h/2\,b = 1$, the corresponding values of n and x_0/b are shown in Table 2.

Table 2. *Neutral Axis and Shear Center for Channel Section*

n	1	3	7	15	31	51	101	∞
$-(x_0/b)_{\beta=90°,\ \alpha_0=90°}$	0.250	0.155	0.095	0.054	0.029	0.018	0.010	0
$(e'/b)_{\beta=0,\ \alpha_0=0}$	0.375	0.350	0.341	0.337	0.335	0.334	0.334	0.333

The analysis in this section has thus far been restricted to the investigation of bending stresses. Now, attention is turned to the determination of the shear stress distribution τ. With reference to Fig. 1, the shear force V and the resulting moment vector M act at an angle β to the y and x axes, respectively. Hence, if the force, equilibrium in the longitudinal direction of the beam is considered, it can be readily shown that

$$\tau = -\frac{V \cos(\alpha_0 - \beta) \int_0^s (y \cos\alpha_0 - x \sin\alpha_0)^{1/n}\, t\, ds}{t \int_A (y \cos\alpha_0 - x \sin\alpha_0)^{1+1/n}\, dA}. \tag{3.8}$$

With the shear stresses determined, the location of the shear center is investigated as the next step. By definition, the shear center is the point through which the shear force V must act so that the resulting shear stresses give rise to a zero net twisting moment. It is well known that in the linear elastic theory the shear center is a purely geometric property of the section i. e., independent of the direction of the force V. Whether this is also the case in the nonlinear elastic theory requires verification.

With the geometrical quantities illustrated in Fig. 1, the condition of zero twisting moment may be written in the form

$$h \int_A^B \tau\, t\, ds + V e_\beta = 0. \tag{3.9}$$

Here, since the twisting moments are taken about the point D, it is apparent that the shear stresses in the lower flange and the web make zero contributions. Eqs. (3.8) and (3.9) determine e_β. Further, it can be shown that e_β and e', the distance on the x axis of the point of intersection of the axis and the shear force V, have the trigonometric relation

$$e_\beta/b = \cos\beta\,[e'/b - (h/2\,b)\tan\beta]. \tag{3.10}$$

Thus, the elimination of e_β/b between Eqs. (3.9) and (3.10) yields

$$e'/b = (h/2\,b)\left\{\tan\beta + 2\,\frac{\cos(\alpha_0-\beta)}{\cos\beta} \times \frac{\cos\alpha_0 \int_0^b \int_0^s [h/2 - (x_0 + b - s)\tan\alpha_0]^{1/n}\, t\, ds\, ds}{\int_A (y\cos\alpha_0 - x\sin\alpha_0)^{1+1/n}\, ds}\right\}. \tag{3.11}$$

If $n = 1$, the Eq. (3.11) reduces to the simple result

$$(e'/b)_{n=1} = 3/2\,(3 + h/2\,b). \tag{3.12}$$

It is apparent from Eq. (3.11) that e' (Fig. 1) is not only a geometric property of the section but also a function of the direction β of the shear force V. In other words, if the channel section of a nonlinear elastic material is to have a zero resultant twisting moment, the distance from the center of the web to the point of intersection of the shear force and the axis of symmetry varies with the direction of the shear force. On the other hand, if the section is of a linear elastic material, Eq. (3.12) shows that this distance is independent of the direction of the shear force and locates the shear center for a given geometry of the cross section. For $n > 1$, the integrals in Eq. (3.11) may be evaluated numerically for given values of $h/2b$ and β (the corresponding values of α_0 and x_0 being evaluated from Eqs. (3.4) and (3.5)). As a typical illustration, the variation of e'/b with β for $n = 3$ and $h/2b = 1$ are shown in Table 1.

It is of interest to point out that the preceding analysis is considerably simplified when $\beta = 0$. For this case Eq. (3.11) yields

$$(e'/b)_{\substack{\beta=0\\ \alpha_0=0}} = \frac{1}{2[1 + h/2b(2 + 1/n)]}. \tag{3.13}$$

Eq. (3.13) gives the variation of e'/b with n for given values of $h/2b$. As an illustration, this variation is shown in Table 2 for $h/2b = 1$.

While the preceding analysis has been concerned specifically with a channel section beam of a nonlinear elastic material, the extension of the method to other types of cross sections would not, as a rule, present any difficulty.

(b) Thin-walled Closed Sections

Having discussed the location of the shear center of thin-walled open sections, a similar investigation may be carried out for thin-walled closed sections. The details of this analysis may be found in Ref. [*8*]. Briefly, the closed section is transformed to an open section by a fictitious cut as in the linear theory. The indeterminate shear flow at the cut is determined by the minimization of the complimentary energy. The shear flow at any point in this fictitious open section is determined by the analysis in the preceding section. The sum of the two shear flows yields the shear distribution for the closed section.

Once the shear flow distribution is determined, the distance from a fixed point can be obtained at which the shear force must act to produce zero resultant twisting moment on the section. As for the open cross section, this requirement is mathematically given by an equation of the type (3.9) which must be solved for the required distance for any given value of n. The method of analysis outlined will be illustrated by the solution of a particular example.

The example considered is the symmetric idealized airfoil section shown in Fig. 2. The cross section is assumed to have a constant wall thickness with the curved portion being a semi-circle and the angle $\vartheta = 15°$. The location of the shear center for three values of n is presented in Table 3.

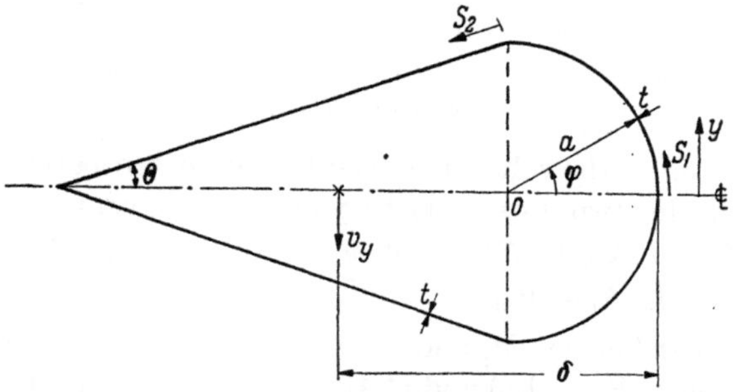

Fig. 2. Idealized airfoil section

Table 3. *Airfoil Section* — n *vs* δ/a

n	1	3	7
δ/a	0.869	1.210	2.361

It is clear from this table that δ/a, the distance of the application of the shear force V_y from a fixed point, shifts considerably if the requirement of a zero resultant twisting moment is to be satisfied.

4. Bending of Annular Plates

(a) Analysis Based upon Eq. (2.19)

The problems of the annular plates subjected to the moments along the edges shown in Fig. 3 are radially symmetric. Thus, the radial and tangential coordinates r and ϑ correspond to principal directions. Therefore, Eq. (2.19) yields the principal curvatures

$$\varkappa_r = -w'' = \left(\frac{M_r^2 - M_r M_\vartheta + M_\vartheta^2}{\mu^2}\right)^m \left(\frac{2 M_r - M_\vartheta}{\mu}\right),$$

$$\varkappa_\vartheta = -w'/r = \left(\frac{M_r^2 - M_r M_\vartheta + M_\vartheta^2}{\mu^2}\right)^m \left(\frac{2 M_\vartheta - M_r}{\mu}\right). \tag{4.1}$$

where

$$\mu = (K\sqrt{3})\,(\sqrt{3})^{\frac{1}{2m+1}}, \tag{4.2}$$

w is the deflection, and the primes denote differentiations with respect to r. Finally, the problem of the annular plates shown in Fig. 3 is governed by the equilibrium equation

$$(r M_r)' - M_\vartheta = 0. \tag{4.3}$$

Eq. (4.1) and (4.3) determine the moments M_r and M_ϑ and the deflection w.

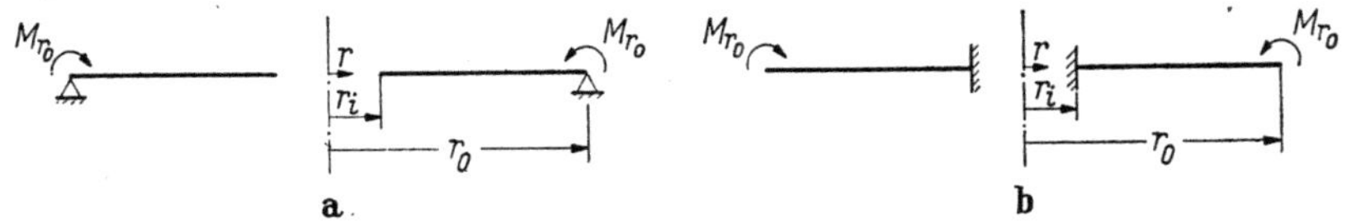

Fig. 3. Annular plates

As a first step in the solution, w can be eliminated between the first and second of the two Eqs. (4.1) to yield the moment relation

$$\frac{m(M_r^2 - M_r M_\vartheta + M_\vartheta^2)'}{(M_r^2 - M_r M_\vartheta + M_\vartheta^2)} = \frac{r(2M_\vartheta - M_r)' + 3(M_\vartheta - M_r)}{r(M_r - 2M_\vartheta)}. \tag{4.4}$$

The moment distributions M_r and M_ϑ may now be obtained by a simultaneous solution of Eqs. (4.3) and (4.4). However, an examination of the equations shows that the following simplification is possible.

With the definition

$$y = M_r / r\, M_r', \tag{4.5}$$

the equilibrium Eq. (4.3) may be introduced into Eq. (4.4) and the following first order differential equations obtained:

$$\frac{dr}{r} = \frac{2(2m+1) + 2(2m+1)y + (m+2)y^2}{2[(2m+1) + (m+2)y](1+y+y^2)}\, dy, \tag{4.6}$$

$$\frac{dM_r}{M_r} = \frac{2(2m+1) + 2(2m+1)y + (m+2)y^2}{2y[(2m+1) + (m+2)y](1+y+y^2)}\, dy. \tag{4.7}$$

A simultaneous solution of Eqs. (4.6) and (4.7) for given boundary conditions yields the radial moment distribution. However, these conditions must be expressed in terms of the variable y in order to facilitate the solution of any given problem. To this end, Eqs. (4.6) and (4.7) are integrated to yield the relations

$$A\, r = f(y)$$
$$= \left(y + \frac{2n}{n+3}\right)^{\frac{a}{2(n+3)}} (1+y+y^2)^{b/4} \exp\left[\frac{2c-b}{2\sqrt{3}} \tan^{-1}\left(\frac{1+2y}{\sqrt{3}}\right)\right], \tag{4.8}$$

$$B M_r = g(y)$$
$$= y\left(y + \frac{2n}{n+3}\right)^{\frac{a_1}{2(n+3)}} (1+y+y^2)^{b_1/4} \exp\left[\frac{2c_1-b_1}{2\sqrt{3}} \tan^{-1}\left(\frac{1+2y}{\sqrt{3}}\right)\right]. \tag{4.9}$$

Here,

$$a = \frac{4n(n+3)}{n^2+3}; \quad b = \frac{(n-1)(n-3)}{n^2+3}; \quad c = \frac{2n(n-1)}{n^2+3},$$
$$a_1 = -\frac{2(n+3)^2}{n^2+3}; \quad b_1 = -\frac{2n(n-1)}{n^2+3}; \quad c_1 = -\frac{(n-1)(n+3)}{n^2+3},$$
$$n = 2m + 1,$$

and A and B are arbitrary constants.

With the use of Eqs. (4.8) and (4.9) the two annular plate problems will now be discussed.

The annular plate on simple supports shown in Fig. 3a is considered first. Since the moment at the inner edge $M_{r_i} = 0$, it follows from equilibrium that $M_{\vartheta_i} = M'_{r_i}$. If $M_{r_i} = M_{\vartheta_i} \neq 0$, then Eq. (4.5) shows that the corresponding value of y_i at the inner edge must be zero. The final solution will verify that $M_{\vartheta_i} \neq 0$ and hence, $y_i = 0$.

The value of y_0 is obtained by proceeding in the following manner. Firstly, $r = r_0$ and the corresponding $y = y_0$ are substituted in Eq. (4.8). A similar substitution of $r = r_i$ and the corresponding $y = y_i$ is then made. A division of the two results yields

$$\frac{r_0}{r_i} = \frac{f(y_0)}{f(y_i)}. \tag{4.10}$$

For given values of the ratio r_0/r_i this equation is now solved for y_0, since $y_i = 0$. With the values of y_i and y_0 known, Eqs. (4.8), (4.9) and the equilibrium Eq. (4.3) can be used to write the nondimensional expressions

$$R = r/r_0 = f(y)/f(y_0), \tag{4.11}$$

$$m_r = M_r/M_{r_0} = g(y)/g(y_0), \tag{4.12}$$

$$m_\vartheta = M_\vartheta/M_{r_0} = (1 + y)\, g(y)/y\, g(y_0). \tag{4.13}$$

Once the moment distributions are known from these equations, the deflection of the plate may be obtained by the use of the moment-curvature relation (4.1). Hence, in nondimensional form,

$$w^* = \frac{w}{r_0^2 (M_{r_0}/\mu)^{2m+1}} = \int_1^R (m_r - 2m_\vartheta)(m_r^2 - m_r\, m_\vartheta + m_\vartheta^2)^m\, R\, dR. \tag{4.14}$$

Therefore, the solution to the problem shown in Fig. 3a is completely determined.

The annular plate fixed at the inner edge is considered next (see Fig. 3b). Since the slope at the inner edge is zero, it follows that $\varkappa_{\vartheta_i} = 0$. Eq. (4.1) then shows that at the inner edge $M_{\vartheta_i} = (1/2)\, M_{r_i}$. This relation can be combined with Eqs. (4.3) and (4.5) to yield $y_i = -2$ which, when substituted in Eq. (4.10) determines y_0 for a given value r_0/r_i. Finally, Eqs. (4.11), (4.12), and (4.13) give the moment distributions. These may be used with the first of the moment-curvature relation (4.1) to yield the deflection

$$w^* = \frac{w}{r_0^2 (M_{r_0}/\mu)^{2m+1}} = \int_{R_i}^R \int_{R_i}^R (m_\vartheta - 2m_r)(m_r^2 - m_r\, m_\vartheta + m_\vartheta^2)^m\, dR\, dR. \tag{4.15}$$

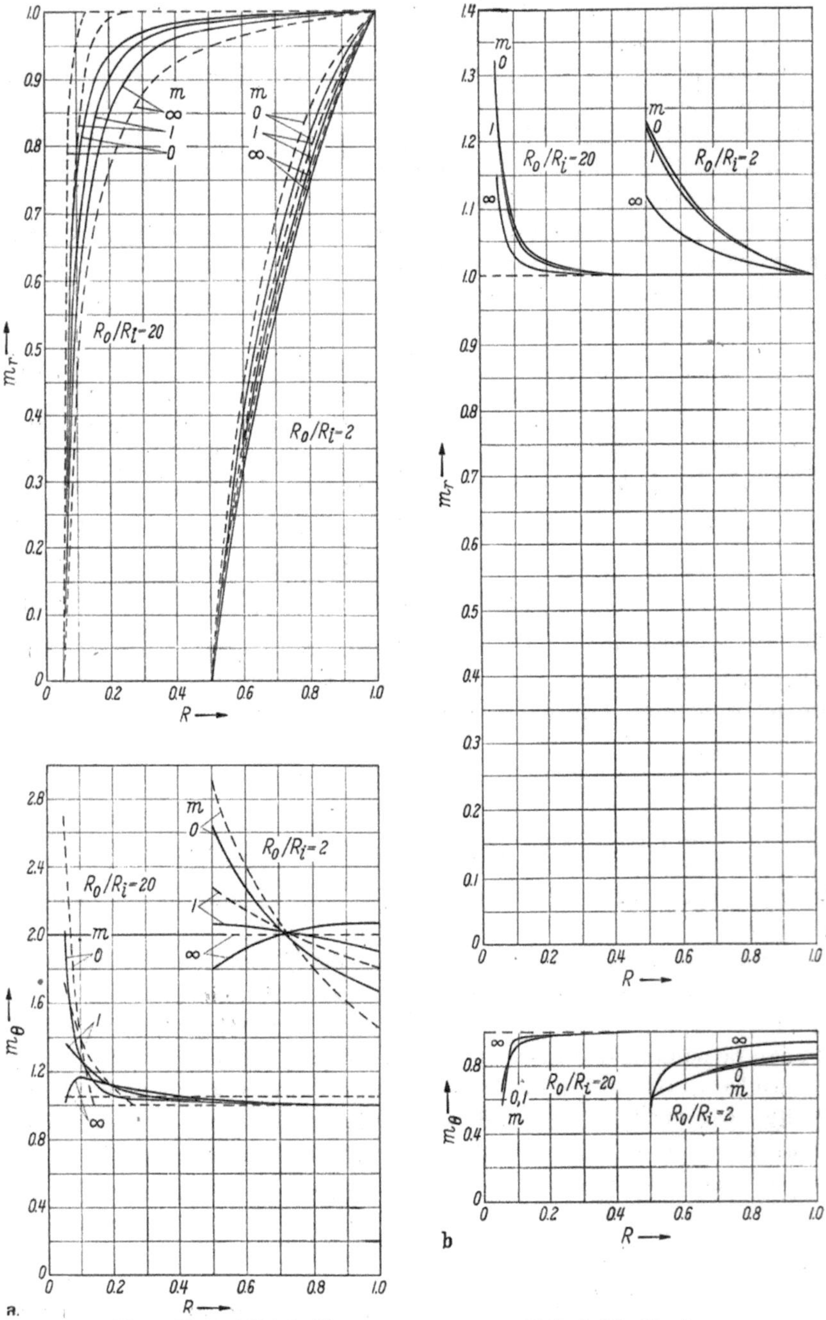

Fig. 4. Moment distribution

------ Method of Section 4a
········ Method of Section 4b

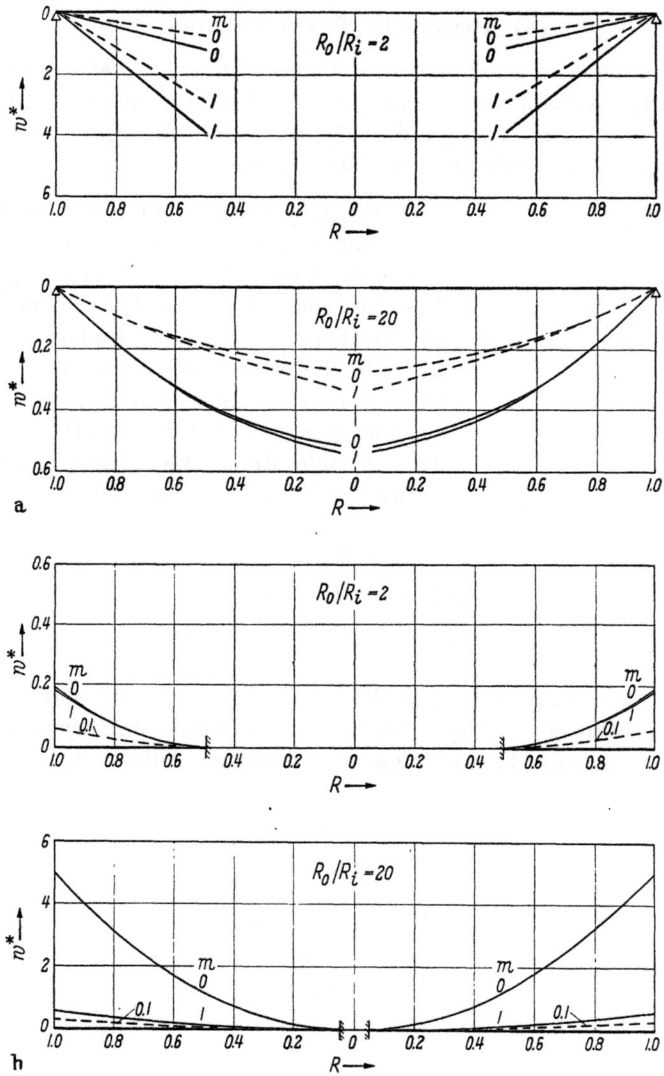

Fig. 5. Deflections

——— Method of Section 4a

········ Method of Section 4b

For values of $m = 0$, 1 and ∞, the moment distributions m_r and m_ϑ for the two problems considered, for two values of $R_0/R_i = 2$ and 20 are shown in Fig. 4. Also, the deflections w^* are plotted in Fig. 5 for values of $m = 0$ and 1.

(b) Analysis by Tresca Criterion

The complete solution of two problems of annular plates subject to pure bending moment has been presented on the basis of Eq. (2.19). The solutions of the same two problems based on the TRESCA criterion and the associated flow rule will now be carried out.

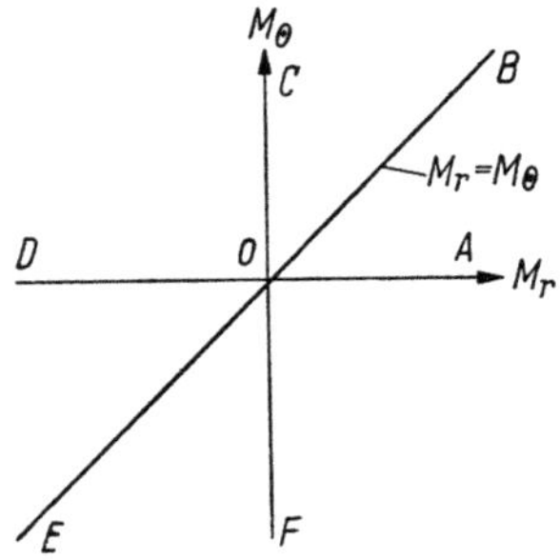

Fig. 6. Stress resultant space

With the discussion of Sec. 2 as basis, the annular plate on simple supports is considered first. At the free edge $M_{r_i} = 0$, thus the inner edge r_i must lie on the vertical FC (Fig. 6). Since the applied moment $M_{r_0} > 0$, the entire plate must lie in regime BOC, if it is assumed that $M_\vartheta > M_r$. The solution must verify this hypothesis. With these considerations, the following mathematical problem is formulated.

$$\varkappa_r = -w'' = 0,$$
$$(r\,M_r)' - M_\vartheta = 0,$$
$$\varkappa_\vartheta = -w'/r = (M_\vartheta/\mu)^{2m+1}. \tag{4.16}$$

The boundary conditions are $M_{r_i} = 0$, $(M_r)_{r=r_0} = M_{r_0}$ and $w_{r_0} = 0$. Hence, the solutions of Eq. (4.16) together with the boundary conditions yield the following dimensionless moments and deflection:

$m = 0$

$$m_r = \frac{\ln(R/R_i)}{R\ln(1/R_i)},$$
$$m_\vartheta = \frac{1}{R\ln(1/R_i)},$$
$$w^* = \frac{1-R}{\ln(1/R_i)}. \tag{4.17}$$

$m > 0$

$$m_r = \frac{(R_i/R)^{\frac{1}{2m+1}} - R_i/R}{R_i^{\frac{1}{2m+1}} - R_i},$$
$$m_\vartheta = \frac{[2\,m/(2\,m+1)]\,(R_i/R)^{\frac{1}{2m+1}}}{R_i^{\frac{1}{2m+1}} - R_i},$$
$$w^* = \left[\frac{2\,m}{(2\,m+1)\left(R_i^{\frac{1}{2m+1}} - R_i\right)}\right]^{2m+1} R_i(1-R). \tag{4.18}$$

The stress solutions presented must now be examined for the validity of the hypothesis $M_\vartheta > M_r$. Thus, the solutions show that the hypothesis is valid, provided

$$\ln(R/R_i) \leq 1,$$

i. e.

$$R \leq e\, R_i, \text{ for } m = 0, \tag{4.19}$$

and

$$(R_i/R)^{\frac{1}{2m+1}} - (R_i/R) \leq \frac{2\,m}{2\,m+1} (R_i/R)^{\frac{1}{2m+1}}$$

i. e.

$$R \leq (2m+1)^{\frac{2m+1}{2m}}\, R_i, \text{ for } m > 0. \tag{4.20}$$

When these inequalities are violated, the hypothesis $M_\vartheta > M_r$ must be modified. One modification would be for a portion of the plate, from r_i to ϱ, to lie in regime BOC and the rest of the plate, from ϱ to r_0, to lie in regime OB. It follows that for values of r from r_i to ϱ, Eq. (4.16) are still valid and for values of r from ϱ to r_0, the governing equations become

$$r\, M_r' = 0,\; M_r = M_\vartheta$$

$$\varkappa_\vartheta + \varkappa_r = -(r\, w')'/r = (M_r/\mu)^{2m+1}. \tag{4.21}$$

The boundary conditions are $(M_r)_{r=r_0} = M_{r_0}$, $w_{r_0} = 0$, and the continuity of M_r, M_ϑ, w, and w' at ϱ must be satisfied.

A solution of Eqs. (4.16) and (4.21) together with the boundary and continuity conditions yields the following dimensionless moments and deflection:

$m = 0$

$$m_r = (\eta/R)\ln(R/R_i),$$

$$m_\vartheta = \eta/R,$$

$$w^* = (1 + 3\eta^2 - 4\eta\, R - 2\eta^2 \ln\eta)/4,\; R_i \leq R \leq \eta. \tag{4.22a}$$

$$m_r = m_\vartheta = 1,$$

$$w^* = (1 - R^2 - 2\eta^2 \ln R)/4,\; \eta \leq R \leq 1. \tag{4.22b}$$

$m > 0$

$$m_r = \frac{2\,m+1}{2\,m} (\eta/R_i)^{\frac{1}{2m+1}} \left[(R_i/R)^{\frac{1}{2m+1}} - (R_i/R)\right],$$

$$m_\vartheta = (\eta/R)^{\frac{1}{2m+1}},$$

$$w^* = (1 + 3\eta^2 - 4\eta\, R - 2\eta^2 \ln\eta)/4,\; R_i \leq R \leq \eta. \tag{4.23a}$$

$$m_r = m_\vartheta = 1,$$

$$w^* = (1 - R^2 - 2\eta^2 \ln R)/4,\; \eta \leq R \leq 1. \tag{4.23b}$$

Here η is the nondimensional radius ϱ/r. From the solutions it is seen that the portion of the plate from ϱ to r_0 must lie at a point on OB.

Finally, for emphasis, it is pointed out that if $R = \eta$ satisfies the inequalities (4.19) or (4.20), then Eqs. (4.17) or (4.18) must be used and if it does not satisfy those inequalities then Eqs. (4.22) or (4.23) must be used.

The annular plate with fixed inner edge is considered next. It can be verified that the entire plate must lie at a point in the regime OB ($M_r = M_\vartheta$). Thus, the mathematical problem for this case is

$$\begin{aligned} &r\, M_r' = 0\,, \\ &\varkappa_r + \varkappa_\vartheta = -\,(r\, w')'/r = (M_r/\mu)^{2m+1}\,. \end{aligned} \tag{4.24}$$

The boundary conditions are $(M_r)_{r=r_0} = M_{r_0}$ and $w_{r_i} = w'_{r_i} = 0$. Finally, the solutions for the moments and deflections, for all values of m, are

$$\begin{aligned} &m_r = m_\vartheta = 1\,, \\ &w^* = (R_i/2)^2\, [1 + 2 \ln (R/R_i) - (R/R_i)^2]\,. \end{aligned} \tag{4.25}$$

Similar results to those in the preceding section are shown by the dashed lines in Figs. 4 and 5.

Solutions for moment distributions and deflections for two annular plate problems have been presented by two different methods of analysis. Fig. 4 shows that for both the simply supported and clamped plates, the two methods of analysis yield moment distribution solutions which are reasonably close. With regard to the deflections, Fig. 5 shows that the two methods yield solutions which differ significantly from each other. The validity of either of these theoretical predictions should, however, be verified by experimental results.

5. Inclusion of Compressibility Effect

In the consideration of the bending problems in the preceding two sections, it was initially assumed that the structures were made of incompressible materials. If it is desired to include the effect of compressibility of the material, it was shown in Sec. 2 that this may be accomplished by replacing the definition of stress-deviation in Eq. (2.3) by the modified definition in Eq. (2.25). To illustrate this effect, a hollow sphere subjected to uniform internal pressure is considered in this section. The discussion includes only a summary of the analysis and results, since the details may be found in Ref. [7].

The analysis of a hollow sphere subjected to uniform internal pressure is essentially a uni-dimensional problem. While the stress field is characterized by the normal stresses σ_r, σ_θ and σ_φ in the respective spherical coordinate directions r, θ, and φ, point symmetry of the problem shows that $\sigma_\varphi = \sigma_\theta$ and the independent variable is the radial coordinate r. The strains $\varepsilon_r = du/dr$ and $\varepsilon_\varphi = \varepsilon_\theta = u/r$ readily follow from the radial displacement u.

On the bases of the preceding discussion, the behavior of the hollow sphere is governed by the equilibrium equation

$$(r/2)\, d\sigma_r/dr = \sigma_\vartheta - \sigma_r = s_\vartheta - s_r \tag{5.1}$$

and the stress-strain law (2.6). Here, it must be noted that s_r, s_ϑ, and $J_2 = (s_r^2 + 2\, s_\vartheta^2)/2$ are obtained from the modified definition of s_{ij} in Eq. (2.25). As a first step in the solution, the stress-strain law (2.6) and the strain-displacement relations for ε_r and ε_ϑ are used to eliminate u and obtain a relation between σ_ϑ and σ_r. This relation is used together with the equilibrium Eq. (5.1) to yield the differential equation in σ_r:

$$4 + r\,\sigma_r''/\sigma_r' = \frac{m[1 + a(\sigma_r/r\,\sigma_r')]}{b(3/2 + m\,a)(\sigma_r/r\,\sigma_r')^2 + [(m+1)\,b + 2\,m\,a\,c](\sigma_r/r\,\sigma_r') + (2\,m+1)\,c}, \tag{5.2}$$

where $a = 2(1-3\alpha)/(1-2\alpha)$, $b = 2(1-3\alpha)^2$, and $c = (1-4\alpha+6\alpha^2)/2$. Eq. (5.2) is of the second order and the boundary conditions are $\sigma_r = \sigma_i$ at $r = r_i$ the internal surface and $\sigma_r = 0$ at $r = r_0$, the external surface.

Before attempting the solution of Eq. (5.2), scrutiny of the equation shows that a simplification similar to that used in the preceding section is possible. Briefly, with the definition

$$x = \sigma_r/r\,\sigma_r', \tag{5.3}$$

Eq. (5.2) is reduced to the following first order differential equations

$$\frac{dr}{r} = \frac{1 + D\,x + E\,x^2}{1 + F\,x + G\,x^2 + H\,x^3}\,dx, \tag{5.4}$$

$$\frac{d\sigma_r}{\sigma_r} = \frac{1 + D\,x + E\,x^2}{x(1 + F\,x + G\,x^2 + H\,x^3)}\,dx, \tag{5.5}$$

equivalent to Eqs. (5.2) and (5.3), respectively. Here, the constants D, E, F, G, and H depend on m, the exponent in Eq. (2.6), and α the compressibility constant defined in Eq. (2.25) (see, Ref. [7]). The transformed conditions for the solution of Eqs. (5.4) and (5.5) are

$$\sigma_r = \sigma_i,\ r = r_i,\ x = x_i \text{ and } \sigma_r = 0,\ r = r_0,\ x = 0.$$

For a given ratio r_0/r_i, the constant y_i may be readily determined as in the preceding section and the solution of σ_r follows from Eq. (5.5). The stress σ_ϑ is then quite simply determined by the use of Eq. (5.1).

As a numerical example, a hollow sphere of radius ratio $r_0/r_i = 1.5$ was considered. For values of the exponent $m = 0$, 1, and ∞ and the compressibility constant $\alpha = 0$, 1/6 and 1/3, the stress distributions are shown in Fig. 7. It should be noted that for $m = 0$, the stress solutions are the LAMÉ solutions which are independent of α. For other values of m, the radial stress distribution is qualitatively the same for all values of

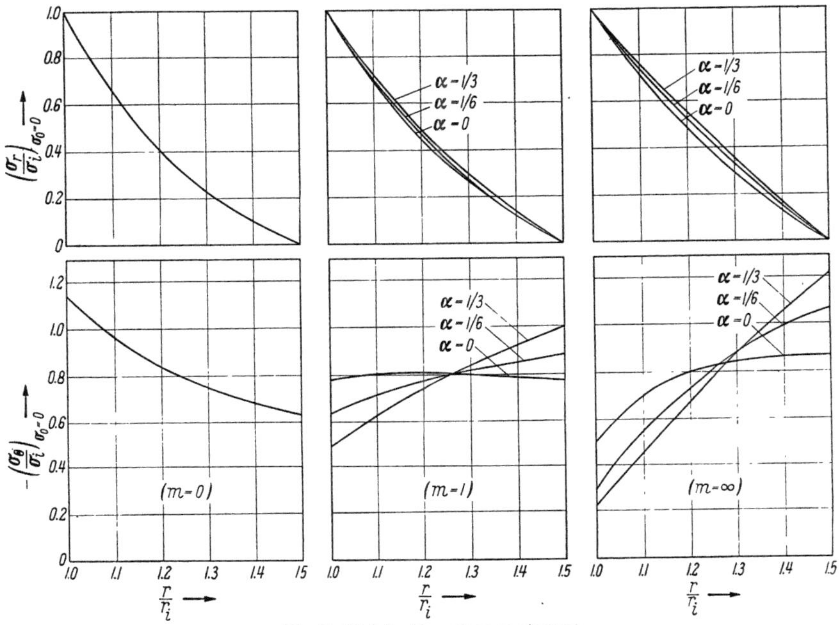

Fig. 7. Distribution of creep stresses

$\alpha (0 \leq \alpha \leq 1/3)$. On the other hand, it is clearly seen that the circumferential stress distribution for $m \geq 1$ differs radically from that of the LAMÉ solution.

References

[1] HOFF, N. J.: Stress Distribution in the Presence of Creep. IX. Congrés International de Mécanique Appliquée, Actes Tome VIII, Université de Bruxelles, p. 211, 1957.

[2] PRAGER, W.: Strain Hardening Under Combined Stresses. J. Appl. Physics, **16**, No. 21, 837 (1945).

[3] HOFF, N. J.: Effets Thermiques dans le Calcul de la Résistance des Structures d'Avions et d'Engins. AGARD Report No. 52, Jan. 1956.

[4] WAHL, A. M., G. O. SANKEY, M. J. MANJOINE and E. SHOEMAKER: Creep Tests of Rotating Disks at Elevated Temperature and Comparison with Theory. J. Appl. Mech., **21**, No. 3, 225 (1954).

[5] VENKATRAMAN, B.: Solutions of Some Problems in Steady Creep. PIBAL Report No. 402, July 1959. AFOSR TN 57—388.

[6] VENKATRAMAN, B., and P. G. HODGE, jr.: Creep Behavior of Circular Plates. J. of the Mech. and Phys. of Solids, **6**, 163 (1958).

[7] PATEL, S. A., B. VENKATRAMAN and W. P. VAFAKOS: Analysis of Compressibility in Creep by the Nonlinear Elastic Analogue. Int. J. Mech. Sci., 2, 1 (1960)

[8] PATEL, S. A., K. A. V. PANDALAI and B. VENKATRAMAN: Creep-Stress Analysis of Thin-Walled Structures. J. Roy. Aero. Soc. 64, No. 599, 673 (1960).

The Shear Center in Creep of Beams of Thin-walled Open Cross Section[1]

By

Aris Phillips

Yale University, New Haven, Connecticut, USA

Abstract

It is shown that in the case of creep bending of beams of thin-walled open cross section, in general, the neutral axes do not intersect at one point and that the concept of the shear center fails. The shear center is replaced by a family of straight lines such that when the plane of bending passes through such a straight line the cross section will suffer no torsion. These torsionfree axes do not intersect at one point except in special cases. The family of the neutral axes and the family of the torsionfree axes correspond to each other and they depend on the magnitudes of the bending moment and shearing force. When the creep law has the form $\sigma = A\,\dot{\varepsilon}^n$ the two families are independent of the magnitudes of the bending moment and shearing force. The implication of this result is that for torsionfree bending of a beam to be possible it is necessary to have the above creep law. Exceptions to this result are discussed. Expressions useful for the calculation of the two families of straight lines are given.

1. Introduction

It is the purpose of this paper to clarify the concept of the shear center for a straight thin walled beam of open cross section under conditions of creep. It will be assumed that plane cross sections remain plane and that the longitudinal stresses are those which follow from pure bending. This means that shearing stresses and shearing forces are neglected when the longitudinal stresses are obtained. Then the longitudinal stresses are used for determining the shearing stresses from

[1] This investigation has been supported by the National Science Foundation of the United States Government.

equilibrium considerations alone. It is assumed that both the longitudinal stresses and the shearing stresses are uniformly distributed over the thickness of the wall and that the shearing stresses are parallel to the middle surface of the wall. Finally it is assumed that the deformation of the cross section is not large enough to influence appreciably the value of the stresses. These assumptions are the standard ones of the elementary theory of bending of beams and have been used by many authors, for example, TIMOSHENKO [1], MAILLART [2], and WEBER [3], in discussing the problem of the shear center when HOOKE's law is valid.

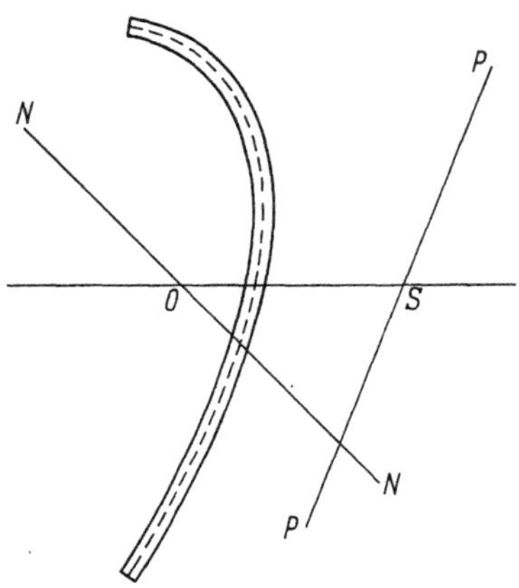

Fig. 1.

Consider the cross section in Fig. 1 where the point 0 is the centroid and the point S is the shear center when HOOKE's law is valid. It is well known that for linear elasticity all neutral axes pass through the centroid. To each neutral axis NN there corresponds a line PP passing through the shear center S such that when the plane of loading passes through PP the neutral axis will be NN and the cross section will suffer no torsion. Therefore, in the linear elastic case the shear center is the point which belongs to all loading planes producing no torsion. We can also say that to the family of neutral axes NN passing through the centroid, there corresponds the family of *torsionfree axes* PP passing through the shear center S.

In the case of creep the concepts of the neutral axis and of the torsionfree axis and of their corresponding families are still valid. In this paper we shall give expressions for determining the two families of axes for creep for any thin-walled cross section. By means of a general proof it will be shown that for an arbitrary cross section the neutral axes do not intesect at one point. It will also be shown that the torsionfree axes do not need to intersect at one point. Hence, the concept of the shear center has to be replaced by the concept of the family of torsionfree axes.

When the cross section has two axes of symmetry then the neutral axes and the torsionfree axes pass through the centroid. When the cross section consists of portions with straight middle lines which — or the extensions of which — all intersect at a point S, this point S will be the intersection of all torsionfree axes; in this case, however, the neutral axes do not need to intersect at one point.

The family of the neutral axes depends on the magnitude of the bending moment. The family of the torsionfree axes depends on the magnitude of the bending moment and on the magnitude of the shearing force. In the special case, however, in which the creep law is of the form

$$\sigma = A\,\dot{\varepsilon}^n \tag{1}$$

the two families remain stationary when the magnitudes of the bending moment and of the shearing force change. These two families depend on the exponent n although not on the coefficient A. Creep law (1) has been proposed by NORTON [4] and is used extensively in the solutions of creep problems.

Consider now a cantilever beam of open cross section loaded by a concetrated force at the end. It is seen that for a given value of the force the bending moment changes with the distance from the force. Hence, unless Eq. (1) is valid, the position of the torsionfree axis changes with the distance of the cross section from the load and consequently torsionfree bending is impossible except in special cases. When, however, Eq. (1) is valid torsionfree bending is possible even if the magnitude of the load changes.

The problem of the shear center in creep is intimately connected with the problem of the shear center in nonlinear elasticity and in strain-hardening because of the validity of the elastic analog (5). The shear center when HOOKE's law is valid has been discussed by TIMOSHENKO [1], MAILLART [2], WEBER [3], and EGGENSCHWYLER [6]. HANDELMAN [7] considers the case of strain-hardening but he assumes that the neutral axis passes through the centroid of the cross section which as it is shown in the present paper is not correct. JOHNSON and MELLOR [8] consider strain hardening but only for the case of symmetrical bending with a powerlaw of the type (1). The case of creep has been considered by DONATH [9] in his doctoral thesis and by PATEL and PANDALAI [10]. DONATH's work is based on some of the results presented in this paper which have been obtained a few years ago. The work presented in Ref. [9] and [10] deals only with the case of a channel cross section with a powerlaw of the type (1). The present paper on the other hand gives the general theory for any thin walled cross section under any creep law.

2. The Neutral Axis

Consider a thin walled beam of arbitrary cross section as in Fig. 2. The system of coordinates XYZ is selected such that the X axis is parallel to the centroidal line of the beam. The assumption that plane cross sections remain plane during creep leads to the expression

$$\varepsilon = \alpha\,y + \beta\,z + \gamma \tag{2}$$

for the strains, where α, β, and γ are functions of time. From Eq. (2) we obtain

$$\dot{\varepsilon} = \dot{\alpha}\, y + \dot{\beta}\, z + \dot{\gamma} \tag{3}$$

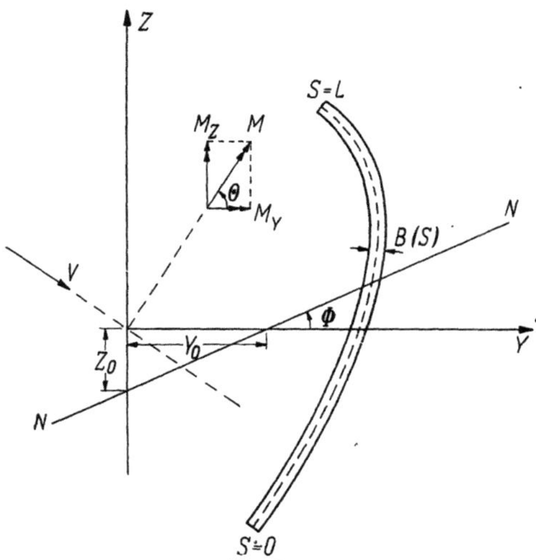

Fig. 2.

for the strain rate, where the dots denote derivatives with respect to time. If

$$\sigma = f(\dot{\varepsilon}) \tag{4}$$

is the stress-rate of strain law of creep we obtain

$$\sigma = f(\dot{\alpha}\, y + \dot{\beta}\, z + \dot{\gamma}) \tag{5}$$

for the stress distribution. The function f is such that for $\dot{\varepsilon} = 0$ we also have $\sigma = 0$.

In elasticity and in plasticity the neutral axis is by definition the line for which $\varepsilon = 0$ and therefore, because of the strain-stress law also $\sigma = 0$. In creep, however, $\sigma = 0$ implies $\dot{\varepsilon} = 0$ but not necessarily $\varepsilon = 0$. Indeed, from

$$\dot{\varepsilon} = \dot{\alpha}\, y + \dot{\beta}\, z + \dot{\gamma} = 0 \tag{6}$$

we obtain by integration

$$\varepsilon = \varepsilon_0 \tag{7}$$

or

$$\alpha\, y + \beta\, z + \gamma = \alpha_0\, y + \beta_0\, z + \gamma_0, \tag{7a}$$

where ε_0, α_0, β_0, and γ_0 are the values of ε, α, β, and γ at $t = 0$ and these values need not be zero. For this reason, we shall define as neutral axis the line of zero strain rate.

The three equations of statics give

$$\int_{s=0}^{s=L} B(s)\, \sigma(s)\, ds = 0, \tag{8}$$

$$\int_{s=0}^{s=L} B(s)\, \sigma(s)\, y\, ds = -M_z, \tag{9}$$

$$\int_{s=0}^{s=L} B(s)\, \sigma(s)\, z\, ds = M_y, \tag{10}$$

where M_z and M_y are the bending moments with respect to the axes Z and Y repectively. The width of the cross section at s is denoted by $B(s)$, and L is the length of the middle line of the cross section. It is assumed, of course, that no axial force is applied.

With

$$\sigma(s) = f(\dot{\alpha}\, y + \dot{\beta}\, z + \dot{\gamma}), \tag{11}$$

Eqs. (8), (9), and (10) can be written in the form

$$\int_{s=0}^{s=L} B(s)\, f(\dot{\alpha}\, y + \dot{\beta}\, z + \dot{\gamma})\, ds = 0, \tag{12}$$

$$\int_{s=0}^{s=L} B(s)\, f(\dot{\alpha}\, y + \dot{\beta}\, z + \dot{\gamma})\, y\, ds = -M_z \tag{13}$$

$$\int_{s=0}^{s=L} B(s)\, f(\dot{\alpha}\, y + \dot{\beta}\, z + \dot{\gamma})\, z\, ds = M_y \tag{14}$$

These three Equations permit, in principle, the determination of $\dot{\alpha}$, $\dot{\beta}$, and $\dot{\gamma}$ when M_z and M_y are known. Once $\dot{\alpha}$, $\dot{\beta}$, and $\dot{\gamma}$ have been determined the position of the neutral axis can be obtained from Eq. (6).

Eq. (6) shows that the position of the neutral axis is determined by the two quantities

$$Y_0 = -\dot{\gamma}/\dot{\alpha}, \quad Z_0 = -\dot{\gamma}/\dot{\beta}. \tag{15}$$

Fig. 2 shows that $(Y_0, 0)$ and $(0, Z_0)$ are the coordinates of the intersections of the neutral axis with the Y and Z axes respectively.

The direction of the neutral axis is given by the ratio

$$Z_0/Y_0 = \dot{\alpha}/\dot{\beta} = -\tan\Phi, \tag{16}$$

where Φ is the angle which the neutral axis makes with the Y axis. The direction of the moment vector is determined by the ratio

$$M_z/M_y = \tan\theta, \tag{17}$$

where θ is the angle of the moment vector with the Y axis.

When $\dot{\gamma} \neq 0$ Eqs. (12) and (16) give

$$\int_{s=0}^{s=L} B(s)\, f\left[\dot{\gamma}\left(-\frac{y}{Y_0} + \frac{z}{Y_0\tan\Phi} + 1\right)\right] ds = 0, \tag{18}$$

$$\frac{-\int_{s=0}^{s=L} B(s) f\left[\dot{\gamma}\left(-\frac{y}{Y_0} + \frac{z}{Y_0\tan\Phi} + 1\right)\right] y\, ds}{\int_{s=0}^{s=L} B(s) f\left[\dot{\gamma}\left(-\frac{y}{Y_0} + \frac{z}{Y_0\tan\Phi} + 1\right)\right] z\, ds} = \frac{M_z}{M_y} = \tan\theta. \tag{19}$$

In principle, we can eliminate Y_0 between these two equations and obtain a relation between Φ and $\dot{\gamma}$ for a given θ. Hence for a given direction of the moment vector the neutral axis changes with the value of $\dot{\gamma}$. However, for $\dot{\gamma}$ to change it is necessary that the bending moment will be changed, if the angle θ remains constant. We conclude that the neutral axis generally will not remain constant when the angle θ is constant but the bending moment changes in magnitude.

When the creep law has the form of Eq. (1), Eqs. (12) to (14) take the forms

$$\int_{s=0}^{s=L} B(s)\left(-\frac{y}{Y_0}+\frac{z}{Y_0\tan\Phi}+1\right)^n ds=0, \tag{20}$$

$$A\dot{\gamma}^n \int_{s=0}^{s=L} B(s)\left(-\frac{y}{Y_0}+\frac{z}{Y_0\tan\Phi}+1\right)^n y\,ds=-M_z, \tag{21}$$

$$A\dot{\gamma}^n \int_{s=0}^{s=L} B(s)\left(-\frac{y}{Y_0}+\frac{z}{Y_0\tan\Phi}+1\right)^n z\,ds=M_y \tag{22}$$

and Eq. (19) takes the form

$$\frac{M_z}{M_y}=\tan\theta=\frac{-\int_{s=0}^{s=L} B(s)\left(-\frac{y}{Y_0}+\frac{z}{Y_0\tan\Phi}+1\right)^n y\,ds}{\int_{s=0}^{s=L} B(s)\left(-\frac{y}{Y_0}+\frac{z}{Y_0\tan\Phi}+1\right)^n z\,ds}. \tag{23}$$

We observe that in Eqs. (20) and (23) the variable $\dot{\gamma}$ is not included and therefore to each θ corresponds now a single value of Φ. This means that to each angle θ now corresponds one neutral axis irrespective of the magnitude of the bending moment.

We shall give now a general proof of the statement that the neutral axes do not intersect at one point, except when the cross section has two axes of symmetry or when HOOKE's law is valid. We shall restrict our proof to a cross section which is symmetrical with respect to the Y axis. Certainly, if the neutral axes do not have the property of passing through the same point for a symmetrical cross section, they will not have this property for non-symmetrical cross sections. Assume that all neutral axes pass through the same point. Then because of the symmetry this point must lie on the Y axis. Therefore Y_0 must be independent of the angle Φ. Hence the derivative of the left hand side of Eq. (18) with respect to Φ must be equal to zero. We obtain

$$\int_{s=0}^{s=L} B(s)\frac{d\sigma}{d\dot{\varepsilon}}\dot{\gamma}\frac{z}{Y_0\cos^2\Phi}ds=0 \tag{24}$$

which can also be written in the form

$$\int_{s=0}^{s=L} B(s)\frac{d\sigma}{d\dot{\varepsilon}}z\,ds=0. \tag{25}$$

Introducing the new variable

$$s'=s-\frac{L}{2} \tag{26}$$

we write Eq. (25) in the form

$$\int_{s'=-\frac{L}{2}}^{s'=\frac{L}{2}} B(s') \frac{d\sigma}{d\dot{\varepsilon}} z\, ds' = 0. \tag{27}$$

We now remark that z is antisymmetrical with respect to the Y axis, that $B(s')$ is symmetrical with respect to the Y axis, and that $d\sigma/d\dot{\varepsilon}$ is always positive. Hence for the left hand member of Eq. (27) to be zero it is necessary that $d\sigma/d\dot{\varepsilon}$ will be a symmetrical function with respect to the Y axis. The quantity $d\sigma/d\dot{\varepsilon}$, however, will not be a symmetrical function with respect to the Y axis *for all neutral axes* except when it has a constant value, or when the cross section has two axes of symmetry and the function $f(\dot{\varepsilon})$ satisfies the condition $f(-\dot{\varepsilon}) = -f(\dot{\varepsilon})$. We conclude that the assumption made above that all neutral axes pass through the same point is correct only when $d\sigma/d\dot{\varepsilon}$ has *a* constant value, or when the cross section has two axes of symmetry and $f(\varepsilon)$ satisfies the condition $f(-\dot{\varepsilon}) = -f(\dot{\varepsilon})$. This proof is valid not only for thin-walled cross sections but for any cross section.

We conclude this section of the paper by adding that when $\gamma = 0$ and $\dot{\beta} \neq 0$, Eqs. (20) and (23) will be replaced by

$$\int_{s=0}^{s=L} B(s)\,(z - y \tan\Phi)^n\, ds = 0, \tag{28}$$

$$\frac{-\int_{s=0}^{s=L} B(s)\,(z - y\tan\Phi)^n\, y\, ds}{\int_{s=0}^{s=L} B(s)\,(z - y\tan\Phi)^n\, z\, ds} = \frac{M_z}{M_y} = \tan\Phi. \tag{29}$$

3. The Shear Flow

From Fig. 3 it follows that the equation of equilibrium for an element of the cross section is

$$\frac{\partial}{\partial s}\,(B(s)\cdot\tau(s)) = -\frac{\partial\sigma(s)}{\partial x}\,B(s). \tag{30}$$

Integrating we obtain

$$\tau(s) = -\frac{1}{B(s)}\int_0^s B(s)\,\frac{\partial\sigma(s)}{\partial x}\,ds + f_1(x). \tag{31}$$

As the edges of the cross section are free from shearing stresses we obtain $\tau(0) = 0$. Hence

$$f_1(x) = 0. \tag{32}$$

Therefore Eq. (31) becomes

$$\tau(s) = -\frac{1}{B(s)} \int_0^s B(s) \frac{\partial \sigma(s)}{\partial x} ds. \tag{33}$$

These shearing stresses produce a shear flow which in turn gives rise to a torque about the X axis. This torque is

$$T = \int_{s=0}^{s=L} B(s)\, \tau(s)\, R(s)\, ds \tag{34}$$

where $R(s)$ is the normal distance from the X axis to the shearing stress $\tau(s)$. We write therefore

$$T = - \int_{s=0}^{s=L} R(s) \int_0^s B(s) \frac{\partial \sigma(s)}{\partial x} ds\, ds. \tag{35}$$

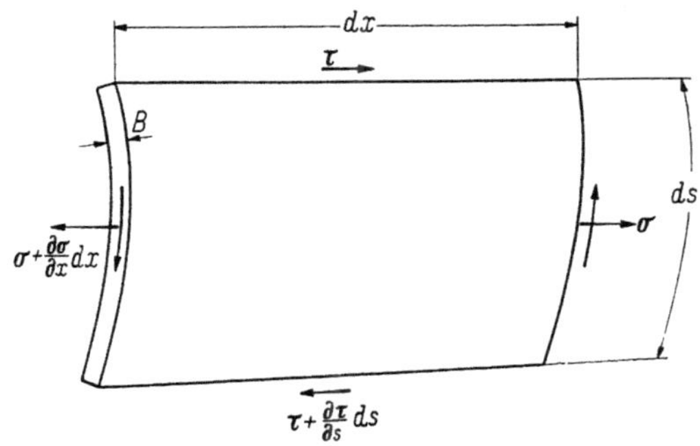

Fig. 3.

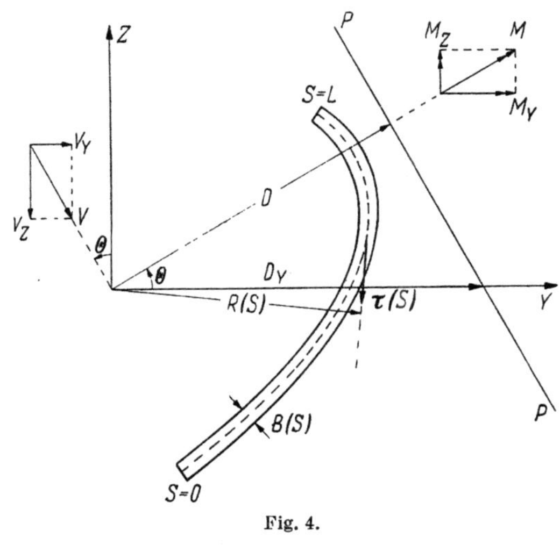

Fig. 4.

In order to eliminate the twist produced by this torque it is necessary to displace the plane of loading parallel to itself from the position passing through the X axis by the amount, Fig. 4,

$$D = T/V, \tag{36}$$

where

$$V = \frac{dM}{dx} = \frac{V_z}{\cos\theta} = \frac{dM_y}{dx} \Big/ \cos\theta \tag{37}$$

is the shearing force. Eq. (10) gives

$$\frac{dM_y}{dx} = \int_{s=0}^{s=L} B(s) \frac{\partial \sigma(s)}{\partial x} z\, ds. \tag{38}$$

Hence

$$D = \frac{-\int\limits_{s=0}^{s=L} R(s) \int\limits_0^s B(s) \frac{\partial \sigma(s)}{\partial x} ds\, ds}{\int\limits_{s=0}^{s=L} B(s) \frac{\partial \sigma(s)}{\partial x} z\, ds} \cos\theta. \tag{39}$$

The motion of the plane of loading in the direction of the Y axis is given by

$$D_y = \frac{D}{\cos\theta} = -\frac{\int\limits_{s=0}^{s=L} R(s) \int\limits_0^s B(s) \frac{\partial \sigma(s)}{\partial x} ds\, ds}{\int\limits_{s=0}^{s=L} B(s) \frac{\partial \sigma(s)}{\partial x} z\, ds}. \tag{40}$$

Eq. (35) can be simplified considerably when $R(s)$ is a stepwise constant function. This occurs when the cross section consists of portions with straight middle lines as, for example, is the case with the cross section in Fig. 5. We use the integration by parts formula

$$\int\limits_{s=0}^{s=L} u\, dv = \left[u\, v\right]_0^L - \int\limits_{s=0}^{s=L} v\, du \tag{41}$$

with

$$u = R(s) \int\limits_0^s B(s) \frac{\partial \sigma(s)}{\partial x} ds, \qquad v = s. \tag{42}$$

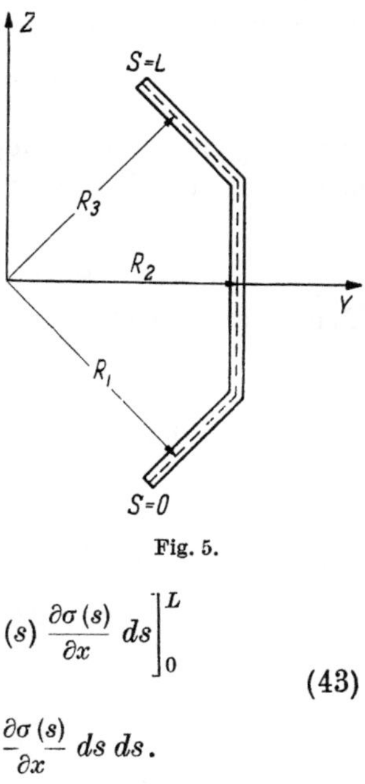

Fig. 5.

Then

$$\int\limits_{s=0}^{s=L} R(s) \int\limits_0^s B(s) \frac{\partial \sigma(s)}{\partial x} ds\, ds = \left[s\, R(s) \int\limits_0^s B(s) \frac{\partial \sigma(s)}{\partial x} ds\right]_0^L$$
$$- \int\limits_0^L s\, R(s)\, B(s) \frac{\partial \sigma(s)}{\partial x} ds - \int\limits_0^L s \frac{dR(s)}{ds} \int\limits_0^s B(s) \frac{\partial \sigma(s)}{\partial x} ds\, ds. \tag{43}$$

But from Eq. (30) it follows that

$$\left[s\, R(s) \int\limits_0^s B(s) \frac{\partial \sigma(s)}{\partial x} ds\right]_0^L = -L\, R(L)\, B(L)\, \tau(L) = 0. \tag{44}$$

Therefore

$$T = \int\limits_{s=0}^{s=L} s\, R(s)\, B(s) \frac{\partial \sigma(s)}{\partial x} ds + \int\limits_{s=0}^{s=L} s \frac{dR(s)}{ds} \int\limits_0^s B(s) \frac{\partial \sigma(s)}{\partial x} ds\, ds. \tag{45}$$

When $R(s)$ is a stepwise constant function the second integral in (45) vanishes and the expression of the torque simplifies to

$$T = \int_{s=0}^{s=L} R(s)\, B(s) \frac{\partial \sigma(s)}{\partial x}\, s\, ds. \tag{46}$$

In this case Eq. (40) can be written in the form

$$D_y = \frac{\int_{s=0}^{s=L} R(s)\, B(s) \frac{\partial \sigma(s)}{\partial x}\, s\, ds}{\int_{s=0}^{s=L} B(s) \frac{\partial \sigma(s)}{\partial x}\, z\, ds}. \tag{47}$$

Eq. (47) is much simpler than Eq. (40).

Consider now the special case in which the creep law is given by Eq. (1) and assume that the direction of the plane of bending is given. Assuming $\dot{\gamma} \neq 0$ we can write

$$\sigma = A\, \dot{\gamma}^n \left(-\frac{y}{Y_0} - \frac{z}{Z_0} + 1\right)^n, \tag{48}$$

where Y_0 and Z_0 are independent of the magnitude of the bending moment and therefore of x also. We obtain

$$\frac{\partial \sigma}{\partial x} = A \left(-\frac{y}{Y_0} - \frac{z}{Z_0} + 1\right)^n \frac{\partial(\dot{\gamma}^n)}{\partial x}. \tag{49}$$

Hence for a beam of constant cross section Eq. (40) gives

$$D_y = -\frac{\int_{s=0}^{s=L} R(s) \int_0^s B(s) \left(-\frac{y}{Y_0} - \frac{z}{Z_0} + 1\right)^n ds\, ds}{\int_{s=0}^{s=L} B(s) \left(-\frac{y}{Y_0} - \frac{z}{Z_0} + 1\right)^n z\, ds}. \tag{50}$$

Similarly Eq. (47) becomes

$$D_y = \frac{\int_{s=0}^{s=L} R(s)\, B(s) \left(-\frac{y}{Y_0} - \frac{z}{Z_0} + 1\right)^n s\, ds}{\int_{s=0}^{s=L} B(s) \left(-\frac{y}{Y_0} - \frac{z}{Z_0} + 1\right)^n z\, ds}. \tag{51}$$

The right hand members of Eqs. (50) and (51) and therefore also D_y, are independent of the magnitude of the bending moment and therefore of the magnitude of the shearing force also.

If the creep law has the general form of Eq. (4), Eq. (47) becomes

$$D_y = \frac{\int_{s=0}^{s=L} R(s)\, B(s) \frac{\partial}{\partial x}\left[f\left\{\dot{\gamma}\left(-\frac{y}{Y_0} - \frac{z}{Z_0} + 1\right)\right\}\right] s\, ds}{\int_{s=0}^{s=L} B(s) \frac{\partial}{\partial x}\left[f\left\{\dot{\gamma}\left(-\frac{y}{Y_0} - \frac{z}{Z_0} + 1\right)\right\}\right] z\, ds}. \tag{52}$$

In Eq. (52) Y_0 and Z_0 are not independent of the magnitude of the bending moment and $\dot{\gamma}$ cannot be eliminated from the right hand member. It follows that, in general, D_y will depend on the magnitudes of the bending moment and shearing force.

We conclude this section by giving the expression for D_y when $\dot{\gamma} = 0$ and $\dot{\beta} \neq 0$. In this case expression (51) will be replaced by

$$D_y = \frac{\int\limits_{s=0}^{s=L} R(s)\, B(s)\, (z - y \tan \Phi)^n\, s\, ds}{\int\limits_{s=0}^{s=L} B(s)\, (z - y \tan \Phi)^n\, z\, ds}. \tag{53}$$

4. The Shear Center

It has been shown previously that when the creep law is given by Eq. (1), D_y is independent of the magnitudes of the bending moment and of the shearing force. In this section it will be shown that the torsionfree axes do not need to intersect at one point when $n \neq 1$.

Consider a cross section with one axis of symmetry. The axis Y is selected to coincide with this axis of symmetry. If the torsionfree axes intersect at one point, this point must be located on the Y axis. Hence in order to prove our proposition it is necesary tos show that D_y does not remain constant when the direction of the plane of bending changes.

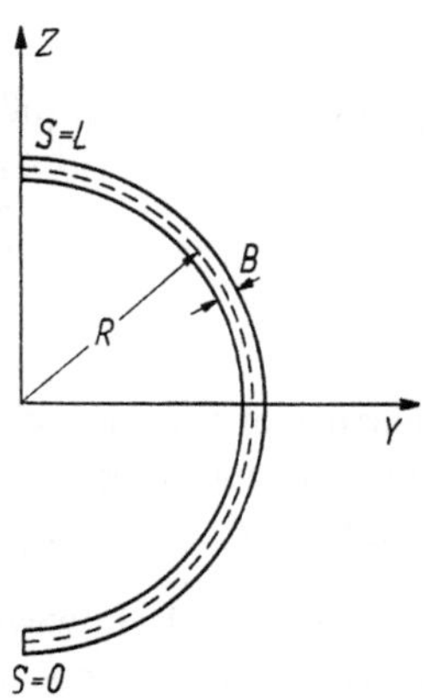

Fig. 6.

We shall consider a semicircular cross section of constant thickness, Fig. 6. Eq. (51) becomes

$$\frac{D_y}{R} = \frac{\int\limits_{s=0}^{s=L} \left(-\frac{y}{Y_0} - \frac{z}{Z_0} + 1\right)^n s\, ds}{\int\limits_{s=0}^{s=L} \left(-\frac{y}{Y_0} - \frac{z}{Z_0} + 1\right)^n z\, ds}. \tag{54}$$

Introducing the new variable s' by means of Eq. (26) we obtain

$$\frac{D_y}{R} = \frac{\int\limits_{s'=-\frac{L}{2}}^{s'=\frac{L}{2}} \left(-\frac{y}{Y_0} - \frac{z}{Z_0} + 1\right)^n s'\, ds' + \frac{L}{2} \int\limits_{s'=-\frac{L}{2}}^{s'=\frac{L}{2}} \left(-\frac{y}{Y_0} - \frac{z}{Z_0} + 1\right)^n ds'}{\int\limits_{s'=-\frac{L}{2}}^{s'=\frac{L}{2}} \left(-\frac{y}{Y_0} - \frac{z}{Z_0} + 1\right)^n z\, ds'}. \tag{55}$$

The second term in the numerator of the right hand member of Eq. (55) is equal to zero. Using the new variable ω which is introduced by means of equations

$$y = R \cos \omega, \; z = R \sin \omega, \; s' = R\,\omega, \; L = R\,\pi, \tag{56}$$

we obtain

$$\frac{D_y}{R} = \frac{\int\limits_{-\frac{\pi}{2}}^{\frac{\pi}{2}} \left(-\frac{R \cos \omega}{Y_0} - \frac{R \sin \omega}{Z_0} + 1\right)^n \omega\, d\omega}{\int\limits_{-\frac{\pi}{2}}^{\frac{\pi}{2}} \left(-\frac{R \cos \omega}{Y_0} - \frac{R \sin \omega}{Z_0} + 1\right)^n \sin \omega\, d\omega}. \tag{57}$$

which can be written in the form

$$\int\limits_{-\frac{\pi}{2}}^{\frac{\pi}{2}} \left(-\frac{R \cos \omega}{Y_0} - \frac{R \sin \omega}{Z_0} + 1\right)^n \left(\omega - \frac{D_y}{R} \sin \omega\right) d\omega = 0. \tag{58}$$

For $n = 1$ Eq. (58) becomes equal to

$$\int\limits_{-\frac{\pi}{2}}^{\frac{\pi}{2}} \left(\omega - \frac{D_y}{R} \sin \omega\right) \sin \omega\, d\omega = 0 \tag{59}$$

because of symmetry and because of the vanishing of the integrals for ω $\sin \omega$, $\omega \cos \omega$, $\cos \omega \sin \omega$. Hence for $n = 1$ the value of D_y/R is independent of Y_0 and Z_0. This means that for $n = 1$ a shear center exists.

For $n \neq 1$, however, Eq. (58) can not be simplified as above, and therefore it does not seem likely that, in general, the result of the integration in Eq. (58) will show that D_y/R is independent of Y_0 and Z_0. In the absence of a general proof we select the value $n = 1/3$ and compute the value of D_y/R by means of numerical integration. For $Z_0/Y_0 = 10$ we obtain the value 1.187; for $Z_0 = Y_0$ we have 1.245; and for $Z_0 = 0$ numerical integration of Eq. (53) gives the value 1.252. It is seen therefore that D_y/R is not constant, and a shear center does not exist.

Another way of showing that a shear center does not exist for $n \neq 1$ is to consider the value $n = 3$ which allows a closed integration of (58). This value of n obviously is not a realistic one for creep. However, it can certainly be assumed that if the integral in (58) gives D_y/R as a function of Y_0 for $n > 1$, the same will be true for $n < 1$. The integration of

Eq. (58) for $n = 3$ gives the following expression for D_y/R:

$$\frac{D_y}{R} = \frac{-\frac{6\,R}{Z_0} + \frac{3\,\pi\,R^2}{2\,Y_0\,Z_0} - \frac{4}{3}\,\frac{R^3}{Y_0^2\,Z_0} - \frac{28\,R^3}{18\,Z_0^3}}{2 - \frac{3\,\pi\,R^3}{8\,Y_0^3} - \frac{3\,\pi\,R^3}{8\,Y_0\,Z_0^2} + \frac{4\,R^2}{Y_0^2} - \frac{3\,\pi\,R}{2\,Y_0} + \frac{2\,R^2}{Z_0^2}}. \tag{60}$$

We remark also that Y_0 and Z_0 are connected with each other by an equation of the form

$$\int_{-\frac{\pi}{2}}^{\frac{\pi}{2}} \left(-\frac{R\cos\omega}{Y_0} - \frac{R\sin\omega}{Z_0} + 1\right)^3 d\omega = 0 \tag{61}$$

giving the position of the neutral axis. Integrating Eq. (61) we obtain

$$\pi - \frac{4\,R^3}{3\,Y_0^3} - \frac{2\,R^3}{Y_0\,Z_0^2} + \frac{3\,\pi\,R^2}{2\,Y_0^2} + \frac{3\,\pi\,R^2}{2\,Z_0^2} - \frac{6\,R}{Y_0} = 0. \tag{62}$$

Eliminating Z_0 by using Eqs. (60) and (62) we obtain the following expression for D_y/R:

$$\frac{D_y}{R} = \left[\frac{\pi - \frac{6\,R}{Y_0} + 4.7124\,\frac{R^2}{Y_0^2} - \frac{4}{3}\,\frac{R^3}{Y_0^3}}{-4.7124 + 2\,\frac{R}{Y_0}}\right]^{1/2}\cdot$$

$$\left[\frac{23.3874 - 24.8733\,\frac{R}{Y_0} + 8.3776\,\frac{R^2}{Y_0^2} - 0.5926\,\frac{R^3}{Y_0^3}}{-56 + 10.5055\,\frac{R}{Y_0} - 11.7810\,\frac{R^2}{Y_0^2} + 5.3333\,\frac{R^3}{Y_0^3} - 0.7854\,\frac{R^4}{Y_0^4}}\right]. \tag{63}$$

We see therefore that D_y/R is not constant but it is a function of the position of the neutral axis. Therefore the torsionfree axes do not need to intersect at one point.

Cases in which the torsionfree axes must intersect at one point are the following.

1. When $n = 1$, as shown previously. This follows also from the validity of the principle of superposition.

2. When the cross section has two axes of symmetry and the function $\sigma = f(\dot{\varepsilon})$ is of the type $f(-\dot{\varepsilon}) = -f(\dot{\varepsilon})$.

3. When the cross section consists of portions with straight middle lines which — or the extensions of which — have a common intersection S. This point S is then the intersection of all torsionfree axes. The correctness of this statement follows from the fact that by properly selecting the system of axes XYZ we can in this case reduce $R(s)$ to be zero for all values of s. Fig. 7 shows an example of such a cross section.

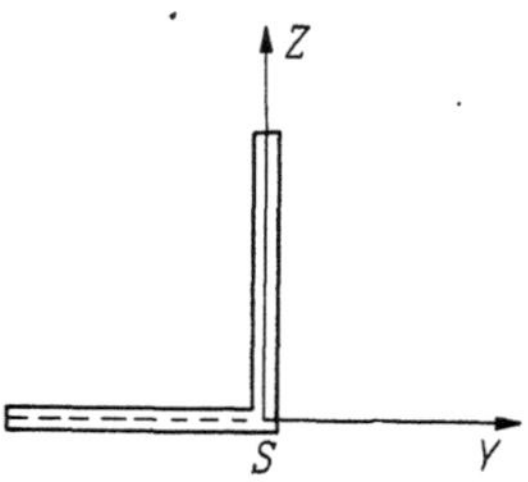

Fig. 7.

References

[1] Timoshenko, S.: Bulletin Institute of Ways of Communication, 1913.
[2] Maillart, R.: Schweizerische Bauzeitung, **77**, 195 (1921).
[3] Weber, C.: ZAMM, **4**, 334 (1924).
[4] Norton, H.: "The Creep of Steel at High Temperatures", New York: McGraw Hill, 1929.
[5] Hoff, N. J.: Quarterly of Applied Mathematics, **12**, 49 (1954).
[6] Eggenschwyler, A.: Schweizerische Bauzeitung, **76**, 206 (1920).
[7] Handelman, G. H.: Journal Aeronautical Sciences, **18**, 749 (1951).
[8] Johnson, W., and P. B. Mellor: Applied Scientific Research, AVI 467 (1957).
[9] Donath, E.: "Non-symmetrical Bending of Thin-walled Beams in Strain-Hardening and Creep", Ph. D. dissertation, Yale University, April 1959.
[10] Patel, S. A., and K. A. V. Pandalai: Stress Distribution in Beams of Thin Walled Sections in the Presence of Creep, Pibal Report 486, February 1959. Also, Patel, S. A., K. A. V. Pandalai and B. Venkatraman: Creep Stress Analysis of Thin-walled Structures, Pibal Report 497, July 1959.

Mechanics of Creep and Combined Stresses

By

Joseph Marin

Pennsylvania State University, State University, Pennsylvania, USA

Summary

A tensile creep strain-time-stress relation is proposed for the basis of creep stress and strain analysis. This relation has the advantages that: (1) it can be used for both short time transient creep as well as long time creep, (2) it gives a good representation of experimental creep data and (3) it leads to relatively simple mathematical creep stress and strain analysis for various types of structurel members and loadings.

The proposed tension creep relation is first discussed and various solutions in the mechanics of creep based upon this relation are summarized to show the relative simplicity of the theoretical solutions and the results obtained. Applications discussed include members subjected to both simple and combined stresses as well as statically indeterminate structural problems. In some cases experimental results are given to support the theories obtained using the proposed tensile creep equation. The paper concludes with a review of problems needing further attention.

1. Introductory Comments

An examination of the literature dealing with mechanics of creep shows that considerable mathematical difficulty arises when the theory is based on certain creep strain-stress-time relations. The current importance of short time transient creep makes it necessary to consider transient creep deformations in creep stress and strain analysis. A consideration of transient creep in stress and strain analysis introduces mathematical difficulties. For these reasons, it becomes important to select a tensile creep strain-stress-time relation which not only gives a good representation of the test data, but also leads to the simplest possible mathematical creep theories for the analysis of structural and machine members. The simple creep relation used in this paper and discussed below complies with these requirements.

2. Proposed Simple Tension Creep Strain-Stress-Time Relation

Many creep strain-stress-time relations have been proposed during the past 50 years. Some of these are reviewed by FINNIE and HELLER [1]. For many years creep relations were suggested for long periods of times, since until recently most applications were for long periods of time. For example, in the design of steam turbines, creep design has been based upon time periods of the order of 10 years. For predicting creep strain-stress-time relations covering intermediate or short periods of time, the initial and transient strains must be considered (Fig. 1). Further compli-

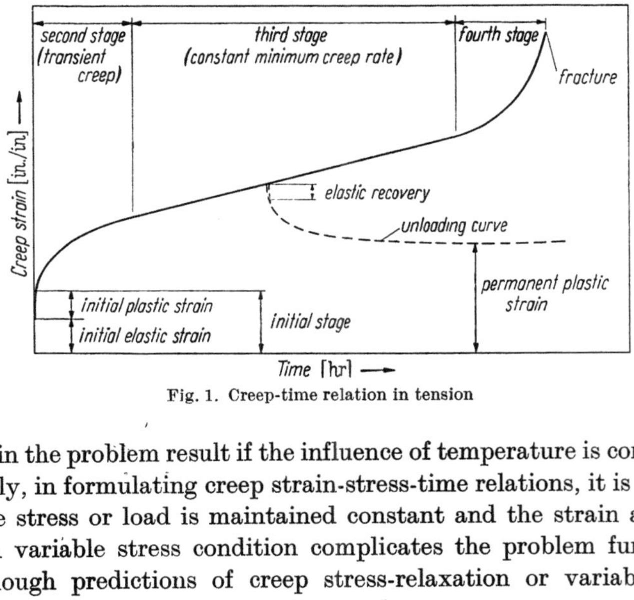

Fig. 1. Creep-time relation in tension

cations in the problem result if the influence of temperature is considered. Generally, in formulating creep strain-stress-time relations, it is assumed that the stress or load is maintained constant and the strain and time vary. A variable stress condition complicates the problem further.

Although predictions of creep stress-relaxation or variable stress conditions can be made theoretically based upon constant stress condition these theoretical results do not always agree with their experimental values.

The proposed creep strain-stress-time relation expressed by Eq. (3) has the following desirable characteristics:

1. It agrees well with available tension creep test results.

2. It leads to relatively simple mathematics in the mechanics of creep and stress and strain analysis of structural and machine members.

3. It is compatible with certain basic physical considerations regarding creep behavior.

The proposed relation first considers the constant stress creep strain (ε_t) of Fig. 1 as made up of four parts (as indicated in Fig. 2): an initial elastic strain ε_e, an initial plastic strain ε_p, a transient creep strain ε_{tr} and a minimum creep rate strain ε_m. That is

$$\varepsilon_t = \varepsilon_e + \varepsilon_p + \varepsilon_{tr} + \varepsilon_m. \tag{1}$$

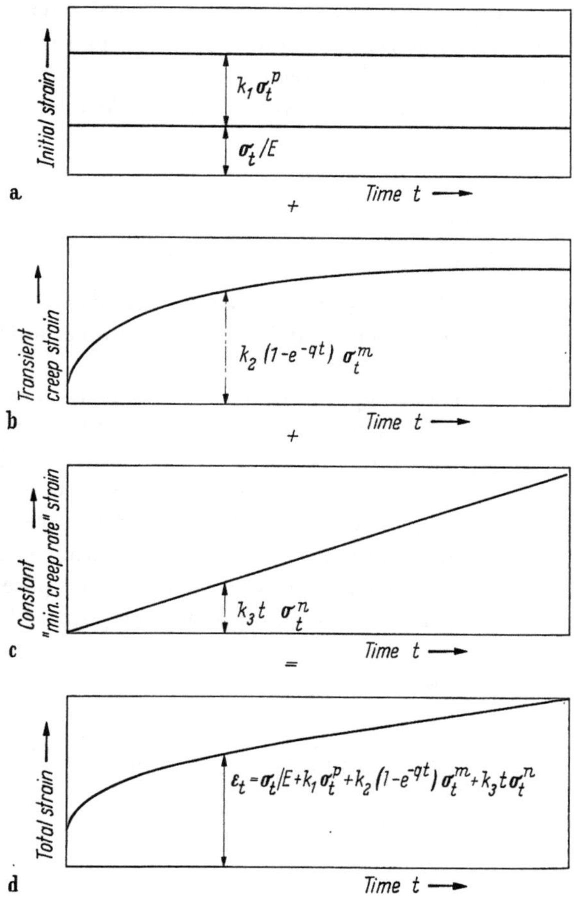

Fig. 2. Superposition method of defining creep

Defining the creep strain ε_t in terms of the components as given in Eq. (1), is in agreement with physical concepts of creep as discussed by OROWAN [2] and others. The next step in the development of a creep relation is the selection of appropriate expressions for each of the strains in Eq. (1). Many relations can be proposed for the various terms in Eq. (1). In Ref. [3] the following expression was proposed for defining the creep strain (ε_t),

$$\varepsilon_t = \sigma_t/E + k_1' \sigma_t^p + k_2 (1 - e^{-qt}) \sigma_t^n + k_3 t \sigma_t^n, \tag{2}$$

where σ_t = the constant tensile stress,
t = the time,
e = the base of natural logarithms
and E, k_1', p, k_2, q, n, k_3 and n = material constants.

Although Eq. (2) can be used as a basis for developing creep stress and strain analyses, in most problems it leads to mathematical difficulties. For this reason, a simplification of Eq. (3) will be made by assuming $p = m = n$ and $\sigma_t/E + k_1' \sigma_t^p = k_1 \sigma_t^n$. Then Eq. (2) can be expressed as

$$\varepsilon_t = B_t (\sigma_t/\sigma_0)^n, \tag{3}$$

where B_t is a time function of value

$$B_t = \sigma_0^n [k_1 + k_2 (1 - e^{-qt}) + k_3 t]. \tag{4}$$

In Eqs. (3) and (4), σ_0 is an arbitrary constant expressed in psi, so that $(\sigma_t/\sigma_0)^n$ in Eq. (3) is dimensionless and B_t has the simple units of in. per in.

It should be noted that Eq. (3) provides a simple creep strain-stress-time relation which at the same time retains flexibility in representing creep test data. This flexibility is provided by the five material creep constants appearing in Eq. (3). Most creep relations that have been proposed have fewer than five constants and for this reason do not generally provide as good a fit of the test data. Another characteristic of the proposed relation is that a mechanistic model, proposed independently by STULEN [4], can be used to represent the proposed creep relation (Fig. 3).

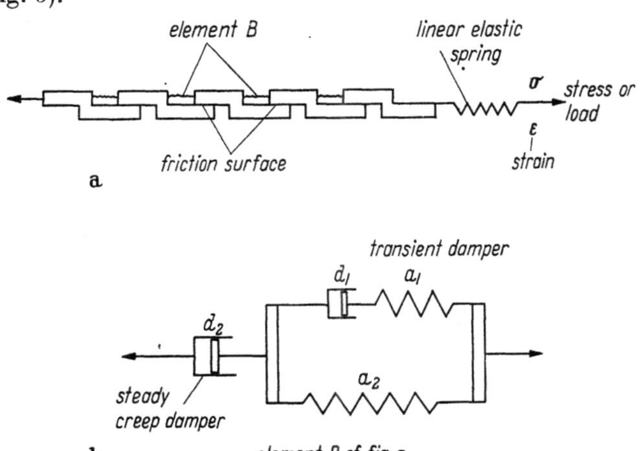

Fig. 3. Mechanistic model for creep behavior

DORN [5] in basic studies of the creep phenomenon has proposed equations expressing the influence of temperature on the creep strain-stress-time relations. These basic studies are based on the concept of activation energy [5]. For a constant temperature and assuming that no change in the structure of the material occurs with time, DORN proposes two relations:

For high stresses

$$d\varepsilon_t/dt = k_t\, e^{k_2}\, \sigma_t. \tag{5}$$

For low stresses

$$d\varepsilon_t/dt = k_1\, \sigma_t^n. \tag{6}$$

Eq. (6) for low stresses can be shown to be a special case of Eq. (3). Eq. (5) for high stresses can also be shown to be similar to Eq. (3), by noting that Eq. (5) may be expresses as

$$\varepsilon_t = k_t\, [\sigma_t + k_2'\, \sigma_t^2 + k_3'\, \sigma_t^3 + \cdots] \tag{7}$$

where

$$k_t = (A + B\,t). \tag{8}$$

Another related type of creep strain-stress-time equation can be found in the Russian literature [6]. A more general creep relation, similar to that proposed in Ref. [6] has been suggested by SHARMA [18].

$$\varepsilon_t = \sum_{i=1}^{i=n} k\, \sigma_t^p + \sum_{i=1}^{i=n} F(t)\, \sigma_t^q + \sum_{i=1}^{i=n} G(t)\, \sigma_t^m. \tag{9}$$

Eq. (9) is a more general relation than the proposed Eq. (3), since it uses many more constants and terms to express the creep strain. If $i = 1$, $p = q = m = n$, and $k = F = G = B_t$, then Eq. (9) reduces to Eq. (3). Although Eq. (9) will naturally provide a better representation of creep test results than the proposed relation as defined by Eq. (3), examination of test results indicated that Eq. (3) is sufficiently accurate for most purposes. In addition, mathematical complications in the use of Eq. (9), when applied to creep stress and strain analysis, gives further support to the uses of the simpler Eq. (3).

The applications of Eq. (3) in uniaxial tension creep stress-relaxation and dynamic creep are given in Ref. [7] and [8].

3. Mechanics of Creep Stress and Strain Analysis for Simple Stresses

Creep stress and strain analysis for various types of simple stress members, including members subjected to bending and torsion, can be obtained with relative simplicity using the tension creep Eq. (3) as a basis. A significant feature of Eq. (3) is that mathematically it is similar to the creep rate relation $C = B\,\sigma_t^n$, which has been frequently used in the past as the basis for creep stress-strain analysis. This means that available solutions based on $C = B\,\sigma_t^n$ can be adapted for solutions to creep problems based upon $\varepsilon_t = B_t\,\sigma_t^n$. Some simple creep stress problems based upon Eq. (3) will be summarized in the following section. These problems are discussed to show the relative simplicity in using Eq. (3) as a basis for mechanics of creep.

(a) Pure Bending of a Beam

Creep stresses in beams subjected to bending can be determined based upon the usual assumptions including: plane cross-sections remain plane, bending deflections are small and tension and compression material creep constants are the same. In evaluating these stresses the tension creep strain-stress-time relation of Eq. (3) is used, (in place of HOOKE's law) and the conditions of equilibrium of the stresses, namely $\Sigma M = M$ and $\Sigma F = 0$ are employed. It can then be shown that the maximum creep stress in a beam of rectangular cross-section subjected to pure bending is defined by

$$\sigma_b = \left(\frac{M\,h}{2\,I}\right)\left(\frac{2\,n+1}{3\,n}\right), \tag{10}$$

where M is the bending moment.

$I = b\,h^3/12$ is the moment of inertia of the cross-section b and h are the width and depth of the beam, respectively.

An examination of Eq. (10) shows that the maximum bending stress is equal to the elastic stress, which is the first term $M\,h/(2\,I)$, times a creep factor $(2\,n+1)/(3\,n)$.

It can also be shown, in a manner similar to that used in the elastic theory, that the differential equation for creep deflections is

$$D\,\frac{d^2y}{dx^2} = M^n, \tag{11}$$

where D corresponds to the flexural rigidity $E\,I$ in the elastic theory. For a beam of rectangular cross-section the value of D is,

$$D = (\sigma_0/B_t)^n\,[(2\,b)^n\,(h/2)^{2n+1}/(2+1/n)^n]. \tag{12}$$

An examination of Eqs. (11) and (12) shows that these equations are analagous to the corresponding elastic equations. However, since M in Eq. (11) is raised to a power n (instead of $n = 1$ in the elastic case) some mathematical complications arise in solving Eq. (11) for the creep deflections. In particular, the presence of n in Eq. (11) makes it impossible to use the simplifying procedure of superposition in creep structural analysis. The suitability of Eq. (11) in predicting creep deflection in specimens subjected to pure bending is illustrated in Fig. 4.

Theories for bending creep stresses and deflections for materials with different tension and compression creep properties can also be developed using an equation similar to Eq. (3) for compression creep. The problem of creep relaxation in bending can be analyzed based upon Eq. (3). Fig. 5 gives a comparison of experimental and theoretical results for bending creep relaxation in Plexiglas.

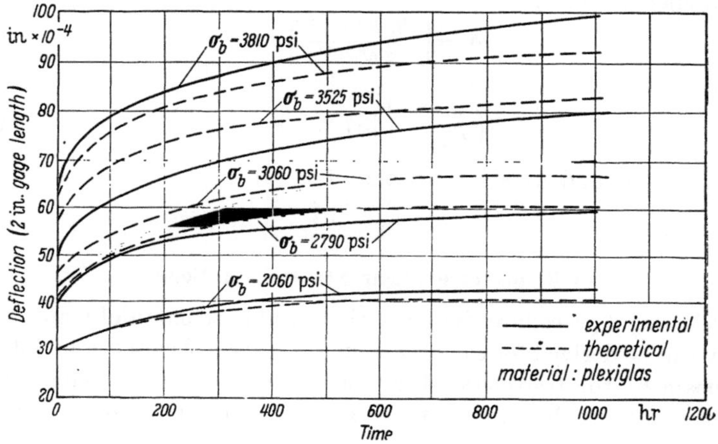

Fig. 4. Comparison of experimental and theoretical values of creep in bending

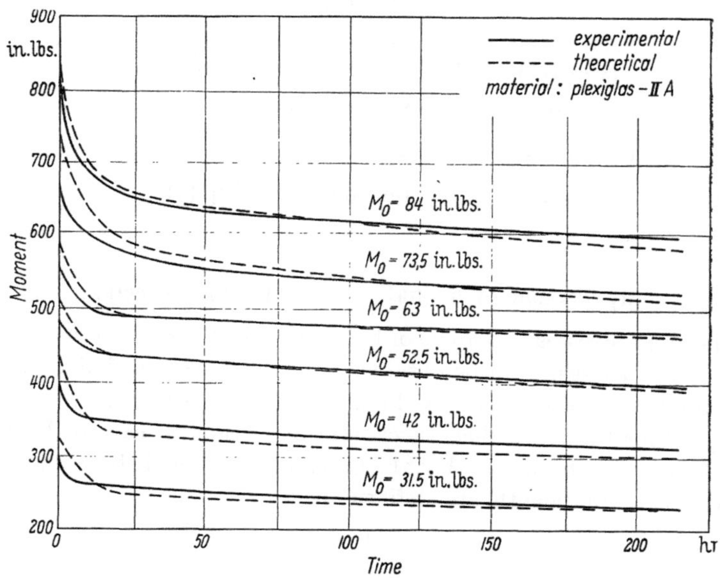

Fig. 5. Comparison of experimental and theoretical values for bending creep relaxation

(b) Creep Deflections in Beams due to Shear

For short beams the creep deflections due to transverse shear stresses cannot be neglected. By following procedures similar to those used in the elastic theory, expressions for shear creep deflections can be obtained.

For the case of a simply supported beam of length L with circular cross-section of diameter d and subjected to a uniformly distributed load F, it can be shown (9) that the maximum creep deflection including the shear deflection is

$$\delta = \left[\frac{B_t F^n L^{n+2}}{2 d^{3n+1}}\right] J, \tag{13}$$

where

$$J = \left(\frac{d}{2L}\right)^{n+1} [\alpha_1^n \alpha_2] + \alpha_3/\alpha_4 \tag{14}$$

and α_1, α_2, α_3 and α_4 are functions of n (9).

c) Transverse Shear Stress in a Beam

Following a procedure similar to that used in strength of materials, a theory can be derived for the maximum transverse shear stress produced in a beam by the transverse shear force. For a beam of circular cross-section it can be shown that the maximum transverse shear stress is,

$$\tau_m = \alpha_1 \frac{V}{A}, \tag{15}$$

where $\alpha_1 \approx \dfrac{\pi(2n+1)(4n+1)(6n+1)(97n^2+36n+3)}{4(n+1)(3n+1)(5n+1)(136n^2+42n+3)}$ (16)

V = the transverse shear force

and A = the cross-sectional area.

Eq. (15) is based upon the tension creep Eq. (3).

(d) Torsion of a Shaft of Circular Cross-Section

The creep stresses and angle of twist in a bar of circular cross-section subjected to twisting can be obtained using a procedure similar to that used in strength of materials and employing an equation corresponding to Eq. (3) for the creep shear stress-strain relation in place of HOOKE's Law. It can then be shown that the maximum torsional stress and angle of twist per unit length are defined by

$$\tau_m = \left(\frac{Tr}{J}\right)\left(\frac{3n_s+1}{4n_s}\right) \tag{17}$$

and

$$\beta = \left(\frac{B_s}{r S_0^{n_s}}\right)\left(\frac{Tr}{J}\right)^{n_s}\left(\frac{3n_s+1}{4n_s}\right)^{n_s}. \tag{18}$$

A creep relaxation theory for torsion based upon Eq. (3) can also be developed (10). Fig. 6 gives a comparison between theoretical and experimental values of torsion creep-relaxation relations.

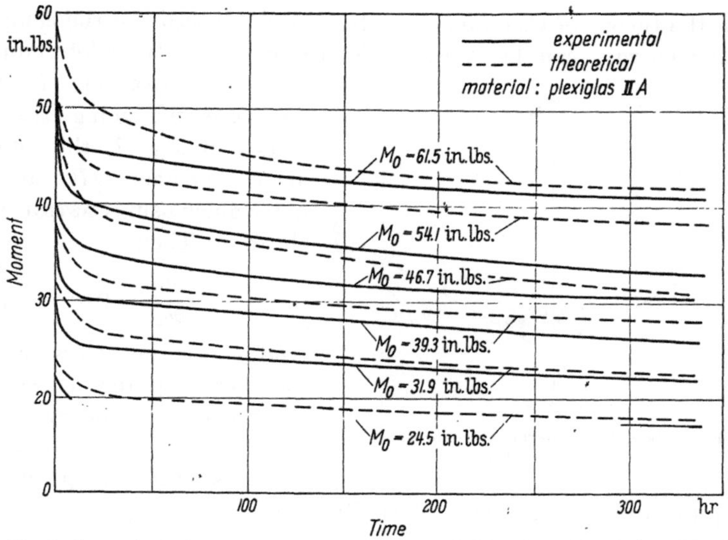

Fig. 6. Comparison of experimental and theoretical values for torsion creep relaxation

4. Mechanics of Creep for Statically Indeterminate Structures

Various conventional methods of structural analysis, based upon HOOKE's law, have been developed for the determination of statically indermediate forces and stresses in structural and machine members. These methods are generally of two types, *variational or energy methods* considering minimum energy of strain or *deformation methods* based upon the deformations produced. To some extent the foregoing methods have been adapted for the analysis of creep problems.

Energy methods. An analogy between steady-state non-linear creep and non-linear elasticity, with applications to creep problems has been made by HOFF [*11*]. Using this non-linear analogy, HOFF points out that the creep stresses based on the steady-state creep relation $\varepsilon = B\,\sigma^n$ can be obtained from the known solutions for non-linear elastic problems based on the tensile stress-strain relation $\varepsilon = B_e\,\sigma^{n_e}$ (where B_e and n_e are elastic constants obtained from a short time tension test). The proposed creep equation $\varepsilon = B_t\,\sigma^n$ [Eq. (3)] is similar to $\varepsilon = B\,\sigma^n$, since B_t simply represents a different time function to provide for small creep strains. For this reason the non-linear analogy proposed by HOFF can be used for short-time creep represented by Eq. (3). Because of the analogy between the non-linear creep equation and the non-linear elastic relation, any variational principle applicable to non-linear elastic systems may be applied to steady-state non-linear creep problems. For this reason the theory of CASTIGLIANO, when stated in terms of the complementary energy, can be used to determine statically indeterminate forces. For example, for the structure undergoing creep as shown in Fig. 7, this stress σ_b in member OB can be determined using CASTIGLIANO theorem. Then,

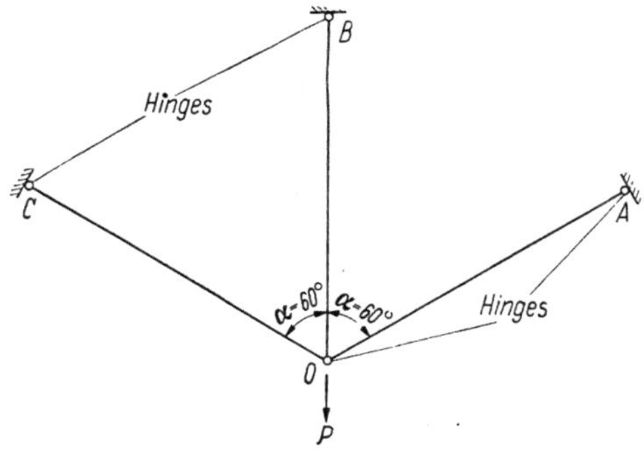

Fig. 7. Statically indeterminate structure subjected to creep

$$\frac{\partial \bar{U}}{\partial \sigma_b} = 0, \tag{19}$$

where σ_b = the stress in the member OB and $\bar{U}$ = the complementary energy which can be expressed in terms of σ_b by using the creep relation of Eq. (3) and the condition of equilibrium.

By the foregoing procedure

$$\sigma_b = \frac{2^{1/n}(P/A)}{1 + 2^{1/n}}. \tag{20}$$

Where A represents the area of each of the three members in Fig. 7. By writing the equation for equilibrium of forces at point 0, the other stresses σ_a in members OA and OC can be determined.

Variational methods of analysis with the strain composed of both elastic and steady creep strains have been developed by WANG and PRAGER [*12*]. Variational methods have also been applied to the more general case of non-steady-state creep combined with elasticity for the case of visco-elastic materials [*13*].

Deformation Methods. Creep deformation methods of structural analysis for statically indeterminate structures were developed by MARIN [*14*]. The various elastic deformation methods of structural analysis including the Slope Deflection, Moment Distribution and Relaxation Methods can be developed for methods of creep analysis. The derivation of these various methods of analysis can be made in a manner similar to that used in elastic theory. The development of the creep equations for the *Slope Deflection Method* of analysis will be given in the following. For this purpose consider the simply supported beam AB shown in Fig. 8, which is subjected to moments M_1 and M_2 at the ends. By the Slope Deflection Method

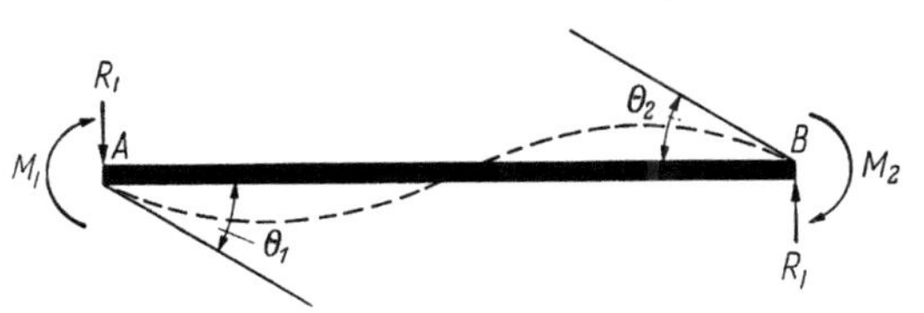

Fig. 8. Creep in statically indeterminate bending

the moments M_1 and M_2 can be evaluated in terms of the slopes or rotations θ_1 and θ_2 at the ends of the beam. To do this the the differential equation for creep deflection in a beam must be used. Based upon the tensile creep Eq. (3) the differential equation for deflection, as given by Eq. (11)., is

$$D\frac{d^2y}{dx^2} = M^n, \tag{a}$$

where D is defined by Eq. (12) for a rectangular cross-section. Since $M = R_1\,x - M_1$, Eq. (11) becomes,

$$D\frac{d^2y}{dx^2} = (R_1 - M_1)^n. \tag{b}$$

Integrating Eq. (b) twice,

$$D\frac{dy}{dx} = \frac{(R_1x - M_1)^{n+1}}{R_1(n+1)} + C_1, \tag{c}$$

$$D_y = \frac{(R_1x - M_1)^{n+2}}{R_1^2(n+1)\,(n+2)} + C_1\,x + C_2. \tag{d}$$

Using the conditions that the deflections are zero at both $x = 0$ and $x = L$, the constants of integration C_1 and C_2 are determined. Then the slopes θ_1 and θ_2 are the values of dy/dx as $x = 0$ and $x =$ L, respectively, as given by Eq. (c). Noting also that, for equilibrium, $R_1 = (M_1 + M_2)/\mathrm{L}$, the slops-deflection equations expressing the slopes in terms of the moments are,

$$D\,\theta_1(M_1 + M_2)^2\,(n+2)\,(n+1) = (M_1 + M_2)\,(-M_1)^{n+1}\,L(n+2) + L(-M_1)^{n+2} - (M_2)^{n+2}\,L \tag{21}$$

and

$$D\,\theta_2(M_1 + M_2)^2\,(n+2)\,(n+1) = (M_1 + M_2)\,(M_2)^{n+1}\,L(n+2) + L(-M_1)^{n+2} - (M_2)^{n+2}\,L. \tag{22}$$

In a particular case, the solution of Eqs. (21) and (22) simultaneously give values of the unknowns M_1 and M_2. However, the actual evaluation of M_1 and M_2 is much more complicated than in the linear elastic case. For most problems the Moment Distribution method is much simpler [*14*]. In considering methods of structural analysis involving the bending of beams, it is necessary to consider the value of n in Eq. (3) as an odd integer. This is necessary in order for the strain ε in Eq. (3) to be negative for compressive stresses and positive for tensile stresses. In an actual problem two solutions can be made for the odd values of n which bracket the actual value of n. Then assuming linear variation between the two answers obtained, the desired value is found by interpolation.

Methods of successive approxation. Numerical procedures have been developed for the analysis of complicated creep problems in which transient creep is considered [*15*]. In these procedures any tensile creep relation can be used.

5. Mechanics of Creep for Combined Stresses

There are many engineering applications in which combined stresses occur. These include pressure vessels and piping used in nuclear power plants and chemical industries and rotating disks used in gas and steam turbines. For this reason the creep properties of materials for combined states of stress must be considered. In considering combined stress creep properties, there are two main properties which need to be evaluated, namely, the stress-time to rupture properties and the creep strain-stress-time properties.

Time to rupture properties. Lack of experimental results on creep stress-rupture properties of materials under combined stresses suggests that a conservative basis be used to predict these strengths. For this reason the maximum shear theory is recommended as a basis for combined stress-creep rupture strength. To express a combined stressrupture theory, a relation defining the uniaxial tensile creep rupture properties must first be selected. Many test data indicate that the tensile rupture strength-time relation for many materials can be expressed by

$$t_r = k_r\, \sigma_{tr}^r, \tag{23}$$

where σ_{tr} = the rupture tensile strength for a temperature θ, t_r = the time to rupture at the stress σ_{t_r} and k_r and r are material constants.

To express a shear theory for stress rupture, the limiting value of the shear rupture stress $\sigma_{s_r} = \sigma_{t_r}/2$ is equated to the three principal shear stresses and the value of σ_{t_r} is replaced by its magnitude as defined by Eq. (23). Then for triaxial stresses the creep rupture-time strength relations become,

$$\left.\begin{aligned} \sigma_{1_r} - \sigma_{2_r} &= (t_r/k_r)^{1/r} \\ \sigma_{2_r} - \sigma_{3_r} &= (t_r/k_r)^{1/r} \\ \sigma_{3_r} - \sigma_{1_r} &= (t_r/k_r)^{1/r} \end{aligned}\right\} \tag{24}$$

Eqs. (24) express the time to rupture strength for combined stresses based upon the conservative shear theory of failure.

Creep Strain-Stress-Time Properties. Several theories have been proposed for combined stresses. All these theories are based upon an assumed tensile creep strain-stress-relation and predict the principal creep strains in terms of the principal stresses and the tension creep material constants. Based upon Eq. (3) for uniaxial tension creep, and assuming that there is a unique relation between the octahedral shear stress- and strain, that the volume remains constant and that the principal shear stresses are proportional to the principal shear strains; it can be shown that the principal creep strains are expressed by

$$\left.\begin{aligned} \varepsilon_1 &= B_t(\sigma_1/\sigma_0)^n\left(R^{\frac{n-1}{2}}\right)(1-\alpha/2-\beta/2)\\ \varepsilon_2 &= B_t(\sigma_1/\sigma_0)^n\left(R^{\frac{n-1}{2}}\right)(\alpha-\beta/2-1/2)\\ \varepsilon_3 &= B_t(\sigma_1/\sigma_0)^n\left(R^{\frac{n-1}{2}}\right)(\beta-1/2-\alpha/2), \end{aligned}\right\} \tag{25}$$

where $R = 1 + \alpha^2 + \beta^2 - \alpha - \alpha\beta - \beta$,

$\alpha = \sigma_2/\sigma_1$, $\beta = \sigma_3/\sigma_1$

and σ_1, σ_2 and σ_3 are the three principal stresses.

In developing Eqs. (25) it is assumed that the stress ratios α and β remain constant. That is, the steady state condition is assumed.

Relatively few experiments have been conducted to determine the validity of combined creep strain relations such as given by Eq. (25). Most of the tests that have been conducted have used thin-walled cylindrical specimens subjected to torsion and axial tension. A few creep experiments have been made using thin-walled cylindrical specimens subjected to internal pressures and axial tension. A study of the available experimental results and their interpretation by the available theories does not provide conclusive results. It appears that further experimental studies must be made.

There are a number of limitations in the available theories for predicting combined creep strains. These include:

1. The principal stress ratios $\alpha = \sigma_2/\sigma_1$ and $\beta = \sigma_3/\sigma_1$ must be maintained constant. That is, the steady state condition is assumed.

2. The entire strain including the initial strain, transient creep strain and constant rate creep strain are not satisfactorily considered. In the theory defined by Eq. (25), the initial elastic and plastic strain is provided by expressing these strains approximately as $\varepsilon = k_1\,\sigma^n$.

3. Available theories do not agree generally with the combined creep stress results. In addition, the amount of available experimental information is meager and inadequate.

4. Available theories generally do not provide for anisotropy and for different creep constants in simple tension and compression. An attempt to provide for this factor is given in a paper by Marin and Pao [*16*].

5. The tensile or compressive creep properties in various directions are the same. The influence of directional effects can be considered as has been provided in plasticity theory.

6. The creep strains are small. In some applications the creep strains may be large or finite. A theory for the creep of a thick-walled cylinder considering large strains was developed by RIMROTT [*17*]. There is a need to develop further mechanics of creep considering large or finite strains.

7. It is assumed that anisotropy does not develop with time during flow. The influence of time on anisotropy has not been determined and presents a complication in the development of the mechanics of creep.

8. The loads are statically applied. Theories based upon fluctuating dynamic stress conditions can be developed. One theory assuming constant creep rate strain only was recently developed by SHARMA [*18*].

A theory for non-steady state of creep relaxation for combined stresses was presented by GRIFFITH and MARIN [*19*]. Some preliminary creep relaxation tests for combined torsion and tension support the theory developed [*19*].

Prediction of Instability Creep Loads. HOFF [*20*] developed a theory for predicting the theoretical value of the time for fracture in a tensile creep specimen, based upon the condition that the creep strain approaches infinity at fracture. This concept can be applied to evaluate theoretical fracture time in members subjected to combined stresses. Based upon the criterion proposed by HOFF, the time for creep fracture in a thick-walled cylindrical pressure vessel subjected to internal pressure has been determined [*21*].

Other Applications of Creep Theories. Steady state solutions for creep of circular plates under various types of lateral loads are summarized by FINNIE and HELLER [*1*]. Non-steady state solutions for circular plates have been made by ODQVIST [*22*]. On the creep buckling of columns several theories have been proposed [*1*], while a few solutions are available on the special problems of creep buckling of plates and shells. The evaluation of stress and strain concentration factors in creep introduces many difficulties. An approximate solution for the strain concentration in a plate with a hole subjected to axial tension was obtained by MARIN [*23*]. An examination of the literature on Mechanics of Creep shows that there are many problems that have not been solved.

6. Needs for further Study on Mechanics of Creep

Although the mechanics of creep has been investigated for over a half century, there are many theoretical and experimental problems needing further investigation. No doubt, some of these problems have not been solved because they have only become important in recent years. Other problems have not been completely resolved because of the complexity of the theoretical analysis or the experimental procedure involved. Only a few of the areas requiring further study will be mentioned in this review. These include:

1. Development of creep strain-time-stress-temperature relations for variable stress and variable temperaturs conditions. Various types of simple and combined stresses need to be investigated experimentally in formulating a creep law for these variable load and temperature conditions.

2. Further experimental study on the steady state creep behavior of materials under combined stresses needs to be made. The few investigations that have been conducted include primarily a few biaxial stress experiments using tubular specimens subjected to torsion and axial tension. No experimental creep studies on biaxial compression and triaxial stresses have been made. Before we can have confidence in a theory for creep under combined stresses much more experimental evidence is needed.

3. Development of new theories for creep stress and strain analysis of structural members for both the steady and non-steady states.

4. A more complete investigation both experimentally and theoretically of dynamic creep for both simple and combined stresses.

7. Concluding Remarks

In this paper various problems in mechanics of creep were discussed to show how relatively simple solutions can be obtained when the tension creep strain-stress-time relation expressed by Eq. (3) is used. In addition to providing simple mathematical solutions, the proposed equation considers the important case of short-time transient creep. Furthermore, the tension creep relation proposed has considerable experimental support. The five material constants appearing in this relation afford considerable flexibility in the representation of creep test results.

References

[1] Finnie, I., and W. R. Heller: Creep of Engineering Materials, McGraw Hill Book Company, 1952.

[2] Orowan, E.: "Creep in Metallic and Non-Metallic Materials", Proc. First US National Congress on Applied Mechanics, 1953, pp. 453—472.

[3] Pao, Y. H., and J. Marin: "An Analytical Theory of the Creep Deformations of Materials", Transactions of ASME, 75, 245—253 (1953).

[4] Stulen, F. B.: Report No. C-2721, Mechanical Properties of Metals, Curtiss-Wright Corporation, Caldwell, N. J., 1957.

[5] DORN, J. E.: "The Spectrum of Activation Energies for Creep", Proc. of a Seminar on "Creep and Recovery", published by American Society for Metals, 1957, pp. 255—284.

[6] RABOTNOV, Y. N.: "The Effect of Changing Loads During Creep", N. P. L. Symposium, p. 221, Her Magesty's Stationary Office, England, 1956.

[7] MARIN, J., and J. E. GRIFFITH: "Creep Relaxation of Plexiglas II-A for Simple Stresses", Journal of Engineering Mechanics Division, American Society of Civil Engineers, **82**, No. E. M. 3, July 1956.

[8] WEISSMANN, G. F., Y. H. PAO and J. MARIN: "Prediction of Creep Under Fluctuating Stress and Damping from Creep Under Constant Stress", Proc. Second US Congress of Applied Mechanics, 577—583 (1954).

[9] MARIN, J.: "Determination of the Creep Deflection of a Rivet in Double Shear", Journal of Applied Mechanics, American Society of Mechanical Engineers, Paper No. 58-A-47, 6 pages, 1958.

[10] MARIN, J., and J. E. GRIFFITH: "Creep Relaxation in Torsion", Proc. 9th International Congress on Applied Mechanics, 8, 316—327 (1957).

[11] HOFF, N. J.: "Approximate Analysis of Structures in the Presence of Moderately Large Creep Deformations", Quarterly of Applied Mathematics, **12**, No. 1, 49 (1954).

[12] WANG, A. J., and W. PRAGER: "Thermal and Creep Effects in Work-Hardedening Elastic-Plastic Solids", Journal of Aeronautical Sciences, **21**, 343 (1954).

[13] ALFREY, T.: "Non-Homogeneous Stresses in Viscoelastic Media", Quarterly of Applied Mathematics, **2**, 113 (1944).

[14] MARIN, J.: "Mechanics of Creep for Structural Analysis", Trans. ASCE, **108**, 453—481 (1943).

[15] MENDELSON, A., M. H. HIRSCHBERG and S. S. MANSON: "A Genreal Approach to the Practical Solution of Creep Problems, Journal of Basic Engineering, December, 1959, Trans. ASME, **81**, Series D, No. 4, 585—598.

[16] MARIN, J., and Y. H. PAO: "A Theory for Combined Creep Stress Relations for Materials with Different Properties in Tension and Compression", Proc. First National US Congress on Applied Mechanics, published by ASME, (1952), pp. 585—595.

[17] RIMROTT, F. P. J.: "Creep of Thick-Walled Tubes Under Internal Pressure Considering Large Strains", Trans. ASME, (1959), p. A-271.

[18] SHARMA, M. G.: "Some Non-Linear Problems in Mechanics of Materials", Ph. D. thesis, The Pennsylvania State University, 1960.

[19] GRIFFITH, J. E., and J. MARIN: "Creep Relaxation for Combined Stresses", Journal of Mechanics and Physics of Solids, **4**, 283—293 (1956).

[20] HOFF, N. J.: "The Necking and Rupture of Rods Subjected to Constant Tensils Loads, Trans. ASME, **75**, 105—109 (1953).

[21] RIMROTT, F. P. J., E. J. MILLS and J. MARIN: "Prediction of Creep Failure for Pressure Vessels", paper presented at June 1960 National Applied Mechanics Meeting of the ASME.

[22] ODQVIST, F. K. G.: "Non-Steady Membrane Creep of Circular Plates", Arkiv. for Fysik, utgivet Av. kungl. Svenska Vetenskapraakademien, **16**, No. 43, 527—531 (1960).

[23] MARIN, J.: "Creep Stresses and Strains in an Axially Loaded Plate with a Hole", Journal of the Franklin Institute, **208**, No. 1, 53—60 (1959).

Discussion

E. H. LEE, Brown University: The consideration of creep problems presented in this paper appears to be limited to applications of Eq. (3), and this applies for stress suddenly applied at $t = 0$ and maintained constant. For situations involving varying stress, or unloading to zero stress, appropriate interpretations must be introduced. For example, in the case of unloading to zero stress, direct application of (3) would determine instantaneous recovery to zero strain, which is unrealistic, and some modification permitting retention of at least some part of the creep strain must be introduced. Presumably, the relaxation curves shown in Fig. 5 are based on such an interpretation. In view of the relatively small stress relaxation shown, however, perhaps it was possible to use (3) as an approximation, and that this is responsible for the somewhat high values of stress deduced theoretically.

The interpretation of (3) for varying stress is extremely important for stress analysis, since even for constant surface tractions, the influence of combined stresses can lead to a varying stress distribution as in viscoelasticity[1]. To make measurements to elucidate this situation, it seems to the discusser that study of recovery after unloading is particularly pertinent, since different possible assumptions such as direct application of Eq. (3), or use of the equation of state for $\dot{\varepsilon}$ deduced from (3) by differentiation, lead to widely contrasting behavior.

F. K. G. ODQVIST, Royal Institute of Technology, Stockholm: It is difficult for me to reconcile Author's Eq. (3) with the statement that "with removal of the load, creep recovery can be predicted" with this equation.

Author's closing remarks: As Professor LEE points out, the paper by the writer is limited to cases where the stress is suddenly applied and remains constant. The writer agrees with Professor LEE that there is a need for a more general theory to provide for variable stress conditions and stress relaxation. Similarly, there is a need for experiments to determine the recovery behavior of materials after unloading.

As far as Professor ODQVIST's comment is concerned, the original paper did have a statement that creep recovery could be predicted. However, this statement was deleted from the final manuscript.

[1] LEE, E. H.: Viscoelastic stress analysis. Proc. First Symposium on Naval Structural Mechanics, edited by J. N. GOODIER and N. J. HOFF, New York: Pergamon Press, (1960) p. 456.

Lifetime of Structures Subjected to Varying Load and Temperature

By

Shuji Taira

Kyoto University, Kyoto, Japan

Summary

For designing structures which are to be subjected to variable loads and temperatures, it is desirable to know the creep properties of materials under such conditions. Formulas are derived to predict creep in the transient and steady stages and creep rupture time under variable loads and temperatures from data obtained in simple creep or creep rupture tests. The concepts of equivalent stress and equivalent temperature are introduced and the validity of the formulas is discussed on the basis of the results of experiments performed with heat resistant alloys.

Formulas for creep in the transient and steady stages under dynamic load are also presented in connection with fatigue strength. They permit the prediction of dynamic creep strains and creep rupture times from data obtained in static creep and creep rupture tests combined with fatigue test data. Examples are given for several alloys. The analysis is extended to the prediction of creep of structural members under combined static tension and dynamic bending and the results are presented together with experimental results.

1. Introduction

In designing structures which are to be subjected to varying loads and temperatures it is important to know the creep properties of materials under such conditions. The influence of varying load and temperature on creep behavior of metallic materials has attracted the interest of many researchers and a number of papers [*1*—*19*] have been published on the subject. However, the majority of these papers deal with the metallographic or metal-physical aspects of the problem and are not intended for application to problems in mechanics.

The author is also interested in the fundamental problem of creep of metals under varying load and temperature, and he has studied the influence of the history of loading and heating by carrying out experiments. Since it is well known that the creep phenomenon is a process composed of work-hardening and recovery, the influence of varying loads and temperatures can be discussed as a combination of the two effects. The author suggests that the influence be interpreted from the viewpoint of viscoelasticity by introducing a new concept, the internal yield stress. The internal yield stress is the stress level at which the material deforms without the aid of thermal activation energy. Work-hardening increases the internal yield stress and recovery decreases it. The influence of a variation in load and temperature on the creep behavior of metals is thus reduced to the influence of this variation on the internal yield stress.

It is reasonable to assume that the internal yield stress is affected by changes in the metallographic structure during the creep process. Changes in the submicroscopic structure should also be an influencing factor. The former effect is likely to prevail over the latter, but it forms part of a study of metallography. Therefore, the present paper can be taken as dealing with the changes in the submicroscopic structure.

According to the papers published so far, including those due to the author, variations in load or temperature do influence the internal yield stress, but the influence is not serious insofar as changes in metallographic structure do not appear. In other words, it can be said that if the metallographic structure is stable the influence of variations in load or temperature is not large from the standpoint of engineering applications.

The following problem is also of importance: do variations in load and temperature during creep accelerate or retard changes in the metallographic structure and the phenomena of precipitation [*20*, *21*], recrystallization [*20*], spheroidization [*16*] and so forth? The study of this problem has not yet progressed sufficiently for a definite answer. Only a vague statement could be found according to which cyclic variations in temperature during creep accelerate the changes in the microscopic structure [*16*, *19*]. A quantitative estimation of this influence is beyond the range of possibilities at the present time.

In spite of these circumstances the engineer needs an estimation of creep under variable load and temperature. For this reason the author has developed formulas permitting an estimation of the lifetime of materials under variable load and temperature on the basis of data obtained from tests carried out with constant load and temperature. The accuracy of the formulas was examined by means of experiments on various metallic materials. Although in some cases the formulas did not lead to a satisfactory prediction of lifetime, the theory proposed in this paper should be of some use in engineering design.

2. Creep under Varying Load

According to the concept of the mechanical equation of state in solids [*22, 23*], the rate of strain is a function of the instantaneous values of strain, stress and temperature, and is independent of the past history of these quantities. One can write

$$\frac{d\varepsilon}{dt} = f(\varepsilon, \sigma, T). \tag{1}$$

Of course, it is known that Eq. (1) is not true for materials in general, and that the past history affects the rate of deformation.

One of the characteristic behavior types of materials at elevated temperatures is the existence of time-dependent recovery. The influence of load variation on creep behavior is connected with the processes of recovery and work-hardening.

It is suggested, therefore, that the creep process be represented by Eq. (2), instead of by Eq. (1):

$$\frac{d\varepsilon}{dt} = f(\sigma, \sigma, T), \tag{2}$$

where $\bar{\sigma}$ is the resistance against further deformation, that is the critical stress at which the material deforms without the aid of additional activation energy [*23, 24*].

It is also assumed that at any moment the rate of change in the internal yield stress is expressed by a function of both the rate of work-hardening $(\partial\bar{\sigma}/\partial\varepsilon)$ and the rate of recovery $(\partial\bar{\sigma}/\partial t)$ as follows:

$$\delta\bar{\sigma} = \left(\frac{\partial\bar{\sigma}}{\partial\varepsilon}\right)\delta\varepsilon + \left(\frac{\partial\bar{\sigma}}{\partial t}\right)\delta t, \tag{3}$$

where $\delta\varepsilon$ and δt are increments of strain and time at any moment.

Adopting these assumptions we have

$$\begin{aligned} \delta\varepsilon &= F(\bar{\sigma}, \sigma, T)\,\delta\bar{\sigma}, & \delta\varepsilon &> 0 \\ \delta t &= [F(\bar{\sigma}, \sigma, T)/f(\bar{\sigma}, \sigma, T)]\,\delta\bar{\sigma}, & \delta t &\geqq 0 \\ F(\bar{\sigma}, \sigma, T) &= 1\Big/\left\{\left(\frac{\partial\bar{\sigma}}{\partial\varepsilon}\right) + \left[\left(\frac{\partial\bar{\sigma}}{\partial t}\right)\Big/ f(\bar{\sigma}, \sigma, T)\right]\right\}. \end{aligned} \tag{4}$$

If $(\partial\bar{\sigma}/\partial\varepsilon)$, $(\partial\bar{\sigma}/\partial t)$ and $f(\bar{\sigma}, \sigma, T)$ were given in analytical form, the relation of $\delta\varepsilon$ to $\delta\bar{\sigma}$ or that of δt to $\delta\bar{\sigma}$ could be found analytically. If this were the case, we would know the relation between ε and t with $\bar{\sigma}$ as parameter, because we could solve the simultaneous Eqs. (4). More details of this theory are not given here because of limitations in space.

According to the theory, the strain in materials which have experienced a successive variation in stress, $\sigma_0, \sigma_1, \ldots$, at a constant temperature

T_0 can be expressed as

$$\varepsilon = \int_{\bar{\sigma}_0}^{\bar{\sigma}_1} F(\bar{\sigma}, \sigma_0, T_0)\, d\sigma + \int_{\bar{\sigma}_1}^{\bar{\sigma}_2} F(\bar{\sigma}, \sigma_1, T_0)\, d\bar{\sigma} + \cdots + \int_{\sigma_{n-1}}^{\bar{\sigma}} F(\bar{\sigma}, \sigma_{n-1}, T_0)\, d\bar{\sigma}. \tag{5}$$

In order to solve Eq. (5), it is necessary to know the analytical forms of $f(\bar{\sigma}, \sigma, T)$, $(\partial\bar{\sigma}/\partial\varepsilon)$ and $(\partial\bar{\sigma}/\partial t)$. However, an analytical derivation of these forms is a difficult task and for this reason the experimental method [25, 26] is adopted in practice. For instance, creep tests at constant temperature and constant load carried out at various stress levels constitute a method of solution [25].

A comparison of analytical and experimental results obtained with carbon steels and low alloy steels has shown that the influence of stress variation on creep in the transient and steady stages is not serious from the standpoint of engineering applications as long as the variation in stress is cyclic and the time τ at the high stress level is not small as compared to the period p of cyclic variation. In these cases the creep strain in a material subjected to periodically varying loads can be predicted satisfactorily by adopting the concept of the mechanical equation of state in a solid on the strain-hardening basis, and by neglecting the influence of the stress history [27, 28].

However, when the time ratio τ/p of periodic variation of stress is small, recovery proceeds during the low stress or zero stress period and it causes a temporary increase in strain at the moment of quick reloading or increase in load. In such a case, a prediction on the basis of the mechanical equation of state is not satisfactory and one has to resort to a prediction according to Eq. (5).

Fig. 1 [29] shows the case of creep of a 0.14% carbon steel under cyclic loading and unloading at 450°C. The full lines show the experi-

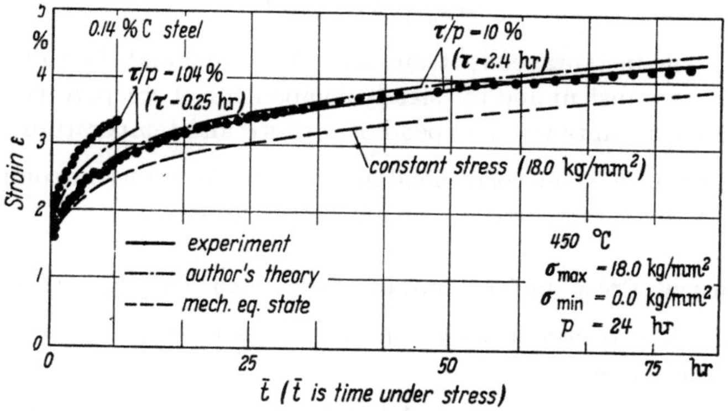

Fig. 1. Creep of 0,14% C steel at 450°C under periodically varying stress (OHJI [29])

mental creep curves for τ/p of 1.04 and 10%. The dotted and chain lines are analytical creep curves corresponding to the author's theory and to the mechanical equation of state in solids, respectively.

With regard to creep in the third stage it was found that periodic stress application initiates the third stage of creep earlier than the constant stress test although the reason is not clear at present.

3. Creep under Varying Temperature

Temperature variation is known to be an influencing factor in the process of creep. When the temperature is raised quickly during the creep test, the strain increases at first more rapidly than predicted by the mechanical equation of state (which disregards the effect of the temperature history) [*31*]. In contrast, when the temperature is lowered quickly, we see a temporary retardation of creep which is known as the incubation or induction period [*14, 15, 31*]. When the temperature variation is periodic, the two effects of temporary increase in creep because of an increase in temperature and the temporary retardation of creep because of a decrease in temperature cancel each other and the combined influence is not serious from the standpoint of application to mechanics problems [*31, 32*].

Thus creep under a periodic variation in the temperature can be estimated satisfactorily with the aid of formulas derived without regard to the influence of the temperature history, at least as far as applications to design are concerned [*22, 23*].

When the mechanical equation of state is assumed to be valid, we have the formula

$$\frac{d\varepsilon}{dt} = f(\varepsilon, \sigma, T). \tag{1}$$

The total creep strain is expressed, as is well-known, by

$$\varepsilon = \varepsilon_0 + \beta\, t^n + \varkappa\, t, \tag{6}$$

where ε_0 is the instantaneous strain, and the second and the third terms represent the transient and the steady components of creep, respectively. β and $\varkappa$ are constants which depend on stress σ and temperature T [*33*].

For simplicity, creep corresponding solely to the transient component will be discussed first under varying temperatures.

3.1. Transient Creep under Constant Stress and Varying Temperature

Creep in the transient stage can be expressed as a function of temperature and time by

$$\varepsilon = \varepsilon_0 + A_0 \exp\left(-\frac{K}{T}\right) t^n, \tag{7}$$

where A_0 and K are constants which depend only in the stress σ [32, 34]. To be exact, the instantaneous strain ε_0 is also dependent on temperature [32], but this dependence will be disregarded here. With the analytical representation of the strain in the form of Eq. (7), Eq. (1) becomes

$$\frac{d\varepsilon}{dt} = A_0^{\frac{1}{n}}\, n \exp\left(-\frac{1}{n}\frac{K}{T}\right)(\varepsilon - \varepsilon_0)^{\frac{n-1}{n}}. \tag{8}$$

When the temperature T varies with a period p, the total strain ε can be obtained by solving Eq. (8) for ε. That is,

$$\int_{\varepsilon_0}^{\varepsilon} (\varepsilon - \varepsilon_0)^{\frac{1-n}{n}}\, d\varepsilon = A_0^{\frac{1}{n}}\, n \int_0^t \exp\left(-\frac{1}{n}\frac{K}{T(t)}\right) dt. \tag{9}$$

From this we have

$$\varepsilon = \varepsilon_0 + A_0\, I^n\, t^n, \tag{10}$$

where

$$I = \frac{1}{p}\int_0^p \exp\left[-\frac{1}{n}\frac{K}{T(t)}\right] dt. \tag{11}$$

The form of the analytical expression in Eq. (10) is the same as that in Eq. (7) for creep under constant temperature. This means that creep under periodic changes in temperatures with a period p is equivalent to creep at the constant temperature T_{ec}, provided T_{ec}, the equivalent constant temperature, is chosen as

$$T_{ec} = -\frac{K}{n}\Big/ \ln I. \tag{12}$$

The constants K and n in Eq. (12) can be determined from creep tests made at constant temperature and under constant stress. Hence creep under periodic temperature changes can be estimated from the data of the ordinary creep test carried out at a constant temperature [34].

3.2. Creep in the Steady Stage under Varying Temperature and Constant Stress

The discussion of the last section can be extended to the case of creep in the steady stage. According to the author's investigations, there exists a relation between the creep rate at any time of the transient stage and that of the steady stage [33]. When the creep strain is expressed in the form of Eq. (7), the creep rate is

$$\frac{d\varepsilon}{dt} = \beta\, n\, t^{n-1} + \varkappa = u\big|_t + \varkappa. \tag{13}$$

The first and the second term are the creep rates in the transient and steady stages, respectively. β and $\varkappa$ are constants dependent on stress σ and temperature T. Surveying a great number of the results of creep tests reported on various metallic materials it is found that there is a general relation between the creep rate at any time in the transient stage and that in the steady stage:

$$u\,|_t\, \varkappa^{-C_1} = C_2\, t^{n-1}, \tag{14}$$

where $C_1 \cong \frac{1}{2}$ and $C_2 = 1 \sim 6$ for a great number of materials [33, 35].

When Eq. (14) is true, we have

$$\varkappa = (A_0\, n/C_2)^{1/C_1} \exp\left(-\frac{1}{C_1}\frac{K}{T}\right). \tag{15}$$

Applying a procedure similar to the reduction of Eq. (10), we obtain

$$\varkappa = (A_0\, n/C_2)^{1/C_1}\, I^{n/C_1} \tag{16}$$

as the creep rate in the steady stage of creep under varying temperature. Comparing Eq. (15) and Eq. (16) we have

$$T_{ec} = -\frac{K}{n}\Big/\ln I. \tag{17}$$

Eq. (17) is the same as Eq. (12).

It follows from the discussion that the creep rate in the steady stage in tests under varying temperature is equal to the rate of creep in the steady stage in tests at a constant temperature T_{ec}, the equivalent constant temperature.

Thus, for creep under periodic stepped variation of temperature the creep strain that includes both the transient and the steady creep is expressed, if Eqs. (10) and (15) are substituted in Eq. (6), as

$$\begin{aligned} \varepsilon = \varepsilon_0 &+ A_0 \exp\left(-\frac{K}{T'_{ec}}\right) t^n \\ &+ \left(\frac{A_0\, n}{C_2}\right)^{1/C_1} \exp\left(-\frac{1}{C_1}\frac{K}{T'_{ec}}\right) t, \end{aligned} \tag{18}$$

where

$$1/T'_{ec} = \frac{1}{T+\Delta T} - \frac{n}{K}\ln\left[\frac{\tau}{p}\exp\left\{-\frac{K}{n}\Big/\left(T_0 + \frac{T_0^2}{\Delta T}\right)\right\} + \left(1 - \frac{\tau}{p}\right)\right].$$

In this formula p is the period of temperature variation and τ is the time elapsed while the specimen is held at a temperature that is higher or lower by ΔT than the standard temperature T_0. It is obvious that for $\tau/p = 1$, $T_{ec} = T_0$, since this is the case of creep under a constant temperature T_0; and for $\tau/p = 0$, $T_{ec} = T_0 + \Delta T$, since this is the case of creep at a constant temperature of $T_0 + \Delta T$.

The validity of the prediction of the creep of materials under periodic stepped variations of temperature from data of creep tests carried out a constant temperatures was examined by experiment. Some of the results obtained are exhibited here. Fig. 2 is the case of 13 Cr steel (oil quenched from 970°C and tempered at 760°C) under a periodic variation of the temperature as shown in the figure, with the period one day. The circles in the figures are experimental points and the full lines represent an analysis based on the data of creep tests on the same material at constant temperatures of 400, 500 and 550°C. It is found that the analytical curves are in a fairly good agreement with the experimental points.

Fig. 3 is the case of AISI 318 type stainless steel (water quenched from 1100°C) under a steady stress of 15.0 kg/mm². A saw-tooth like

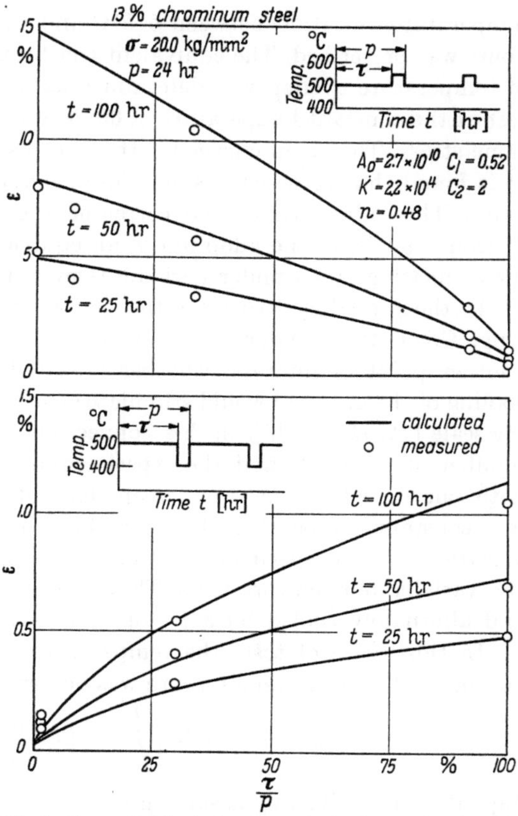

Fig. 2. Creep of 13 Cr steel under periodically varying temperature (NISHIHARA, TAIRA, TANAKA and OHNAMI [33])

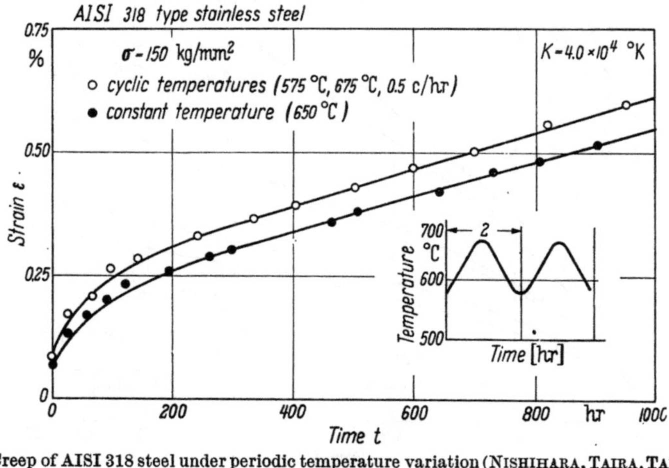

Fig. 3. Creep of AISI 318 steel under periodic temperature variation (NISHIHARA, TAIRA, TANAKA and OHNAMI [33])

temperature variation between 575°C and 675°C with a period of two hours was prescribed. The equivalent constant temperature for this case of temperature cycling was determined as 650°C and therefore a creep test at the constant temperature of 650°C was also carried out under the same stress. The creep curves for the two tests are shown in the figure. It is found that these curves are close even at an elapsed time of 1000 hours. This substantiates the correctness of the present mode of prediction, that is, of the adoption of an equivalent constant temperature for estimating creep under periodic temperature variations.

In the preceding two cases the period of temperature variation was long. Creep tests under quick cyclic variations of temperature have another point of interest, whether repeated quick heating and quick cooling of the material would stimulate structural changes on a microscopic scale [*16, 19, 36, 37*]. In this connection, tests were performed on a number of materials and the experimental creep curves were compared with curves calculated from creep data of ordinary tests made at a few constant temperature levels within the range of the temperature variation. Hollow specimens of 1 mm wall thickness were employed in the variable temperature tests. They were heated by an electric current and alternately cooled by a jet of compressed air.

In this series of tests the temperature was not uniform along the specimen. The mean temperature was determined from the formula [*34*]

$$T^* = -K/\ln\left[\int_0^{l_0} \exp\left(-\frac{K}{T(l)}\right) dl/l_0\right]. \tag{19}$$

Eq. (19) was derived by assuming that the creep strain in a specimen of uneven temperature distribution is equivalent to that in a specimen of uniform temperature of T^* [*34*]. In the experiments of this series, the

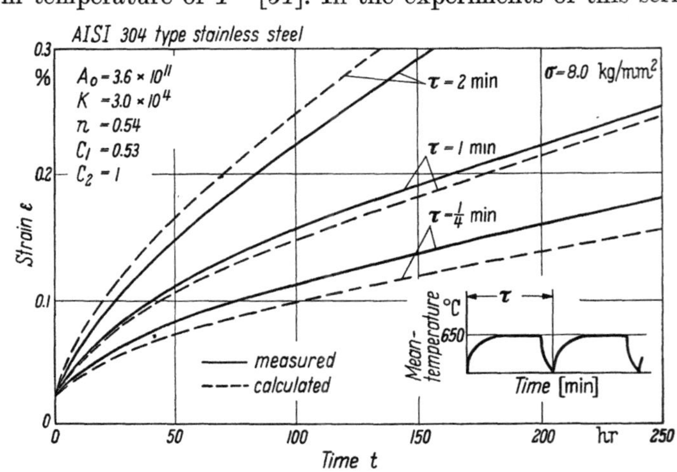

Fig. 4. Creep of AISI 318 steel under quick temperature variation (TAIRA and OHNAMI [*34*])

mean temperature T^* was made to fluctuate according to a prescribed program. The equivalent steady temperature was calculated from the mean temperature T^*. Fig. 4 [34] shows, as an example, the case of AISI 304 type stainless steel. The full lines are experimental creep curves and the dotted lines are the analytical ones.

3.3. Creep under Combined Variation of Temperature and Load

Materials are most often used in practice under conditions in which the temperature and the load vary simultaneously. Unfortunately, the creep problem is very complex under such conditions [1].

On the basis of the author's experiments the following conclusions can be drawn [32]: if the variation in temperature and load is cyclic and the two cycles are synchronized in such a way that the temperature rises when the load increases, it is likely that the experimental creep curves under simultaneous variation of temperature and load are close to the analytical curves derived from the data of standard creep tests with the aid of the mechanical equation of state. If the two cycles are not synchronized in such a manner, the validity of the conventional method of prediction is in question.

If the stress were markedly lowered during the high temperature period, after having subjected the material to high stress at a lower temperature, evidently recovery would proceed without noticeable work-hardening. This would cause an excessive increase in creep strain during the next period of variation. This is the reason why the conventional method of predicting creep fails.

3.4. Creep Rupture under Varying Temperature

The prediction of life in creep rupture under cyclic variation of temperature from the data of creep rupture under constant temperature has been discussed by E. L. Robinson [18] and J. Millers [1].

If the period of temperature cycling is denoted as p, the temperature during an infinitesimal time interval $(\Delta t)_i$ as T_i, and the rupture life for a constant temperature T_i as L_i, then the life consumption during a fragment of time $(\Delta t)_i$ is represented by

$$E_i = \frac{L}{L_i} \frac{(\Delta t)_i}{p} . \tag{20}$$

L is the life for the test temperature T.

Assume that the material subjected to periodic temperature variations fractures when

$$\sum_{i=1}^{s} E_i = 1, \; s = p/(\Delta t)_i . \tag{21}$$

In addition the formula

$$L_i = B_0 \exp(Q/T_i), \tag{22}$$

is assumed to hold, with B_0 and Q constants. Then the life is represented by

$$L = B_0 p \Big/ \int_0^p \exp\left[-\frac{Q}{T(t)}\right] dt. \tag{23}$$

Thus, the equivalent constant temperature for creep rupture under the temperature variation $T(t)$ is represented by

$$T_{er} = -Q/\ln\left[\frac{1}{p}\int_0^p \exp\left\{-\frac{Q}{T(t)}\right\} dt\right]. \tag{24}$$

T_{er} is equal to T_{ec}, the equivalent constant temperature for creep under varying temperature, when $Q = K/n$.

Fig. 5 shows the results of experiments on AISI 318 steel in the case of a saw-tooth type cycling of the temperature. The range of temperature

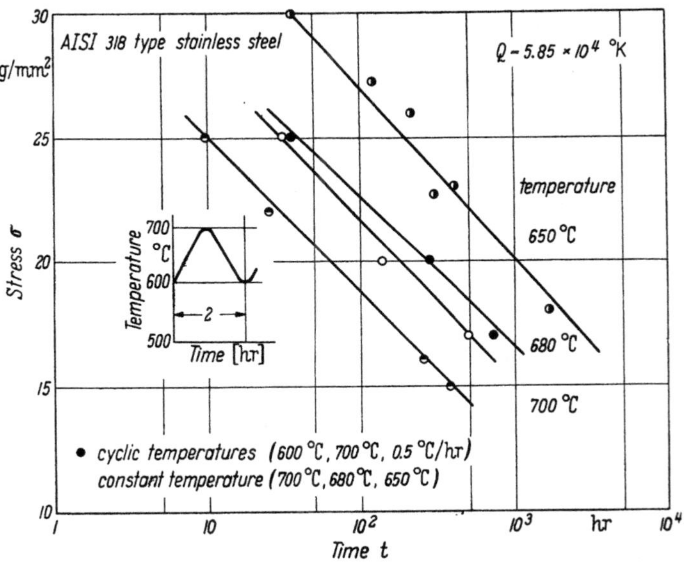

Fig. 5. Creep rupture of AISI 318 steel under quick variation of temperature (TAIRA and OHNAMI [37])

variation was between 600°C and 700°C, and the equivalent constant temperature T_{er} was determined as 680°C. The temperature was cycled once per two hours. Besides, creep rupture tests were performed by fixing the test temperature at 680°C under various stress levels. The closed circles are the results for the cyclic temperature variation. The stress-rupture time relation for fixed temperatures of 650°C and 700°C is also shown in the figure.

It can be seen from the figure that the life for periodic cycling is a little longer than that for the constant temperature of 680 °C, which is equal to the equivalent steady temperature for cyclic temperature. The difference can probably be attributed to differences in the changes in the metallographic structure during creep in the two series of tests.

The example shows that the estimation of creep rupture life of materials under periodic temperature variations is less reliable than that of the creep strain. It is possible that fracture is greatly affected by changes in the structure during the tertiary stage of creep.

4. Strength under Dynamic Loading

Structures are often subjected to dynamic loading and the resistance of structural members to dynamic stress is another point of interest in the design of structures. The deformation of a material under dynamic stress is called dynamic creep, and fracture under dynamic stress is known as dynamic creep rupture or fatigue.

Concerned with the strength of materials under dynamic stress are the prominent papers of B. J. Lazan and others [*38—49*], who have investigated the characteristic behavior of various metallic materials under dynamic stress.

The purpose of the present paper is to discuss a practical procedure for predicting the strength of materials under variable loads or temperatures from the data of tests obtained under steady conditions. From this point of view, it is intended to interrelate dynamic creep with creep under static loading. Dynamic creep rupture is also treated analytically and is related to creep rupture under static loading combined with fatigue.

In the discussion which follows, the working stress is taken as composed of the dynamic component σ_a and the static component σ_m, the mean stress; and the ratio of the two components is called the stress ratio A.

4.1. Dynamic Creep

It has been described in the preceding section that the concept of the mechanical equation of state is applicable approximately for the prediction of creep under stresses of slow periodic variation. This is now extended to the analysis of dynamic creep [*50*]. When the temperature is constant, Eq. (1) is

$$\frac{d\varepsilon}{dt} = f(\sigma, \varepsilon)\,. \tag{1'}$$

The creep strain under a static load is again represented as

$$\varepsilon = \varepsilon_0 + \beta\, t^n + \varkappa\, t\,. \tag{6}$$

Under a constant temperature, ε_0, β and $\varkappa$ are constants which depend on stress only, being independent of time. The strain rate is again

$$\frac{d\varepsilon}{dt} = n\,\beta\,t^{n-1} + \varkappa. \tag{7}$$

By eliminating t from Eqs. (6) and (7), we would have a formula corresponding to Eq. (1'). But this cannot be given in a simple analytical form. If the creep strain were given by a single term of either the transient or the steady component, a formula would be obtained in a simple form. For the case of dynamic creep when the strain is taken as composed of the steady component only, the solution has been given by LAZAN [*40*]. The dynamic creep in the transient stage is therefore discussed here.

If the steady component of creep is neglected, Eq. (6) becomes Eq. (25), where β is expressed as a power function of the stress σ:

$$\varepsilon = a\,\sigma^{\alpha} t^{n}, \tag{25}$$

where a and α are constants independent of stress and time. Then we have an equation corresponding to Eq. (1') in the form

$$\frac{d\varepsilon}{dt} = n\,a^{\frac{1}{n}}\,\sigma^{\frac{\alpha}{n}}\,\varepsilon^{\frac{n-1}{n}}. \tag{26}$$

If the stress were given as a function of time t, we could solve Eq. (26). When the stress variation is cyclic with a short period, the problem becomes the problem of dynamic creep.

Under ordinary service conditions the stress wave in dynamic loading would be close to a sinusoidal fluctuation superposed on a static component, such as

$$\sigma = \sigma_m + \sigma_a \sin\omega\,t, \tag{27}$$

where ω is the angular velocity of the alternating stress component. Substituting Eq. (27) in Eq. (26) and integrating, we have

$$\varepsilon = a\,\sigma_m^{\alpha}\left[\int_0^t (1 + A\sin\omega\,t)^{\frac{\alpha}{n}}\,d(\omega\,t)\right]^n,$$

where A is the stress ratio. Denoting the number of stress cycles as N, we obtain

$$\varepsilon = a\,\sigma_m\left[\frac{N}{\omega}\int_0^{2\pi} (1 + A\sin\omega\,t)^{\frac{\alpha}{n}}\,d(\omega\,t)\right]^n.$$

Using the relation $t = 2\pi N$, we have

$$\varepsilon = a\,\sigma_m^{\alpha}\,t^n\,I^n, \tag{28}$$

where

$$I = \frac{1}{2\pi}\int_0^{2\pi} (1 + A\sin\omega\,t)^{\frac{\alpha}{n}}\,d(\omega\,t). \tag{29}$$

If we put

$$\sigma_{ed} = \sigma_m\,I^{\frac{n}{\alpha}} \tag{30}$$

the following formula is obtained for the strain in dynamic creep reached at time t:

$$\varepsilon = a\,\sigma_{ed}^{\alpha}\,t^{n}. \tag{31}$$

Eq. (31) has a form similar to that of Eq. (25). This means that the creep curve in dynamic creep under the stress of Eq. (27) is similar to the creep curve in static creep under the stress σ_{ed} of Eq. (31). Thus, σ_{ed} is called the equivalent static stress.

It follows from Eqs. (29) and (30) that σ_{ed} is determined if the ratio α/n is known from data obtained in static creep tests of the material and if the stress condition in dynamic loading is prescribed by stipulating the stress ratio A and the mean stress σ_m or the stress amplitude σ_a. This is the principle which underlies the prediction of transient strain in dynamic creep from data on static creep.

If we take the logarithm of Eq. (25), which is the formula of creep under static conditions, we have

$$\log\sigma + \frac{n}{\alpha}\log t = \frac{1}{\alpha}\log\frac{\varepsilon}{a}. \tag{32}$$

This means that the ratio α/n is determined by the inclination of the straight line in the log σ — log t diagram for static creep. The stress range diagram for the combination of alternating stress and mean stress component is shown in Fig. 6 in non-dimensional form with α/n as parameter. Any combination of σ_a and σ_m on a line of the figure gives the same amount of strain, provided σ_{ed} is the same. On straight lines through the origin the stress ratio is constant.

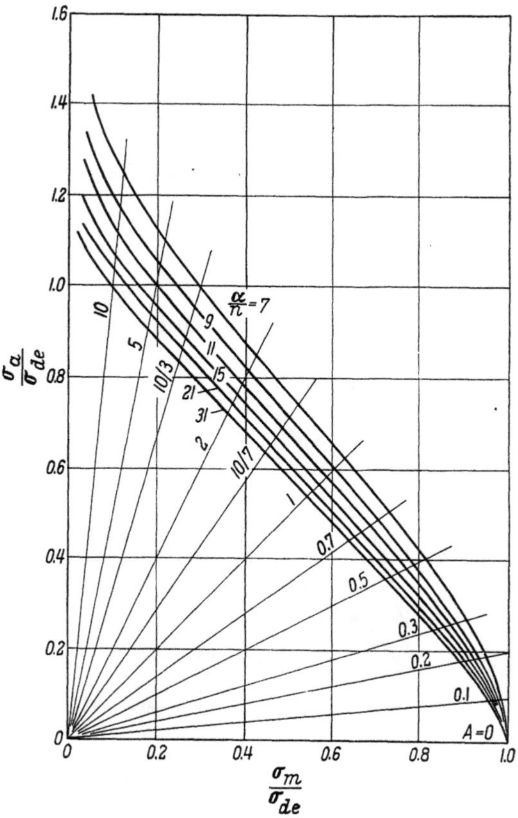

Fig. 6. Theoretical stress range diagramm for dynamic creep in non-dimensional expression (TAIRA, TANAKA and KOTERAZAWA [50])

Fig. 6 provides a convenient means to determine the equivalent static stress σ_{ed} when data on static creep are available and the nature of the dynamic stress is known. First we determine the value of α/n from the static creep data and find the value of the stress ratio A. Next we can determine the value of σ_m/σ_{ed} by locating the intersection point of the two lines in the figure. As soon as the value of σ_{ed} is known, we can draw a creep curve for the dynamic stress condition using the data of static creep.

The validity of the analysis was checked by experiments [50]. Fig. 7 shows the example of a 0.15% carbon steel. Fig. 7(a) contains the creep

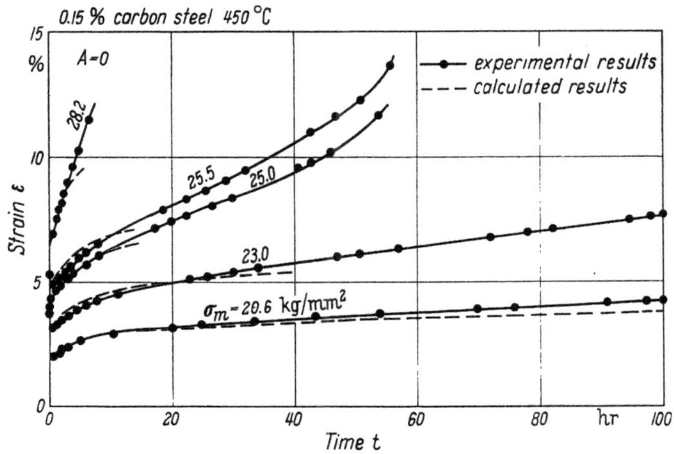

Fig. 7a. Static creep curves for 0,15% carbon steel at 450 °C (TAIRA, TANAKA and KOTERAZAWA [50])

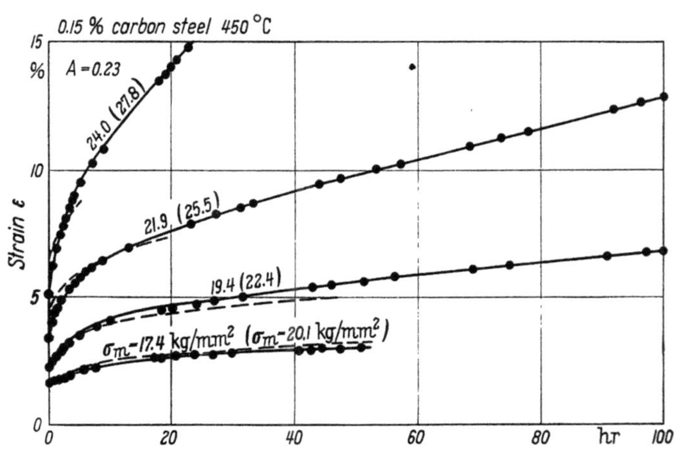

Fig. 7b. Dynamic creep curves for 0,15% carbon steel at 450 °C (TAIRA, TANAKA and KOTERAZAWA [50])

curves for static creep and (b) those for dynamic creep with a stress ratio $A = 0.23$. In the figures the full lines are experimental curves and the dotted lines are due to analysis. It is found that the analytical curves are in fairly good agreement with the experimental curves for a wide range of the stress σ_{ed}, especially when the stresses are low. Fig. 8 is a stress range diagram for the same material derived from Fig. 7. Fig. 9 is the

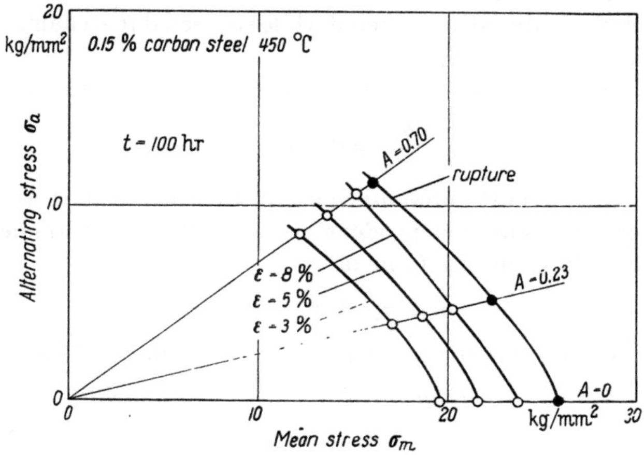

Fig. 8. Stress range diagramm for dynamic creep and dynamic creep rupture of 0,15% carbon steel at 450°C (TAIRA, TANAKA and KOTERAZAWA [50])

non-dimensional stress range diagram. The full line is determined by analysis and the circles are plotted from Fig. 9. Satisfactory agreement is found in the figure when the stress ratio is less than 0.70. In the range of stress ratios over 0.70, the influence of fatigue should be taken into account necessitating additional considerations [51]. These will be described later.

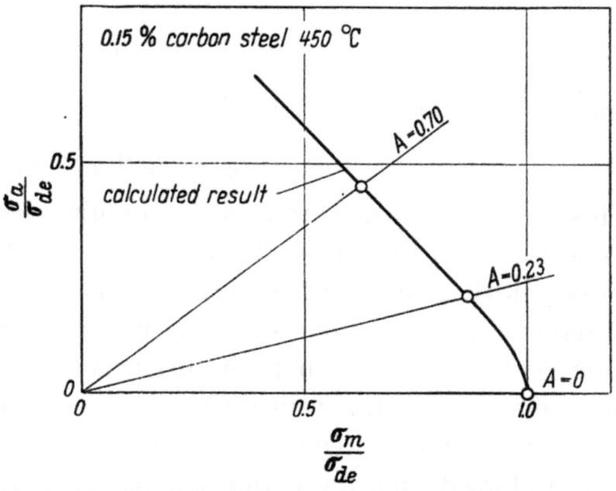

Fig. 9. Non-dimensional stress range diagramm for dynamic creep of 0,15% carbon steel at 450°C (TAIRA, TANAKA and KOTERAZAWA [50])

It is of interest to note that if the equivalent static stress σ_{ed} is determined by utilizing the creep curves in the early period in accordance with the method described, this value of σ_{ed} can be applied to the range of later stage of creep by simply substituting it in Eq. (6) [51].

4.2. Dynamic Creep Rupture

The basic idea of the estimation of dynamic creep rupture from the data of static creep rupture tests [*50*, *52*] will be briefly discussed. It consists of the idea of cumulative damaged which causes materials to fracture. The assumptions made are as follows:

1. Fracture occurs when the total amount of damage Φ_c due to creep, which the material has suffered while subjected to dynamic loading, reaches a critical value Φ_{cc}.

2. The rate of increase of creep damage is represented by

$$\frac{d\Phi_c}{dt} = a_c(|\sigma|)^{\alpha_c} \tag{33}$$

as a power function of the absolute value of the stress applied. In Eq. (33) a_c and α_c are constants independent of stress. When the applied stress is given, as before, in the form of

$$\sigma = \sigma_m + \sigma_a \sin \omega t \tag{27}$$

we have a formula for the initiation of fracture in the form

$$\Phi_{cc} = a_c \sigma_m^{\alpha_c} t_c I_c, \tag{34}$$

where

$$I_c = \frac{1}{2\pi} \int_0^{2\pi} (|1 + A \sin \omega t|)^{\alpha_c} d(\omega t). \tag{35}$$

If we put

$$\sigma'_{ed} = \sigma_m (I_c)^{\frac{1}{\alpha_c}}, \tag{36}$$

the critical damage Φ_{cc} is given as

$$\Phi_{cc} = a_c(\sigma'_{ed})^{\alpha_c} t_c. \tag{37}$$

This means that fracture under dynamic stress occurs at the same time as static creep rupture under the stress σ'_{ed}. Thus, σ'_{ed} is called the equivalent static stress for dynamic creep rupture. σ'_{ed} can be determined for dynamic stressing under any combination of the mean stress σ_m and the alternating stress σ_a when the material constant α_c is known from the data of static creep rupture tests. α_c is determined from the $\log \sigma - \log t$ diagram for static creep rupture tests of the material as the inclination of the plotted line. As in the case of dynamic creep, we can establish a stress range diagram for dynamic creep rupture. Fig. 10 is the stress range diagram in terms of non-dimensional quantities, derived with the aid of Eq. (36). In the figure, α_c is taken as a parameter and curves are drawn for several values of α_c that are probable for engineering materials.

Examples are shown for a number of materials. Fig. 11 shows the comparison of the analytical stress range diagram with the results of experiments on a 0.15% carbon steel at 450°C [*50*, *51*]. The analytical curve is the full line inclined at about forty-five degrees to the abscissa through 1.0, and the experimental results are shown by circles. It is seen from the figure that the analysis is in good agreement with experiment when the stress ratio A is less than about 1.5. When the stress ratio is greater than this value, the validity of the analysis is not established. Thus we see the limitation of the analysis. For stress ranges involving large stress ratios it would be necessary to take into account the damage due to fatigue besides the creep damage [51].

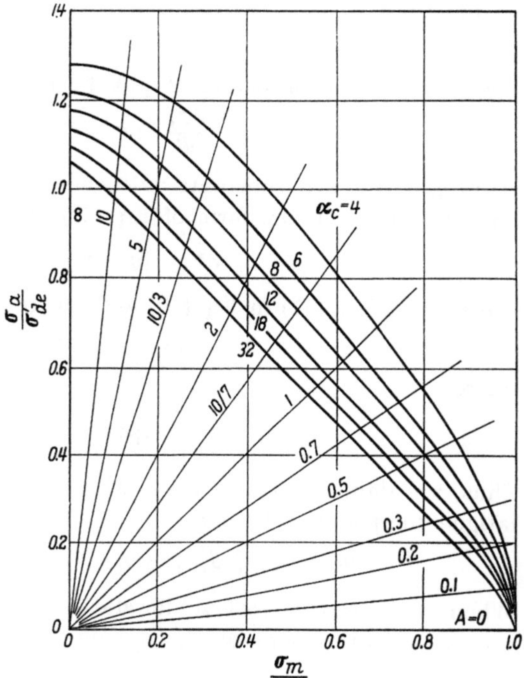

Fig. 10. Theoretical stress range diagramm for dynamic creep rupture in non-dimensional expression (NISHIHARA, TAIRA, TANAKA and KOTERAZAWA [52])

In the case of dynamic stressing with large stress ratios, the damage suffered by the material is taken as composed of the creep damage Φ_c and the fatigue damage Φ_f. It is assumed that the material fractures when the total damage Φ_s reaches a critical value:

$$\Phi_s = \Phi_c + \Phi_f. \qquad (38)$$

As to the creep damage Φ_c, Eqs. (33)—(37) are applicable for this case also.

With respect to the fatigue damage Φ_f, it is reasonable to assume that the difference between instantaneous and mean stress σ_m is effective in creating fatigue damage. Thus, denote the absolute value of

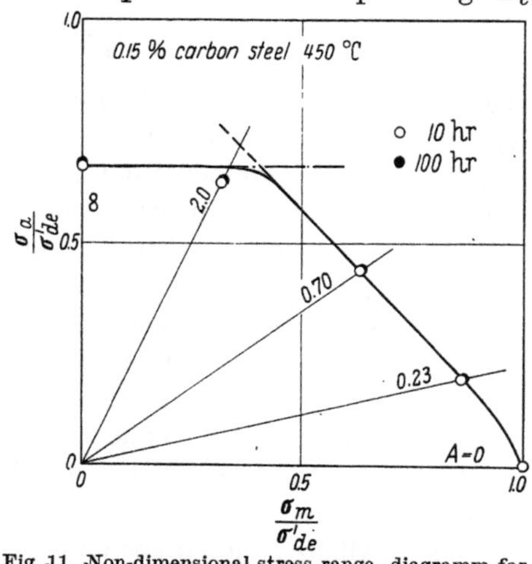

Fig. 11. Non-dimensional stress range diagramm for dynamic creep rupture of 0.15% carbon steel at 450 °C (TAIRA and KOTERAZAWA [51])

the difference as σ^*:

$$\sigma^* = |\sigma - \sigma_m|. \tag{39}$$

In the same manner as before, we put the increasing rate of the damage as

$$\frac{d\Phi_f}{dt} \doteq a_f(\sigma^*)^{\alpha_f}, \tag{40}$$

where a_f and α_f are constants independent of σ. If the dynamic stress applied is given in the form of Eq. (27), we have σ^* as

$$\sigma^* = |\sigma_a \sin \omega\, t|. \tag{41}$$

Substituting Eq. (41) in Eq. (40) and integrating, we have

$$\Phi_f = a'_f\, \sigma_a^{\alpha_f}\, t_c \tag{42}$$

where

$$a'_f = \frac{a_f}{2\pi} \int_0^{2\pi} (|\sin \omega\, t|)^{\alpha_f}\, d(\omega\, t). \tag{43}$$

Thus, the damage at any time, which is composed of the creep and fatigue components, becomes

$$\Phi_s = a_c\, \sigma_a^{\alpha_c}\, I'_c\, t_c + a'_f\, \sigma_a^{\alpha_f}\, t_c, \text{ where } I'_c = \frac{I}{A^{\alpha_c}}. \tag{44}$$

When we define a stress level for static creep rupture that leads to fracture at the same time as the dynamic stress σ'_{ed}, we can write

$$\Phi_s = a_c(\sigma'_{ed})^{\alpha_c}\, t_c. \tag{45}$$

Comparing Eq. (44) and Eq. (45), we have

$$\frac{\sigma_a}{\sigma'_{ed}} = \left(I_a + \frac{k_f}{k_c}\, \sigma_a^{\alpha_f - \alpha_c}\right)^{-\frac{1}{\alpha_c}}, \tag{46}$$

where

$$k_c = \frac{a_c}{\Phi_s} \text{ and } k_f = \frac{a'_f}{\Phi_s}.$$

The right hand side of Eq. (46) is known when the values of α_c, α_f and k_f/k_c are known for any combination of σ_a and σ_m of the dynamic stress applied. α_c is determined from static creep rupture tests and α_f from fatigue data obtained in tests under completely alternating stress. The value of k_f/k_c is given when α_c and α_f are determined.

The full line in Fig. 11 including the whole range of the stress ratios is based on the present analysis. The analytical curve is close to the experimental results for the entire range of the stress ratio. The dotted and the chain lines in the figure show the stress range for fracture based on an independent consideration of the influences of the creep damage and of the fatigue damage, respectively. It is again emphasized that for rupture under dynamic stress the creep damage is important when the stress ratio is small, and the fatigue damage is important when the stress ratio is large.

Fig. 12 demonstrates the application of the author's analysis to the experimental results obtained by B. J. LAZAN with 24 S-T 4 at 500°C [*44*].

Applying the analysis to various sorts of materials, one finds that there are many cases where the equivalent static stresses σ_{ed} for dynamic creep and σ'_{ed} for dynamic creep rupture are the same [*51*]. In these cases it is recommended that the two stress range diagrams for dynamic creep and dynamic creep rupture be drawn on the same sheet. An example is given in Fig. 13, which is the case of a 13 Cr steel at 450°C. It is to be noted that the stress range curves are common for the case of dynamic stress when the stress ratio is less than unity; for the dynamic stress at large stress ratios only the curve of the stress range for fracture (dotted line) is drawn.

One can conclude from these examples, that the analyses of dynamic creep and dynamic creep rupture described are useful in predicting the lifetime of materials under dynamic loading.

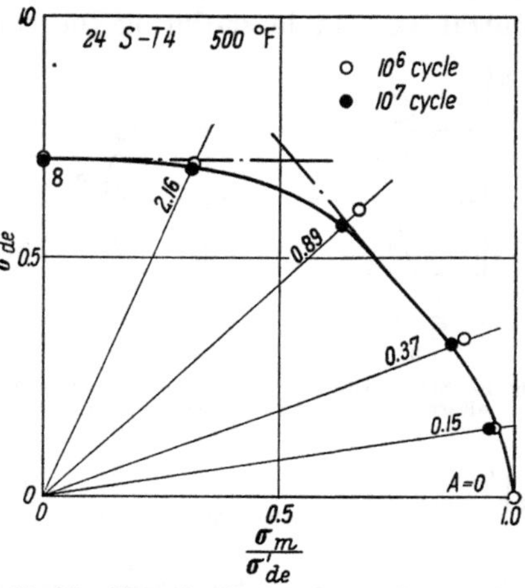

Fig. 12. Non-dimensional stress range diagram for rupture of 24S—T 4 aluminium alloy (TAIRA and KOTERAZAWA [51]) (experimental points are due to B. J. LAZAN [*44*])

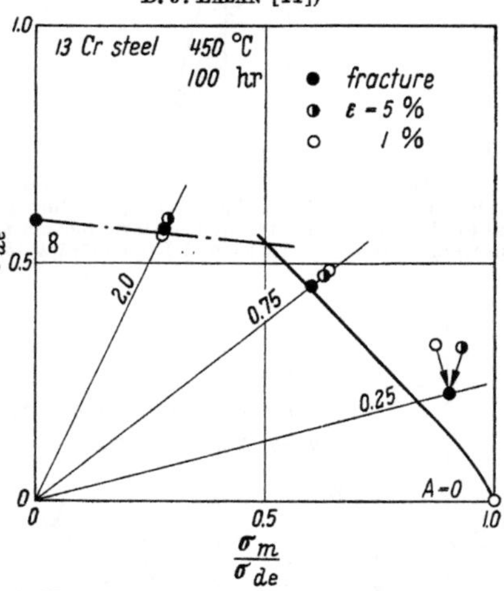

Fig. 13. Non-dimensional stress range diagramm for dynamic creep and dynamic creep rupture of 13 Cr steel at 450°C (TAIRA and KOTERAZAWA [*55*])

4.3. Application

Creep under combined stressing in static tension and alternating bending [*53, 54*] is discussed as an example of application of the analysis described in Section 4.1.

Within the cross section of a specimen subjected to alternating bending superimposed on static tension, the stress distribution may be as shown in Fig. 14a as long as the deformations are elastic. Consequently the material must creep more rapidly in the outer layer than near the neutral axis, at least during the early periods of stressing.

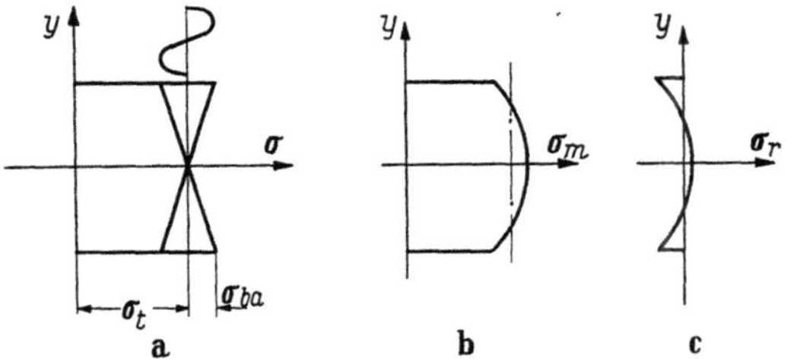

Fig. 14. Stress within a specimen under combined stress of alternating bending and axial tension (TAIRA and KOTERAZAWA [53])

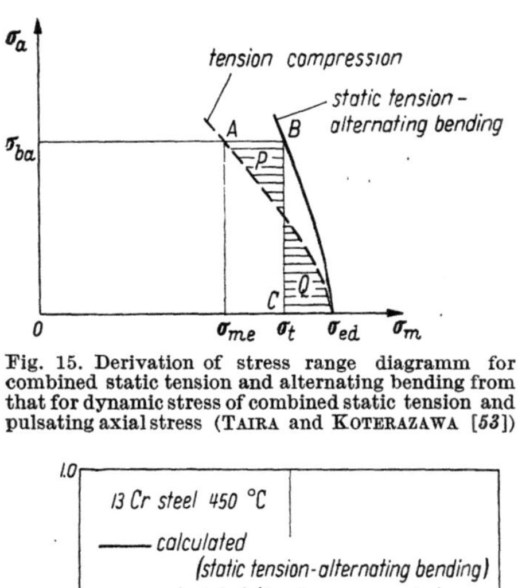

Fig. 15. Derivation of stress range diagramm for combined static tension and alternating bending from that for dynamic stress of combined static tension and pulsating axial stress (TAIRA and KOTERAZAWA [53])

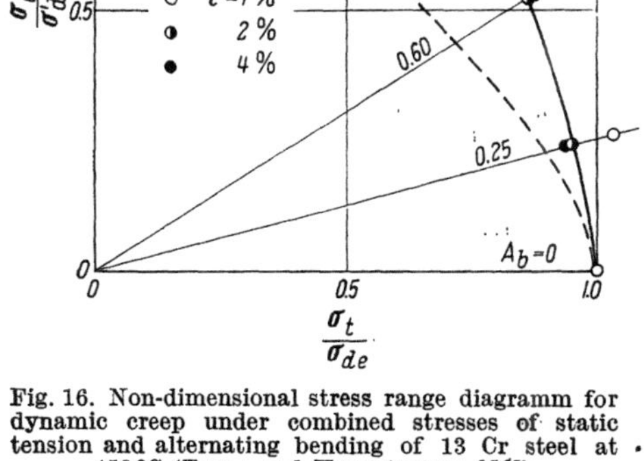

Fig. 16. Non-dimensional stress range diagramm for dynamic creep under combined stresses of static tension and alternating bending of 13 Cr steel at 450°C (TAIRA and KOTERAZAWA [53])

However, the cross section must remain plane even after deformations. This means that the equivalent static stress should be equal for every layer of the section. Therefore, the mean stress is re-distributed as shown in Fig. 14b, causing a distribution of internal stresses as shown in Fig. 14c. The stress distribution (c) develops early during the stress alternations; this has been proved by an observation of the residual stresses.

Under these conditions the material creeps in the axial direction. The stress range diagram for this case of dynamic stress is derived simply from the stress range diagram for alternating axial stress superimposed on static axial stress.

In Fig. 15, let the dotted line be a stress range curve for alternating tensile-compressive stresses combined with a mean axial stress. Consider the case that the axial stress σ_t is given and the magnitude of the alternating stress σ_{ba} is wanted. Take point C on the abscissa, $OC = \sigma_t$. Then draw a vertical line through C. Next draw a horizontal line AB so as to make the area P equal to the area Q. Then point B is a point on the stress range curve for the dynamic stress of combined alternating bending and axial tension. Successive applications of this procedure in taking different values of σ_t yield the continuous curve shown by the full line. The dotted line represents the distribution of the mean stresses.

Fig. 16 shows, by the way of example, a stress range diagram of a 13 Cr steel at 450°C under combined alternating bending and axial tensile stress in non-dimensional form [*53, 54*]. The full line is derived from the dotted line, which is drawn from the data of tests under combined tensile-compressive and static axial stress. The figure shows that the analytical curve is close to the experimental points.

5. Conclusions

The characteristics of creep and creep rupture of metallic materials under varying load and temperature are briefly discussed. The following observations can be listed as the conclusions of the present study:

1. The deformations of materials at elevated temperatures are influenced by the history of loading and heating. As far as creep is concerned, the influence of stress cycling is much more noticeable than the influence of temperature cycling, as long as changes in the metallographic structure are small.

2. Recovery plays an important role in the influence of stress cycling on creep. If the time of recovery during stress cycling is long, that is, if the material is kept at the lower stress level long enough, it clearly shows the effect of the stress history. For this case, the author's theory is applicable for the prediction of creep under cyclic stress from data on creep under steady stress.

3. Temperature cycling does not have a serious influence on creep. The creep under cyclic temperature can be predicted by using the equivalent steady temperature σ_{ec} for creep, which is determined from the data of the usual creep tests at steady temperature together with the prescribed program of temperature cycling.

4. The equivalent steady temperature for creep rupture σ_{er} is also introduced. Prediction of creep rupture life is less accurate than the prediction of creep strain. The poorer agreement is probably due to changes in the metallographic structure during the tertiary stage of creep.

5. Dynamic creep under alternating stress combined with static stress is discussed. The equivalent static stress σ_{ed} for dynamic creep is defined and it is proved that the prediction of dynamic creep from data on static creep by use of the equivalent static stress is in good agree-

ment with experimental results for various materials. Theoretical stress range diagrams for dynamic creep are shown.

6. Dynamic creep rupture under alternating stress combined with static stress is also discussed. The prediction of dynamic creep rupture life from data on static creep rupture and fatigue is described. Theoretical stress range diagrams for dynamic creep rupture and fatigure are derived and compared with experiments on a number of metallic materials at elevated temperatures.

7. The creep of metals at elevated temperature subjected to alternating bending combined with static tension is discussed.

References

[1] ASTM STP No. 165 (1954), "Symposium on Effect of Cyclic Heating and Stressing on Materials at Elevated Temperature".

[2] ROBERTS, I.: Proc. ASTM, **51**, 811 (1951).

[3] FINDLEY, W. N., and G. KHOSLA: J. Appl. Phys., **26**, 821 (1955).

[4] ODQVIST, F. K. G.: J. Mech. Rev., **7**, 517 (1954).

[5] WEISSMAN, G. F., VOH-HAN PAO and J. MARIN: Proc. 2nd US Nat. Cong. Appl. Mech. 577 (1955).

[6] KENNEDY, A. J.: Proc. Roy. Soc., **213**, 492 (1952).

[7] KENNEDY, A. J.: Proc. Phys. Soc., **68**-B, 257 (1955).

[8] SHERBY, O. D., T. A. TROZERA and J. E. DORN: Proc. ASTM **56**, 207 (1957).

[9] LUBAHN, J. D.: Trans. Metallurg. Soc. AIME, **212**, 557 (1958).

[10] RABOTNOV, Y. N.: Proc. Symposium on the Creep and Fracture of Metals at High Temperature, (NPL 1955), p. 221.

[11] TANENBAUM, M., and W. KAUZMANN: J. Appl. Phys. **25**, 451 (1954).

[12] COTTRELL, A. H.: Dislocation and Plastic Flow in Crystals, Oxford, (1953) p. 209.

[13] DORN, J. E.: Proc. Symposium on the Creep and Fracture of Metals at High Temperature, (NPL 1955), p. 221.

[14] LESCHEN, J. G., P. R. CARREKER and J. H. HOLLOMON: Transa. AIME, **180**, 131 (1949).

[15] CARREKER, P. R., J. G. LESCHEN and J. D. LUBAHN: ibid., p. 139.

[16] BROPHY, G. R., and D. E. FURMAN: Transa. ASM, **30**, 1115 (1942).

[17] OLSSEN, K. G.: Archiv. für Eisenhüttenwesen, **28**, 11, 679 (1957).

[18] ROBINSON, E. I.: Transa. ASME, **60**, 253 (1938); **74**, 777 (1952).

[19] BALDWIN, E. E.: Transa. ASME, **78**, 517 (1956).

[20] Proc. "Symposium on the Creep and Fracture of Metals at High Temperature", Section 2, (NPL 1955).

[21] KOCH, W., A. KRISH and H. ROHDE: Stahl und Eisen, **78**, 1251 (1958).

[22] LUDWICK, P.: Elemente der Technologischen Mechanik, Berlin: Springer 1909.

[23] NISHIHARA, T., S. TAIRA, K. TANAKA and K. OHJI: Tech. Rep. Engg. Res. Inst. Kyoto Univ., **6**, 7 (1956) (Rep. No. 33).

[24] NISHIHARA, T., S. TAIRA, K. TANAKA and K. OHJI: Transa. JSME (Japan Soc. Mech. Engrs.), **24**, 452 (1958).

[25] NISHIHARA, T., S. TAIRA, K. TANAKA and K. OHJI: Transa. JSME, **25**, 159, 1048 (1959).

[26] Ibid., p. 1056.

[27] TAIRA, S., K. TANAKA and K. OHJI: Transa. JSME, **25**, 155 (1959).
[28] NISHIHARA, T., S. TAIRA, K. TANAKA and K. OHJI: J. JSTM, **7**, 254 (1958).
[29] OHJI, K.: J. JSTM (Japan Society for Testing Materials), **8**, 209 (1959).
[30] TAIRA, S., K. TANAKA and K. OHJI: J. JSTM, **8**, 664 (1959).
[31] NISHIHARA, T., S. TAIRA, K. TANAKA and M. OHNAMI: Proc. lst Japan Congr. Testing Mater., (1958), p. 50.
[32] Ibid., Proc. 2nd Japan Congr. Test. Mater., (1959), p. 48.
[33] Ibid., Tech. Rep. Engg. Res. Inst. Kyoto Univ., **8**, 2 (1958) (Rep. No. 42).
[34] TAIRA, S., and M. OHNAMI: Proc. 3rd Japan Congr. Test. Mater., (1960), p. 77.
[35] OHNAMI, M.: J. JSTM, **8**, 199 (1959).
[36] TAIRA, S., and M. OHNAMI: J. JSTM, **9**, 89 (1960).
[37] Ibid., Preprint for the Annual Meeting of the Japan Soc. for Mech. Engrs., April 1960.
[38] HEMPEL und TILLMANNS: Mitt. Kais. Wilh. Inst. Eisenforsch., **18**, 163 (1936).
[39] HEMPEL und KRUG: Mitt. Kais. Wilh. Inst. Eisenforschung, **24**, 71 (1942).
[40] LAZAN, B. J.: Proc. ASTM, **49**, 757 (1949).
[41] MANJOINE, M. J.: Proc. ASTM, **49**, 788 (1949).
[42] TAPSELL, H. J., P. G. FORREST and G. R. TREMAIN: Engg., **170**, 189 (1950).
[43] LAZAN, B. J., and E. WESTBERG: Proc. ASTM, **52**, 837 (1952).
[44] DEMONEY, F. W., and B. J. LAZAN: Proc. ASTM, **54**, 769 (1954).
[45] TRAPP, W. J., and R. T. SCHWARZ: Proc. ASTM, **53**, 825 (1953).
[46] HOFFMANN, C. A.: Proc. ASTM, **56**, 1063 (1956).
[47] NACA Subcommittee on Power Plant Materials, NACA Report 1288 (1956).
[48] VITVEC, F., and B. J. LAZAN: "Symposium on Metallic Materials above 1600°F" ASTM STP No. 173 (1956).
[49] HYLER, W. S., and W. F. SIMMONS: Trans. ASME, **78**, 339 (1956).
[50] TAIRA, S., K. TANAKA and R. KOTERAZAWA: Proc. 2nd Japan Congr. Test. Mater. (1960), p. 54.
[51] TAIRA, S., and R. KOTERAZAWA: Tr. JSME, **26**, 167 (1960).
[52] NISHIHARA, T., S. TAIRA, K. TANAKA and R. KOTERAZAWA: Proc. 6th Japan Nat. Congr. Appl. Mech., (1956), p. 221.
[53] TAIRA, S., and R. KOTERAZAWA: Proc. 3rd Japan Congr. Test. Mat., (1960), p. 168.
[54] TAIRA, S., R. KOTERAZAWA and M. INOUE: Bulletin of JSME, **2**, 508 (1959).

Discussion

E. H. LEE, Brown University: The significance of the influence of varying temperature shown in Fig. 3 could be more easily appreciated if the creep curves for temperatures steadily maintained at the upper and lower limits were given. Could Professor TAIRA indicate approximate points on creep curves for temperature maintained at 575 °C and 675 °C, respectively?

M. ŻYCZKOWSKI, Technical University of Kraków: As far as the problems of periodic change in stress or temperature are concerned, the idea of a reduced stress or reduced temperature may be very useful in practice. For instance, for the simplest creep law

$$\dot{\varepsilon} = \alpha\,\sigma^n$$

we have

$$\varepsilon = \int_0^t \alpha\,\sigma^n(\tau)\,d\tau;$$

for a constant temperature and variable load we can write, introducing the reduced stress σ_r,

$$\varepsilon = \int_0^t \alpha\, \sigma^n(\tau)\, d\tau = \alpha\, t\, \sigma_r^n,$$

whence

$$\sigma_r = \sqrt[n]{\frac{1}{t}\int_0^t \sigma^n(\tau)\, d\tau}\ .$$

If, for example, the stress variation is described by $\sigma = \sigma_1$ for $\frac{m-1}{2} T < t < \frac{m}{2} T$, and $\sigma = \sigma_2$ for $\frac{m}{2} T < t < \frac{m+1}{2} T$, $m = 1, 2, 3, \ldots$ we get

$$\sigma_r = \sqrt[n]{\frac{\sigma_1^n + \sigma_2^n}{2}}\ .$$

Similar formulae may be obtained for other creep laws, for other types of change of stress as well as for variable temperature. Complicated formulae may be often replaced by very simple approximate ones, which is especially important for practical applications.

N. J. HOFF, Stanford University: The discussor wishes to call attention to his calculations of the effects of variable temperatures and variable loads on the creep process as presented in "Mechanics Applied to Creep Testing", the William M. Murray Lecture 1958, Proceedings of the Society for Experimental Stress Analysis, Vol. 17, No. 2, p. 1, 1960.

OLEG D. SHERBY, Stanford University, and ALFRED GOLDBERG, US Naval Postgraduate School, Monterey: The "equivalent steady temperature", T_{ec}, developed in this extensive and excellent presentation by Professor TAIRA, is analogous to a temperature-compensated time parameter, θ [1]. We wish to point out that although the use of an equivalent temperature or time does meet with frequent success, microstructural changes, normally expected to be absent under isothermal conditions, may be induced by temperature cycling; the creep strength of the material may be substantially altered during such a change. This possibility was recently brought to our attention when we attempted to predict the creep strain from variable temperature-time data of Armco iron under constant load [2]. Armco iron normally contains 0.03% carbon. The temperature cycling covered a range in which iron transformed from a body-centered-cubic (α) to a face-centered-cubic structure (γ).

It has been shown [1] that if the creep-strain-time relation is known at a given stress at high temperatures the strain-time curve can be predicted for any other temperature or any variable temperature history by the use of a temperature-compensated time parameter, θ, where $\theta = \int_0^t e^{-Q/RT}\, dt$ under a variable temperature history, or $\theta = t\, e^{-Q/RT}$ at constant temperature. Here, t = time, Q = activation energy for creep, R = gas constant and T = absolute temperature. Graphically, this means that tests at various temperatures under a given creep stress will superimpose when plotted as creep strain, ε, versus θ. If a material exhibits only steady state creep characteristics predictions based on the θ concept are simplified. In this case, ε equals $\dot{\varepsilon}/t$. Multiplying and dividing by $e^{Q/RT}$ one obtains $\varepsilon = \dot{\varepsilon}\, e^{Q/RT}\, \theta$. Thus, if the value of $\dot{\varepsilon}\, e^{-Q/RT}$ is known for a given stress the strain can be evaluated for any variable temperature history by determining θ from a graphical integration of an $e^{-Q/RT}$ versus time plot.

In the investigation on Armco iron, by LOZINSKY and SIMEONOVA [2], the specimen experienced creep in both the alpha and gamma phases, hence, in this more general case the total strain becomes

$$\varepsilon_t = (\dot{\varepsilon}\ e^{Q/RT})_\alpha\ \theta_\alpha + (\dot{\varepsilon}\ e^{Q/RT})_\gamma\ \theta_\gamma \tag{1.}$$

Data were analyzed for the temperature cycling of from 800 to 1000 °C at the center of the specimen under a stress, σ, of 0.33 kg/mm²; this is the only test for which sufficient information was available [2]. The temperature history of cycling at the center of the specimen is shown in Fig. 1.

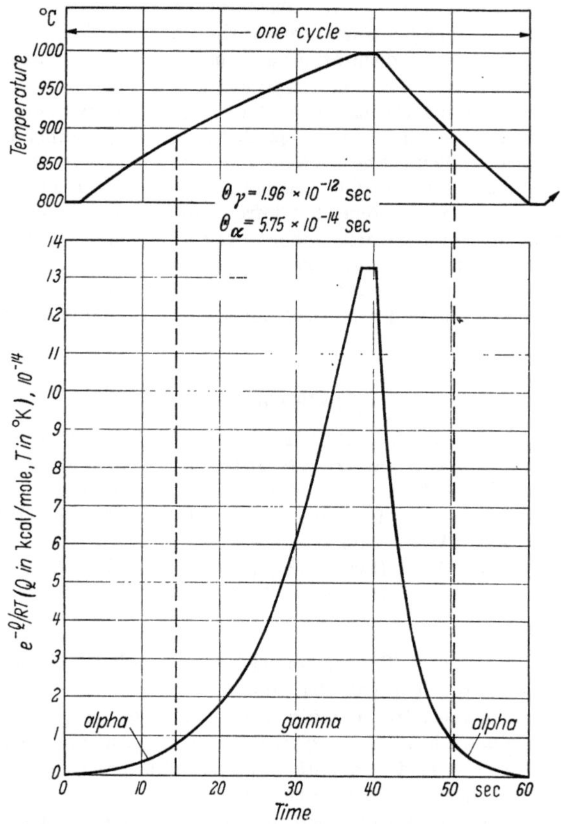

Fig. 1. Determination of temperature-compensated time, Θ, for creep of Armco iron at a stress of 0.33 kg per mm² where specimen creeps in both alpha and gamma phases (Temperature-time data obtained from Ref. [2])

With reference to a solution of ε_t in Eq. (1), $\dot{\varepsilon}_\gamma$ was obtained by extrapolations of data, reported by FELTHAM [3], to $\sigma = 0.33$ kg/mm² at 910 °C and found equal to 2.8×10^{-7} per second. The activation energy for creep of alpha and gamma iron was assumed about equal at 75.0 Kcal/mole [4]. Hence, the value of $\dot{\varepsilon}_\gamma\ e^{Q/RT} = 2.0 \times 10^{7}\ \text{sec}^{-1}$. The creep rate of alpha iron has been shown to be about 200 times faster than gamma iron [4], giving $\dot{\varepsilon}_\alpha\ e^{Q/RT} = 4.0 \times 10^{9}\ \text{sec}^{-1}$. The values of θ_α and θ_γ were obtained graphically from Fig. 1 over one cycle (60 seconds). The transformation temperature of $\alpha \rightleftarrows \gamma$ was estimated at 890 °C [5] where alpha and gamma exist in about equal amounts for a composition of 0.03% C.

A total strain of 0.027% per cycle is predicted. An average strain of 0.10% per cycle was obtained experimentally by LOZINSKY and SIMEONOVA (their Fig. 6). The predicted strain is approximately one-fourth of the observed strain. This discrepancy may be explained by the fact that strain induced by the transformation per se under stress adds to the normally expected strain as strongly implied by LOZINSKY and SIMEONOVA [2] and supported by other investigators [6, 7]. Although no explanations have been presented for this added strain, we would like to suggest that moving interfaces (boundaries between old and new phase) associated with transformation result in sweeping away obstacles responsible for the back stresses at piled-up dislocation regions. Furthermore, the misfit at the interface may act both as a source for additional dislocations and, perhaps more important, facilitates dislocation climb necessary for high temperature deformation.

Thus, when considering the use of an "equivalent steady temperature" or a "temperature-compensated time" for the prediction of creep strains under variable temperature conditions one must be concerned with unusual microstructural changes induced by such temperature cycling. Although working in regions where phase transformations occur may be somewhat infrequent, other possibilities of moving interfaces induced by temperature changes exist; for example, the variation in solubility with temperature in multicomponent systems may well be a cause for interfacial movement. Another complicated system would be that of creep of cold-worked metals which would undergo recrystallization and subgrain formation during deformation.

References

[1] SHERBY, O. D., and J. E. DORN: Trans. AIME, **197**, 324 (1953).
[2] LOZINSKY, MICHAEL G., and IZINA S. SIMEONOVA: Acta Met. **1**, 709 (1959).
[3] FELTHAM, P.: Proc. Phys. Soc. London, B **66**, 865 (1953).
[4] SHERBY, O. D., and J. L. LYTTON: Trans. AIME, **206**, 928 (1956).
[5] Metals Handbook, ASM, 1182 (1948).
[6] DE JONG, M., and J. W. RATHENAU: Acta Met. **7**, 46 (1959).
[7] FOLEY, F. B.: Trans. AIME, **188**, 845 (1950).

Author's closing remarks: The author wishes to thank Professor SHERBY and Professor GOLDBERG for their valuable comments.

As is described in the introduction of the paper, the "equivalent steady temperature" was originally introduced to treat practical problems under isothermal conditions which are the most frequently experienced in practice. The "equivalent steady temperature" does not include the influence of temperature history in its idea of predicting creep strain or creep rupture time under varying temperature conditions from information on creep or creep rupture under steady temperatures at various temperature levels covering the variational conditions. This means that the analysis is made by neglecting the influence of microscopic or submicroscopic structural changes during the variations in temperature; these changes are suspected to differ from those occurring under steady temperature conditions. In this sense the "equivalent steady temperature" and the "temperature-compensated time parameter" are analogous.

The author agrees with Professor SHERBY and Professor GOLDBERG in recognizing the importance of temperature variations encompassing a phase transformation temperature. In this case, a prediction by means of the "equivalent steady temperature" should be analogous to a prediction by means of the "temperature-compensated time parameter", as described in the comments.

It should be noted, however, that even when the treatment is limited to isothermal conditions there is a possibility of the occurrence of submicroscopic structural changes, as is emphasized by Professor SHERBY and Professor GOLDBERG. The author, as many other researchers, is making every effort to clarify this question. Direct observation of submicroscopic structural changes in crept materials is tedious and very difficult, and we are not in a position at the present time to make a conclusive statement about it. However, it is worthwhile to mention that a partial proof is available in an experiment in which the development of creep damage is compared for materials crept under steady temperature and those crept under variable temperature conditions.

Creep damage is directly related to a decrease in shock impact value, as is described in a paper by G. SACHS and W. F. BROWN, Jr.[1]. The author has conjectured that shock impact values of crept materials are very sensitive even to submicroscopic structural changes during creep. Fig. 17 gives an example of a creep damage study, which we expect to publish in the near future. Creep damage is represented by a decrease in the notch impact value $\left(1-\frac{U}{U_0}\right)$, where U_0 and U are the notch impact values of specimens before and after creep deformation. In this experiment, the development of creep damage was studied with AISI 318 type stainless steel, by applying series of temperature conditions of steady as well as varying temperatures.

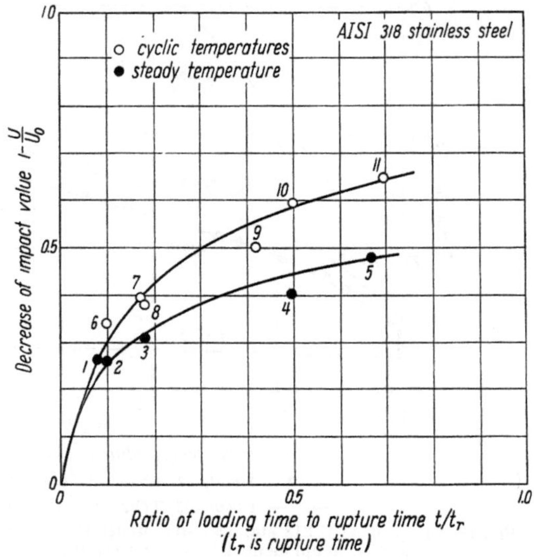

Fig. 17. Creep damage during creep process expressed in term of decrease in shock impact value (TAIRA and OHNAMI)

The figure shows evidence of a clear distinction in the development of creep damage for the two series of tests, although no distinct difference in microscopic structure could be found from observations by means of an electron microcsope. Creep damage is more pronounced in the case of varying temperature than in the case of steady temperature. This suggests that the submicroscopic structure is changed at a more accelerated rate under varying temperatures than under steady temperatures.

There arises another question how this difference in the development of creep damage affects the predicted creep strain or creep rupture life for varying temperature as calculated from information obtained from steady temperature tests. At present, the author wishes to state tentatively that the temperature variation does not noticeably affect the predicted creep strain, but it affects to some extent the

[1] SACHS, G., and W. F. BROWN, Jr.: ASTM, STP 128 6 (1952).

predicted creep rupture life. This is not contradictory to the experimental results shown in Fig. 17 when we take into account the following facts.

When we take creep strain as the basis for design, the analysis is limited to the transient and steady stages of creep of the material. However, when rupture life is discussed, the tertiary stage is included together with the transient and steady stages of creep. For instance, in Fig. 17 the tertiary stage is initiated at the approximate time ratio $\frac{t}{t_r} = 0.2$ and the differences in the development of creep damage are noticeable in the tertiary stage. From this one can deduce that until the beginning of the tertiary stage the differences in the changes in the submicroscopic structure between materials crept at steady temperatures and those under varying temperature are not so pronounced as to affect the predicted creep strain. However, the significant differences in the submicroscopic structure during the tertiary stage of creep render the prediction of creep rupture life more or less uncertain.

Professor SHERBY and Professor GOLDBERG point out considerable deviations between predicted creep strain and the experimental values in the case of a temperature cycling that encompasses the phase transformation temperature and attribute its cause to some unrevealed phenomenon. The author would like to note that this phenomenon might be a particular one accompanying the phase transformation.

The author also wishes to thank Professor LEE for his comments. Fig. 3 is complemented by adding two creep curves for constant temperatures of 575 °C

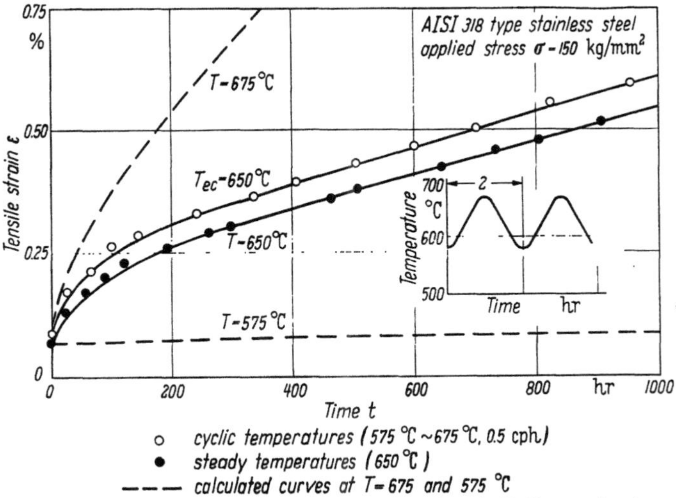

Fig. 18. Creep of AISI 318 stainless steel under steady and periodically varying temperature (Fig. 3 is complimented by adding creep curves for constant temperature)

and 675 °C. These curves were determined from calculations using the same values of the constants as the derivations of the creep curve for cyclic temperature. The revised figure is presented as Fig. 18.

The Effect of Incremental Loading on Creep Behavior of Metals[1]

By

E. T. Onat and T. T. Wang

Brown University, Providence, Rhode Island, USA

Summmary

In structural analysis it is of basic importance to know whether small deviations (in stress and strain) from a basic state obey a linear visco-elastic law for a given metallic material. The paper starts with a discussion of a proposed series of tests by which the existence and the form of such linear relations may be determined. In the following section preliminary results of the proposed type of tests with lead wires are discussed. Finally the importance of the rational determination of time operators in the visco-elastic law is illustrated in the case of creep buckling of initially straight columns. The buckling criteria of RABOTNOV-SHESTERIKOV and SHANLEY are critically examined. The importance of the critical load based on the inelastic modulus is stressed.

1. Introduction

In many important problems of stress analysis in structures the analyst is interested in the following aspect of the creep behavior: A material element is subjected to constant uniaxial stress σ_0 during the time interval $0 < t < t_0$. Suppose that for $t > t_0$ the stress is varied according to the relation

$$\sigma(t) = \sigma_0 + s(t), \tag{1}$$

where $\left|\frac{s}{\sigma_0}\right|$ is small compared with unity (see Fig. 1). The strain response of the element can be written in the form

$$\varepsilon(t) = \varepsilon_0(t) + e(t), \tag{2}$$

where $\varepsilon_0(t)$ is the time dependent strain which would have occurred under the constant stress σ_0 for $t > t_0$. The determination of the rela-

[1] The results obtained in this paper were obtained in the course of an investigation conducted under Contract Nonr 562(20) with the office of Naval Research.

tionship between $e(t)$, $s(t)$ and the parameters σ_0 and t_0 which define the history of loading is many times the key to the rational stress analysis in a structure subjected to transient loading.

Since $\frac{s}{\sigma_0}$ is assumed to be small it may be hoped that this relationship can be represented, for $\frac{t-t_0}{t_0} \ll 1$ by a linear operator equation of the type

$$P(s) = Q(e), \tag{3}$$

where P and Q are linear operators of the form

$$P = \sum_0^n p_r \frac{\partial^r}{\partial t^r}, \quad Q = \sum_0^n q_r \frac{\partial^r}{\partial t^r}$$

and p_r and q_r are functions of the parameters σ_0 and t_0[1]. The inclusion of time derivatives of both s and e in (3) is necessitated by the complex strain behavior usually found in creeping materials which involves an instantaneous response to stress and the phenomenon of primary creep and finally the viscous flow.

The simplest relation of the type (3) which exhibits all three phenomena is the following one which contains the quadratic time operators:

$$p_2 \ddot{s} + p_1 \dot{s} + p_0 s = q_2 \ddot{e} + q_1 \dot{e}. \tag{4}$$

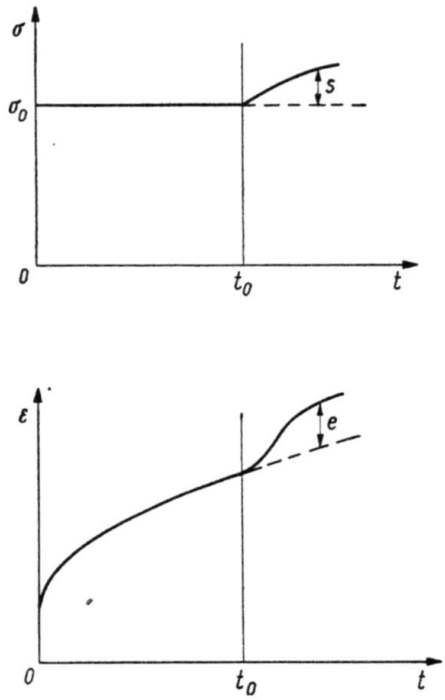

Fig. 1. Stress and strain histories

The response to a constant incremental stress s_0, suddenly applied and maintained to $t > t_0$ is given in this case by

$$e(t) = s_0 e^*(t), \tag{5}$$

where

$$e^*(t) = \frac{1}{q_2/p_2} + \frac{t-t_0}{q_1/p_0} + C\left(1 - e^{\frac{-(t-t_0)}{\tau}}\right)$$

and

$$\tau = \frac{q_2}{q_1}, \qquad C = \frac{p_1}{q_1} - \frac{p_2}{q_2} - \frac{p_0 q_2}{q_1^2}.$$

[1] The present discussion is restricted to isothermal creep, so that the temperature is not a significant variable.

The response consists of an instantaneous response $OA = \frac{s_0}{q_2/p_2}$ (Fig. 2) followed by "primary creep" and "viscous flow" defined by the characteristic values

$$\tan\alpha = \frac{q_1}{q_2}\left(\frac{p_1}{q_1} - \frac{p_2}{q_2}\right) s_0, \qquad \tan\beta = \frac{s_0}{q_1/p_0}$$

and

$$0A' = \left(\frac{p_1}{q_1} - p_0 \frac{q_2}{q_1^2}\right) s_0 .$$

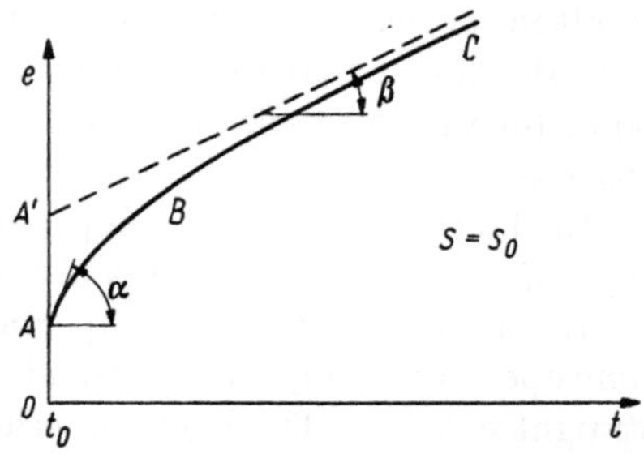

Fig. 2. Response to constant incremental stresses

For an actual material, while the response to a pulse of constant incremental stress may be similar qualitatively to Fig. 2, higher order operators than (4) may be required to represent the behavior in detail [*1*][1]. It is of interest to note here that RABOTNOV and SHESTERIKOV used a linear relationship of the type (4) in their theory of creep buckling [*9*]. These authors' starting point is the "mechanical equation of state"

$$\Phi(\sigma, \varrho, \dot{\varrho}) = 0, \tag{6}$$

where ϱ is the creep strain defined as

$$\varrho = \varepsilon - \frac{\sigma}{\overline{E}}, \tag{7}$$

and $\overline{E}$ is the inelastic modulus. One obtains by taking the time derivative of (6)

$$\Phi_{,\sigma}\dot{\sigma} + \Phi_{,\varrho}\dot{\varrho} + \Phi_{,\dot{\varrho}}\ddot{\varrho} = 0, \tag{8}$$

where comma denotes partial differentiation and the dot the time derivative. Now in the neighborhood of a state point (σ_0, ϱ_0) and t_0 the partial derivatives in (8) may be considered as constants so that (7) and (8) yield for s and e a relation similar to (4) with coefficients

$$p_0 = 0, \ \frac{q_2}{p_2} = \overline{E}, \ \frac{q_1}{p_1} = \frac{\overline{E}\, E_c}{E_c + \overline{E}}$$

and

$$\frac{q_2}{q_1} = \frac{\Phi_{,\dot{\varrho}}}{\Phi_{,\varrho}} = -\frac{\dot{\varepsilon}_0}{\ddot{\varepsilon}_0} . \tag{9}$$

where the creep modulus $E_c = -\frac{\Phi_{,\varrho}}{\Phi_{,\sigma}}$ and $\dot{\varepsilon}_0$ and $\ddot{\varepsilon}_0$ are first and second time derivatives of $\varepsilon_0(t)$ at $t = t_0$.

[1] Numbers in square brackets refer to the Bibliography at the end of the paper.

The main purpose of the present paper is to propose a series of tests by which one may be able to determine whether a linear relation of the type (3) exists for a given material and whether this relation (if it exists) has the simple form predicted by the "mechanical equation of state".

The authors believe that the establishment of these points will lead to a better understanding of the creep phenomenon and it may even result in the construction of a rational phenomenological theory of creep. It is also noted that the existence of (3) may enable one to employ the powerful analytic tools of visco-elasticity in many problems of structural analysis.

In the third section of the paper preliminary results of the proposed type of tests with lead wires are discussed.

In the final section the importance of the rational determination of time operators in (3) is illustrated in the case of creep buckling of initially straight columns. The buckling criteria of RABOTNOV-SHESTERIKOV and SHANLEY are critically examined. The importance of the critical load based on the inelastic modulus is stressed.

2. Proposed Tests

A wide variety of experimental techniques is available to determine whether a linear relationship of the type (3) exists [*1*]. Of these, the dynamic methods (forced vibrations, wave propagation) are preferable in general [*2*]. However for metallic materials the creep behavior in the presence of unloading is, in general, quite different from that in loading so that interpretation of dynamic test results (see, for example, [*3*]) is difficult for such materials. Therefore, the static test methods will have to be employed until a better understanding of the unloading behavior is available.

In the present paper we propose two types of tests[1]. The first type involves application of the constant incremental step loading s_0 at $t = t_0$. The magnitude and sign of s_0 is to be varied from test to test (while σ_0 and t_0 are kept constant) to determine whether the incremental strain response $e(t)$ (see Fig. 1) obeys the linear law

$$e(t) = s_0\, e^*(t) \quad \text{for} \quad t > t_0, \tag{10}$$

where $e^*(t)$ is the incremental creep compliance corresponding to unit incremental step loading. If (10) holds then the coefficients of the linear operators in (3) can be determined (at least in principle) from the measured function $e^*(t)$ [*1*].

---

[1] The proposed tests are concerned with uniaxial stressing. However the proposal can be generalized to tests where multi-axial states of stress are present.

In the second type of tests, s is to be varied as a given function of time to verify experimentally whether the strain response satisfies the superposition principle which is a direct consequence of (3)

$$e(t) = \int_{t_0}^{t} e^*(t - t') \frac{ds}{dt'} dt', \tag{11}$$

where $e^*(t)$ is the previously measured incremental creep compliance.

It goes without saying that the two types of tests proposed here must be repeated at various stress levels and various times t_0 in order to bring out the dependence of linear operators in (3) on σ_0 and t_0.

It must be noted that recent works on creep behavior under varying stress[1] by Carreker, Leschen and Lubahn [5], Lubahn [6], Nishihara, Taira, Tanaka and Ohji [7] treat some aspects of the general program outlined in this section. However to the present authors' knowledge no systematic studies of the type proposed here are available in the literature.

3. Tests with Lead Wires

Exploratory tests of the type proposed in the previous section were conducted with commercially pure lead wires. The test specimens were of $\frac{1}{4}''$ diameter and $16''$ gage length. They were held by serrated grips of length $2\frac{1}{2}''$ at both ends. The specimens had a slightly larger diameter than $\frac{1}{4}''$ when purchased so that some cold stretching was first applied to bring the diameter to $\frac{1}{4}''$ and then the specimens were annealed at 250°F for one hour.

The basic stress was $\sigma_0 = 755$ psi and the time at which incremental loading was applied was $t_0 = 8$ hrs (Fig. 1) for all tests.

Elongations were measured by a dial gage sensitive to $10^{-4}{}''$. The tests were conducted at room temperature and during the phase of incremental loading the temperature was kept within 72 ± 0.5°F.

The incremental strains were obtained from the measured elongations with the aid of (2) where the following linear approximation was employed for $\varepsilon_0(t)$:

$$\varepsilon_0(t) = \varepsilon_0(t_0) + \dot{\varepsilon}_0(t_0)(t - t_0) \qquad \text{for } t > t_0,$$

where $\dot{\varepsilon}_0(t_0)$ denotes the slope of the basic $\varepsilon_0(t)$ curve at $t = t_0$. $\dot{\varepsilon}_0(t_0)$ was measured in each test and the measured value (not an average value) was used in determining the incremental strains. The above linear approximation was permissible for test durations of one hour in view of the fact that $\ddot{\varepsilon}_0(t_0)$ was of the order of -0.15×10^{-4} hr^{-2} while $\dot{\varepsilon}_0$ varied between 3.12×10^{-4} hr^{-1} and 4.86×10^{-4} hr^{-1}.

---

[1] For a recent literature survey see [4].

In Fig. 3 time dependent incremental elongations are shown for various incremental step loadings. In order to assess the importance of

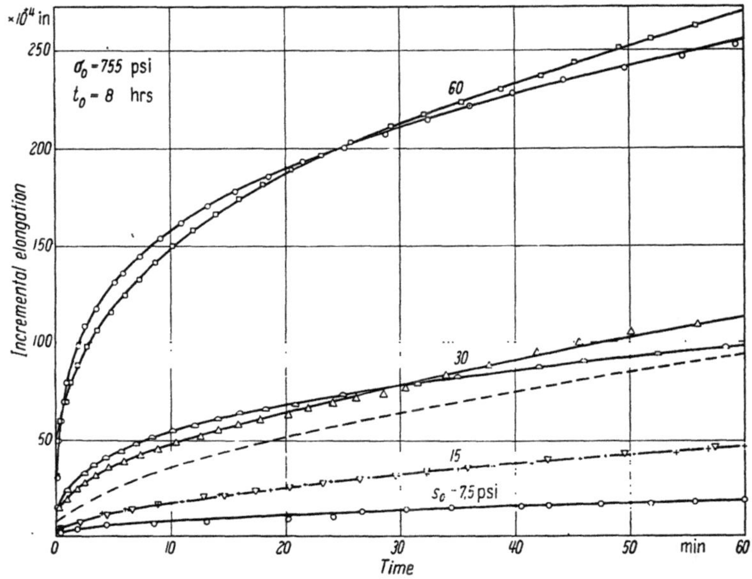

Fig. 3. Incremental elongations following incremental step loading

scatter (which may be considerable since each test is carried out with a different specimen) each test was repeated twice except the one corresponding to $s_0 = 7.5$ psi. The dashed curve introduced for comparison purposes has twice the ordinates of the curve corresponding to $s_0 =$ 15 psi $\left(\frac{s_0}{\sigma_0} = 2\%\right)$. It is seen from Fig. 3 that for incremental step loads of the order of 4% or less of the basic load, there is a clear indication of the linearity in the sense of Eq. (10). However for incremental loads of higher intensity such as $s_0 = 60$ psi, $\frac{s_0}{\sigma_0} = 8\%$, the linearity does not appear to hold.

It is interesting to note that the incremental creep compliance $e^*(t)$ determined on the basis of the strain response to $s_0 = 15$ psi can be represented with good accuracy by the analytical expression

$$e^*(t) = 0{,}020 \times 10^{-4} + 0{,}204 \times 10^{-4}\left(1 - e^{-(t-t_0)/37.5}\right), \qquad (12)$$

where t is measured in minutes. Comparison with (5) shows that the compliance corresponds to an equation of the type (4) where

$$p_0 = 0, \frac{q_1}{p_1} = 4{,}44 \times 10^4 \text{ psi}, \frac{q_2}{p_2} = 50 \times 10^4 \text{ psi}$$

and the retardation time $\tau = \frac{q_2}{q_1} = 37.5$ min.

We observe that $\frac{q_2}{p_2} = 50 \times 10^4$ psi has the nature of an *inelastic* modulus since it is approximately one fifth of the elastic modulus of lead. We also note that the retardation time according to the RABOTNOV-SHESTERIKOV linearization of the mechanical equation of state is 30 hrs. in view of the previously mentioned values of $\dot{\varepsilon}_0$ and $\ddot{\varepsilon}_0$ and (9). The retardation time according to (12) is, however, 37.5 min. The wide discrepancy between the two retardation times is remarkable and is indicative of the further work needed in this area. Another point of interest is that $p_0 = 0$ according to (12) which seems to support the RABOTNOV-SHESTERIKOV linearization. However this result must be accepted with caution since in a similar test (with different σ_0 and t_0) on a lead wire [5], the incremental strain versus time curve had a non-horizontal asymptote.

Before discussing the incremental unloading tests we mention briefly some results of the tests of the second type. In Fig. 4 the results of two such tests are shown. In these tests s was increased linearly in time and with the aid of a water loading device to 30 and 61.1 psi respectively within 25 minutes and was kept constant at the respective values for further times. Indications of linearity for small increments of loading are noted in Fig. 4. Moreover the curve OA which is obtained from (11) and (12) provides a strong indication of the validity of the superposition principle.

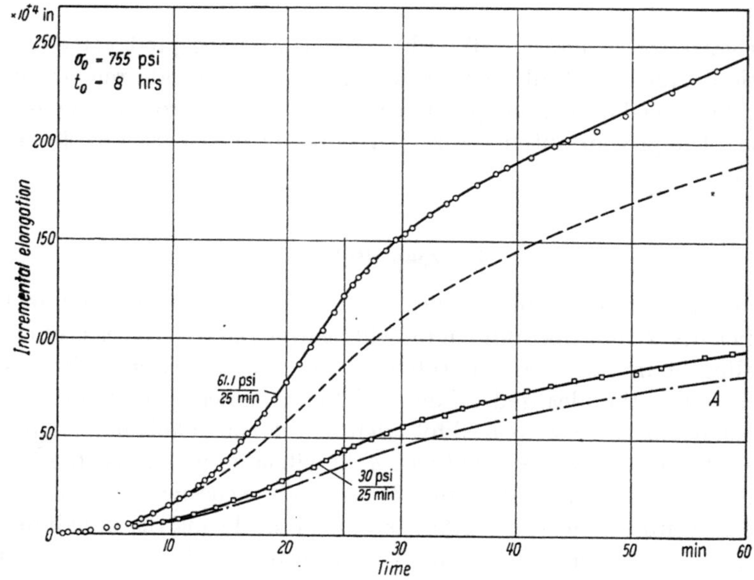

Fig. 4. Incremental elongations caused by linearly increasing loads

In Fig. 5 the incremental unloading behavior is shown for $s_0 = -30$ and -60 psi. In this figure absolute elongations are plotted versus time. For simplicity the origin of ordinates is taken at $\varepsilon_0(t_0)$. The straight lines with labels 1 and 2 indicate the slopes of the $\varepsilon_0(t)$ curve at $t = t_0$ for two tests. It was observed that upon the application of incremental unloading the strains were decreased by an amount $\frac{s_0}{E}$ where $E = 240 \times 10^4$ psi.

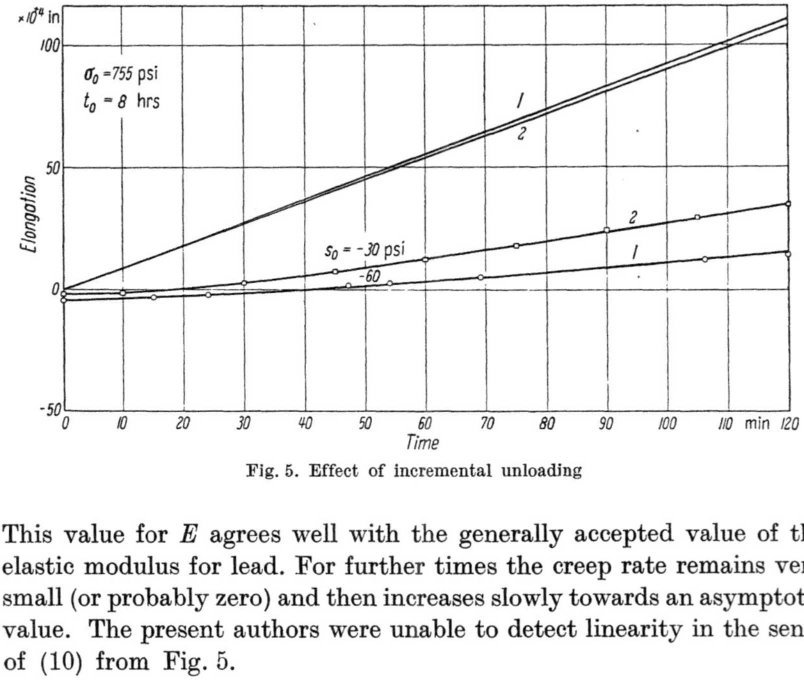

Fig. 5. Effect of incremental unloading

This value for E agrees well with the generally accepted value of the elastic modulus for lead. For further times the creep rate remains very small (or probably zero) and then increases slowly towards an asymptotic value. The present authors were unable to detect linearity in the sense of (10) from Fig. 5.

4. Creep Stability

In this section the importance of the rational determination of the coefficients of time operators in (3) will be illustrated in the case of creep buckling of initially straight columns. For brevity we shall limit the discussion to the idealized column[1] of Fig. 6. which consists of a rigid portion of length $l/2$ supported by a deformable cell of height h composed of two flanges of cross sectional area A. It will be assumed that the flange material obeys a law of the type (3) for small excursions from a steady state in compression. It is further assumed that during the time interval $0 < t < t_0$ the column has carried a vertical load P_0 with no accom-

[1] For a complete discussion of the questions of creep buckling see [8].

panying angular displacement θ. Neglecting the inertia terms in the equations of motion we have for $0 < t < t_0$

$$\sigma_L = \sigma_R = \sigma_0 = -\frac{P_0}{2A} \text{ and } u = u_0(t), \tag{13}$$

where σ_L and σ_R are the stresses in the flanges and $u_0(t)$ is the time dependent vertical displacement caused by creep in the flanges under the compressive stress σ_0.

Suppose that at time $t = t_0$ a dynamic disturbance is applied to the column to cause deviations from the vertical position. If the resulting motion is again regarded as quasi-static we have the equations of equilibrium

$$\begin{aligned} P - P_0 + A(s_L + s_R) &= 0, \\ l\,P\,\theta + b\,A(s_L - s_R) &= 0, \end{aligned} \tag{14}$$

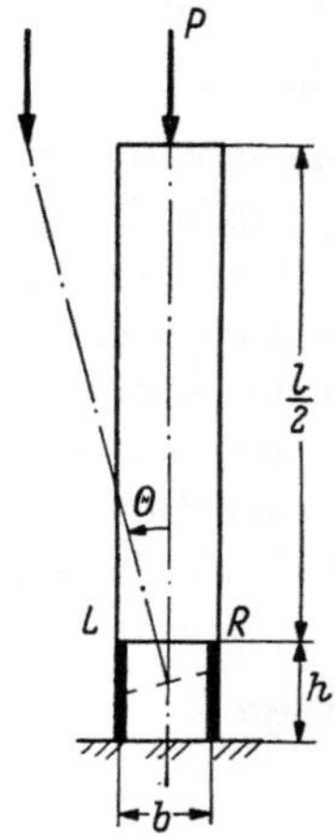

Fig. 6. Model of column

where s_L and s_R denote deviations from the basic state of stress and P the time dependent load. Since $e_L - e_R = -\frac{b}{h}\,\theta$, we obtain from the second equation of (14) and the constitutive equation (3)

$$\Sigma\, p_r \frac{d^r(P\,\theta)}{dt^r} - D\,\Sigma\, q_r \frac{d^r\theta}{dt^r} = 0, \tag{15}$$

where $D = \frac{b^2}{l}\frac{A}{h}$.

In the special case of (4) and $P = P_0 + \lambda(t - t_0)$, (15) reduces to

$$p_2\left(P - D\frac{q_2}{p_2}\right)\ddot{\theta} + p_1\left(2\lambda\frac{p_2}{p_1} + P - D\frac{q_1}{p_1}\right)\dot{\theta} + (\lambda\,p_1 + p_0\,P)\,\theta = 0. \tag{16}$$

It is required to investigate the boundedness of the solutions of (16) with the initial conditions $\theta = 0$, $\dot{\theta} = \dot{\theta}_0$ at $t = t_0$.

The case of $p_0 = 0$ and $\lambda = 0$ was investigated by Rabotnov and Shesterikov [*9*]. In view of the inequality $\frac{q_2}{p_2} > \frac{q_1}{p_1}$ (which is equivalent to $\tan\alpha > 0$ in Fig. 2) one finds that the solutions in this case are bounded if $P_0 < P_1 = D\frac{q_1}{p_1}$.

On the other hand if $\lambda = 0$ but $p_0 > 0$, i. e. if "viscous flow" is present in incremental straining the solutions of (16) are unbounded even for $0 < P_0 < P_1$. This point (which was also noticed in [*10*]) illustrates the importance of an experimental determination of p_0. If the tests indicate that $p_0 \neq 0$, then the validity of the Rabotnov-Shesterikov criterion will be doubtful.

On the other hand we observe that in both cases considered above the power r of the unbounded solution e^{rt} is proportional to $(P_0 - P_2)^{-1}$ where $P_2 = D\frac{q_2}{p_2} = D\,\overline{E}$ so that when P_0 is in the neighborhood of P_2 the structure will lose its usefulness after a short time following the dynamic disturbance. Thus $P_2 = D\,\overline{E}$ has a real significance.

The importance of P_2 or equivalently the inelastic modulus $\overline{E}$ may be seen more clearly if the response of a column to a dynamic disturbance is investigated under gradually increasing loads. In this case (16) has time dependent coefficients and it is singular at $P = P_2$. The nature of the solutions of (16) in the neighborhood of $P = P_2$ is easily investigated with the help of the indicial equation [*11*]. It will be found that

$$\theta \sim (P - P_2)^{1-\gamma},$$

where

$$\gamma = 2 + \frac{D}{\lambda}\frac{q_1 q_2}{p_2^2}\left(\frac{p_1}{q_1} - \frac{p_2}{q_2}\right) > 2.$$

The above result shows that θ will increase to infinity when P reaches the critical value P_2.

It is of interest to compare the critical load P_2 with the critical load $P_S = D\,E_S$ proposed by SHANLEY [*12*] where E_S is iso-tangent modulus. In order to compare E_S with $\overline{E}$ let us reconsider Fig. 2 and assume (with some justification) that the asymptote in this figure represents a portion of the creep curve corresponding to the finite step load $\sigma_0 + s_0$. Under this assumption and by the definition of the iso-tangent modulus we have

$$E_S = \frac{s_0}{0\,A'}.$$

It is seen from Fig. 2 that $E_S < \overline{E}$ and therefore $P_2 > P_S$. The last inequality is significant and it may explain why the SHANLEY theory underestimates the critical loads and the critical times in the creep buckling of columns [*13*]. In fact it would be desirable to compare experimentally obtained buckling times in [*13*] with critical times based on $P_2 = D\,\overline{E}$ where $\overline{E}\,(\sigma_0, t_0)$ is the inelastic modulus obtained in tests of the first type proposed in this paper.

References

[*1*] STUART, H. A.: Editor, Die Physik der Hochpolymeren, Berlin/Göttingen/Heidelberg: Springer 1956, Vol. IV, Ch. I.

[*2*] LEE, E. H.: Stress Analysis for Visco-elastic Materials, Conf. on the Properties of Materials at High Rates of Strain, The Institute of Mechanical Engineers, 1957.

[*3*] BODNER, S. R., and H. KOLSKY: Stress Wave Propagation in Lead, Proc. Third US Nat. Cong. of Appl. Mech., ASME, 495—501 (1958).

[4] MANSON, S. S., G. SACHS and W. F. BROWN, Jr.: Literature Survey on the Effect of Cyclic Heating and Loading of Metals at Elevated Temperatures, The Joint ASTM-ASME Committee on Effects of Temperature on the Properties of Metals, June 1957.
[5] CARREKER, R. P., J. G. LESCHEN and J. D. LUBAHN: Transient Plastic Deformation, Institute of Metal Division, American Institute of Mining and Metallurgical Engineers, Transactions **180**, 139—146 (1949).
[6] LUBAHN, J. D.: Creep Properties and Creep-Stress-Time Relations Mechanics of Creep, Proceedings for a short course on Mechanics of Creep conducted by the Pennsylvania State University, 43—64, July 1954.
[7] NISHIHARA, T., S. TAIRA, S. TANAKA and K. OHJI: Creep of Low Carbon Steel under Interrupted Stressing, Technical Reports of the Engineering Research Institute, Kyoto University, Kyoto, Japan, VI, No. 7, 127—145 (1956).
[8] HOFF, N. J.: A Survey of the Theories of Creep Buckling, Proc. Third US Nat. Cong. of App. Mech., ASME, 29—49 (1958).
[9] RABOTNOV, G. N., and S. A. SHESTERIKOV: Creep Stability of Columns and Plates, Jour. Mechanics and Physics of Solids, **6**, 27 (1957).
[10] ROSENTHAL, D., and H. W. BAER: Elementary Theory of Creep Buckling of Columns, Proc. First US Nat. Cong. App. Mechanics, ASME, 603 (1952).
[11] JEFFREYS, H., and B. S. JEFFREYS: Methods of Mathematical Physics, Cambridge, (1950), p. 481.
[12] SHANLEY, F.: Weight-Strength Analysis of Aircraft Structures, New York: McGraw Hill (1952), Chap. 19.
[13] JAHSMAN, W. E., and F. A. FIELD: Creep Instability and Collapse of Initially Straight Columns, Lockheed Aircraft Corporation, Missile Systems Division, Report No. 2458, June 1958.

Discussion

E. H. LEE, Brown University: I would like first to congratulate the authors on this incremental approach to creep analysis, which appears to offer the possibility of the consideration of general loading situations within the framework of the mathematically relatively simple theory of linear viscoelasticity.

In connection with the discussion of the quadratic differential operator relation:

$$p_2 \frac{d^2\sigma}{dt^2} + p_1 \frac{d\sigma}{dt} + p_0 \sigma = q_2 \frac{d^2\varepsilon}{dt^2} + q_1 \frac{d\varepsilon}{dt} \tag{1a}$$

it is perhaps helpful to observe that it is represented by the viscoelastic model shown in Fig. 1a with the spring constants and dashpot viscosities prescribed in terms of the constant coefficients in Eq. (1a).

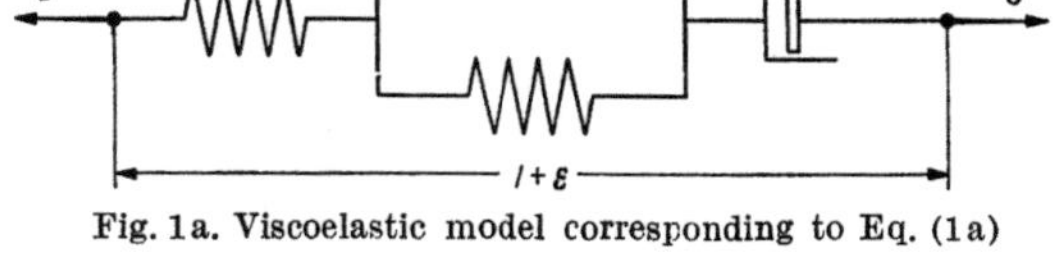

Fig. 1a. Viscoelastic model corresponding to Eq. (1a)

For $p_0 = 0$, integration of Eq. (1a) with respect to t gives:

$$p_2 \frac{d\sigma}{dt} + p_1 \sigma = q_2 \frac{d\varepsilon}{dt} + q_1 \varepsilon \tag{2a}$$

with integration constants satisfying zero initial values. Eq. (2a) correspond to the standard linear solid corresponding to the viscoelastic model shown in Fig. (2a).

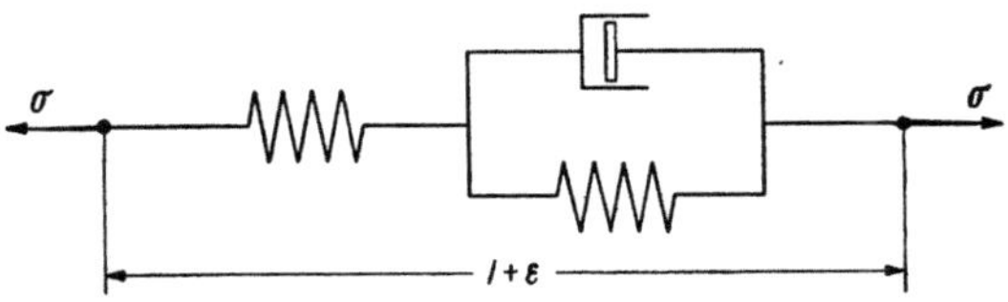

Fig. 2a. Viscoelastic model corresponding to Eq. (2a)

These models assist in the visualization of the response of the laws defined in Eqs. (1a), (2a) respectively to prescribed stress variation. For example, under suddenly applied and maintained constant stress, both exhibit instantaneous and delayed elastic response, but (1a) includes viscous or steady creep strain while (2a) does not. In addition to solutions of Eq. (2a), Eq. (1a) with $p_0 = 0$ allows σ and ε to maintain arbitrary constant values, but this situation is not consistent with Eq. (2a) for zero initial conditions, which represents the physically reasonable situation in this case.

Applicability of the Elastic Analogue to Creep Problems of Plates, Membranes and Beams

By

Folke K. G. Odqvist

Royal Institute of Technology, Stockholm, Sweden

1. Introduction. The Elastic Analogue

The analogy between the phenomena of small elastic-plastic deformations and those of plastic flow from a purely theoretical point of view seems to have been first called attention to by A. A. ILYUSHIN in a paper of 1945 [*1*, *2*]. In 1954 N. J. HOFF showed that, in order to find the asymptotic stress distribution in a body subject to creep under the action of constant forces, it is possible to use an elastic analogue [*3*, *4*]. In many cases it is thus possible to neglect transient phenomena connected with the initial stages in problems of incipient creep. In this way a state of stationary flow is considered, corresponding to secondary creep in uniaxial tension. Such theory is entirely within the Eulerian conception of the flow of matter.

Under other conditions, to be specified in the sequel, the elastic analogue is no longer valid. In certain cases it is necessary to take account also of second order terms in the expressions of the strains and strain rates. This assumption then changes the whole picture of the phenomenon. In this case, we may use with advantage the Lagrangian representation of the flow of matter and hence consider a cartesian strain tensor of the type [*5*]

$$\varepsilon_{ij} = \frac{1}{2}\left[\frac{\partial u_i}{\partial x_j} + \frac{\partial u_j}{\partial x_i} + \frac{\partial u_k}{\partial x_i}\frac{\partial u_k}{\partial x_j}\right], \tag{1.01}$$

where u_i is the displacement vector and x_i the original position of a point in the undeformed state. In (1.01) the summation convention is used.

For the strain rate tensor $\dot{\varepsilon}_{ij}$, we shall introduce the corresponding dotted quantities

$$\dot{\varepsilon}_{ij} = \frac{d\varepsilon_{ij}}{dt} = \frac{1}{2}\left[\frac{\partial \dot{u}_i}{\partial x_j} + \frac{\partial \dot{u}_j}{\partial x_i} + \frac{\partial \dot{u}_k}{\partial x_i}\frac{\partial u_k}{\partial x_j} + \frac{\partial u_k}{\partial x_i}\frac{\partial \dot{u}_k}{\partial x_j}\right] \tag{1.02}$$

using fluxions for the time derivatives.

The following treatment shall be limited to incompressible materials, i. e. we shall have

$$\varepsilon_{kk} = 0 \text{ and also } \dot{\varepsilon}_{kk} = 0. \tag{1.03}$$

We shall use the second order strain invariants (ε_e, "effective strain", and $\dot{\varepsilon}_e$, "effective strain rate") defined by

$$\varepsilon_e^2 = 2\,\varepsilon_{ij}^2/3, \tag{1.04}$$

$$\dot{\varepsilon}_e = 2\,\dot{\varepsilon}_{ij}^2/3. \tag{1.05}$$

In considering time-dependent phenomena, we shall always neglect inertia forces.

2. Subdivision of Creep Problems

We further introduce the stress tensor σ_{ij}, the stress deviation tensor s_{ij}

$$s_{ij} = \sigma_{ij} - (\delta_{ij}\,\sigma_{kk}/3), \tag{2.01}$$

where δ_{ij} is Kronecker's delta, and finally the second order stress invariant ("effective stress") σ_e, defined by

$$\sigma_e^2 = 3\,s_{ij}^2/2 = \frac{3}{2}\,[\sigma_{ij}^2 - \sigma_{kk}^2/3]. \tag{2.02}$$

Higher invariants than those of the second order will be left out of consideration. All stresses will be referred to the undeformed dimensions of the body.

We shall start from stress-strain relations of a type studied previously [*6*] and covering the stage of secondary creep as well as, at least approximately, that of primary creep, *viz.*

$$\dot{\varepsilon}_{ij} = \frac{3}{2}\,F(\sigma_e)\,s_{ij} + \frac{3}{2}\,\frac{d}{dt}\,[G(\sigma_e)\,s_{ij}], \tag{2.03}$$

where the last term has to be dropped when the quantity $\sigma_e\,d\sigma_e \leq 0$; when the converse is true, the term must be retained. This condition should be understood in such a way that retention of the second term in (2.03) shall be connected with the first loading process with ever increasing values of the invariant σ_e^2. Should unloading take place from a certain value σ_{e1}^2 of the invariant, then the second term in (2.03) should be dropped for all intermediate stages until the value σ_{e1}^2 is again exceeded. Eq. (2.03) may be formally integrated

$$\varepsilon_{ij} = \frac{3}{2}\int_0^t F(\sigma_e)\,s_{ij}\,dt + \frac{3}{2}\,G(\sigma_e)\,s_{ij}. \tag{2.04}$$

If loading begins from a stressfree and strainfree state of the body the first stages of deformation (when t is small) may be represented by the second term to the right in (2.04), while the first term may be neglected. On the other hand, during the later stages of the motion, the first term may predominate and the second term may be neglected. We shall treat problems of both kinds in the sequel and shall call them problems of the type *E* and *C*, respectively.

The functions $F(\sigma_e)$ and $G(\sigma_e)$ express the individual behaviour of the material. In general they will depend also on temperature, metallurgical variables etc. Incidentally, we shall use in the sequel the forms

$$F(\sigma_e) = (\sigma_e/\sigma_c)^{n-1}/\sigma_c, \tag{2.05}$$

$$G(\sigma_e) = (\sigma_e/\sigma_0)^{n_0-1}/\sigma_0. \tag{2.06}$$

Here n and n_0 are constants, the former being the well-known "NORTON's exponent". σ_c is the stress required to maintain a specified unit creep rate in uniaxial tension ("limiting creep stress") and σ_0 the stress necessary to produce a specified permanent set in a uniaxial tension test ("proof stress"). If the units are selected suitably, the stresses σ_c and σ_0 can be identified with well-known material constants encountered in the literature of testing materials.

In the case of type *C* problems when the second term in the right-hand member of (2.03) is so small as to be neglected in comparison with the first term, we have, irrespective of the sign of the product $\sigma_e\, d\sigma_e$, the relation

$$\dot{\varepsilon}_{ij} = \frac{3}{2} F(\sigma_e)\, s_{ij}, \tag{2.07}$$

where $F(\sigma_e)$ may be taken as given in (2.05). This is obvious from the fact that, utilizing (1.05), we have

$$\dot{\varepsilon}_e = F(\sigma_e)\, \sigma_e = (\sigma_e/\sigma_c)^n.$$

In this case it may be advantageous to solve (2.07) for s_{ij}, so that we obtain

$$s_{ij} = \frac{2}{3} f(\dot{\varepsilon}_e)\, \dot{\varepsilon}_{ij}. \tag{2.08}$$

Here, incidentally, we shall use the equivalent of (2.05), *viz.*

$$f(\dot{\varepsilon}_e) = \sigma_c\, \dot{\varepsilon}_e^{-(n-1)/n}. \tag{2.09}$$

In the case of type *E* problems when t is small and the first term in the right-hand member of (2.04) may be neglected we may use this relation in the integrated ("finite") form

$$\varepsilon_{ij} = \frac{3}{2} G(\sigma_e)\, s_{ij} \tag{2.10}$$

or in the special case of (2.06)

$$\varepsilon_{ij} = \frac{3}{2} \frac{s_{ij}}{\sigma_0} \left(\frac{\sigma_e}{\sigma_0}\right)^{n_0-1} \tag{2.11}$$

which has the same form as (2.07) except that strain is substituted for strain rate.

Finally from (2.11), utilizing (1.04) we may obtain

$$\varepsilon_e = (\sigma_e/\sigma_0)^{n_0} \tag{2.12}$$

and therefore, in analogy to (2.09) as an alternative to (2.10)

$$s_{ij} = \frac{2}{3}\, g(\varepsilon_e)\, \varepsilon_{ij}, \tag{2.13}$$

where

$$g(\varepsilon_e) = \sigma_0\, \varepsilon_e^{-(n_0-1)/n_0}. \tag{2.14}$$

The Eqs. (2.11) or (2.13) together with the equations of equilibrium and appropriate boundary conditions form the basis of a problem of primary creep or of nonlinear elasticity, E, respectively. The Eqs. (2.07) or (2.08) together with the equations of equilibrium and appropriate boundary (and initial) conditions define a problem of secondary creep, C. If Eqs. (2.05) or (2.09) are used in this connection, the theory becomes a generalization of "NORTON's law" for secondary creep.

When the functions $F(\sigma_e)$ and $f(\dot{\varepsilon}_e)$ as defined by (2.05) and (2.09) are used, it is possible to pass to the limit $n = \infty$ and thus formally to arrive at the theory of rigid-plastic bodies, as pointed out by HOFF who gave the procedure the name "plastic analogue".

The possibility of taking solutions of problems of type E and of substituteing in them the strain rate tensor $\dot{\varepsilon}_{ij}$ for the strain tensor ε_{ij} in order to obtain solutions of problems of type C forms the content of the elastic analogue.

3. Plates and Membranes

We shall from now on limit the treatment to the case of plane plates and membranes, the middle surface of which in the undeformed state coincides with the $x\,y$-plane, with the z-axis as the normal of the surface. The plate thickness is taken as $2\,h$. As usual, we assume a plane state of stress with $\sigma_z = 0$, $\sigma_{xz} = \sigma_{yz} = 0$, so that the strain tensor takes the form

$$\begin{matrix} \varepsilon_x & \varepsilon_{xy} & 0 \\ \varepsilon_{xy} & \varepsilon_y & 0 \\ 0 & 0 & -(\varepsilon_x + \varepsilon_y). \end{matrix} \tag{3.01}$$

A corresponding matrix with fluxions is used for the strain rate tensor.

Further we introduce BERNOULLI's hypothesis of conservation of normals. If u, v, w be the displacement components of a point of the middle surface we obtain from (1.01)

$$\begin{aligned} \varepsilon_x &= \frac{\partial u}{\partial x} + \frac{1}{2}\left(\frac{\partial w}{\partial x}\right)^2 - z\,\frac{\partial^2 w}{\partial x^2}\,, \\ \varepsilon_y &= \frac{\partial v}{\partial y} + \frac{1}{2}\left(\frac{\partial w}{\partial y}\right)^2 - z\,\frac{\partial^2 w}{\partial y^2}\,, \\ \varepsilon_{xy} &= \frac{1}{2}\left(\frac{\partial u}{\partial y} + \frac{\partial v}{\partial x} + \frac{\partial w}{\partial x}\frac{\partial w}{\partial y} - 2\,z\,\frac{\partial^2 w}{\partial x\,\partial y}\right) \end{aligned} \tag{3.02}$$

and alternatively from (1.02)

$$\begin{aligned} \dot\varepsilon_x &= \frac{\partial \dot u}{\partial x} + \frac{\partial w}{\partial x}\frac{\partial \dot w}{\partial x} - z\,\frac{\partial^2 \dot w}{\partial x^2}\,, \\ \dot\varepsilon_y &= \frac{\partial \dot v}{\partial y} + \frac{\partial w}{\partial y}\frac{\partial \dot w}{\partial y} - z\,\frac{\partial^2 \dot w}{\partial y^2}\,, \\ \dot\varepsilon_{xy} &= \frac{1}{2}\left(\frac{\partial \dot u}{\partial y} + \frac{\partial \dot v}{\partial x} + \frac{\partial \dot w}{\partial x}\frac{\partial w}{\partial y} + \frac{\partial w}{\partial x}\frac{\partial \dot w}{\partial y} - 2\,z\,\frac{\partial^2 \dot w}{\partial x\,\partial y}\right). \end{aligned} \tag{3.03}$$

In the Eqs. (3.02) and (3.03), following VON KÁRMÁN, the only second order terms retained are the ones containing the normal deflection w.

Further we have

$$\varepsilon_e^2 = \frac{4}{3}\left(\varepsilon_x^2 + \varepsilon_y^2 + \varepsilon_x\,\varepsilon_y + \varepsilon_{xy}^2\right) \tag{3.04}$$

and

$$\dot\varepsilon_e^2 = \frac{4}{3}\left(\dot\varepsilon_x^2 + \dot\varepsilon_y^2 + \dot\varepsilon_x\,\dot\varepsilon_y + \dot\varepsilon_{xy}^2\right). \tag{3.05}$$

The stress tensor and stress deviation tensor will be denoted

$$\sigma_{ij} = \begin{pmatrix} \sigma_x, & \sigma_{xy}, & 0 \\ \sigma_{xy}, & \sigma_y, & 0 \\ 0, & 0, & 0 \end{pmatrix} \qquad s_{ij} = \begin{pmatrix} s_{11}, & s_{12}, & 0 \\ s_{12}, & s_{22}, & 0 \\ 0, & 0, & s_{33} \end{pmatrix} \tag{3.06}$$

and may be obtained from one another according to

$$\sigma_{ij} = \begin{pmatrix} 2s_{11} + s_{22}, & s_{12}, & 0 \\ s_{12}, & s_{11} + 2s_{22}, & 0 \\ 0, & 0, & 0 \end{pmatrix} \qquad s_{ij} = \begin{pmatrix} (2\sigma_x - \sigma_y)/3, & \sigma_{xy}, & 0 \\ \sigma_{xy}, & (2\sigma_y - \sigma_x)/3, & 0 \\ 0, & 0, & -(\sigma_x + \sigma_y)/3 \end{pmatrix} \tag{3.07}$$

Finally we have

$$\sigma_e^2 = \sigma_x^2 + \sigma_y^2 - \sigma_x\,\sigma_y + 3\,\sigma_{xy}^2. \tag{3.08}$$

Introduction of (3.02) and (3.04) into (2.13) and (3.07) yields

$$\begin{aligned} \sigma_x &= 2g(\varepsilon_e)\,(2\varepsilon_x + \varepsilon_y)/3\,, \\ \sigma_y &= 2g(\varepsilon_e)\,(\varepsilon_x + 2\varepsilon_y)/3\,, \\ \sigma_{xy} &= 2g(\varepsilon_e)\,\varepsilon_{xy}/3\,. \end{aligned} \tag{3.09}$$

Equations corresponding to (3.09) will be given the number (3.10); they will contain the dotted quantities of (3.03) and (3.05) and the function f, since (2.08) is used instead of (2.13). We may now calculate resultant forces and moments per unit length of the normal sections of the median surface parallel to the coordinate axes using (3.09) [or (3.10) respectively] for the stresses as functions of z, the distance from the undeformed middle surface by way of Eqs. (3.02), (3.04) and (2.14) for problems of type E [or (3.03), (3.05) and (2.09), respectively, for problems of type C], *viz.*

$$\begin{aligned} N_x &= \int \sigma_x \, dz, \quad N_y = \int \sigma_y \, dz, \quad N_{xy} = \int \sigma_{xy} \, dz, \\ M_x &= \int \sigma_x \, z \, dz, M_y = \int \sigma_y \, z \, dz, M_{xy} = \int \sigma_{xy} \, z \, dz. \end{aligned} \tag{3.11}$$

All integrals are extended from $z = -h$ to $z = +h$. The external forces may be given as p_x, p_y, p_z per unit area of the undeformed middle surface. The following equations of equilibrium for the deformed plate element are thus obtained:

Projectional equations:

$$\begin{aligned} \frac{\partial N_x}{\partial x} + \frac{\partial N_{yx}}{\partial y} &= -p_x, \\ \frac{\partial N_{xy}}{\partial x} + \frac{\partial N_y}{\partial y} &= -p_y, \\ \frac{\partial T_x}{\partial x} + \frac{\partial T_y}{\partial y} &= -p_z. \end{aligned} \tag{3.12}$$

Moment equations:

$$\begin{aligned} \frac{\partial M_x}{\partial x} + \frac{\partial M_{yx}}{\partial y} + N_x \frac{\partial w}{\partial x} + N_{yx} \frac{\partial w}{\partial y} - T_x &= 0, \\ \frac{\partial M_{xy}}{\partial x} + \frac{\partial M_y}{\partial y} + N_{xy} \frac{\partial w}{\partial x} + N_y \frac{\partial w}{\partial y} - T_y &= 0, \\ N_{xy} - N_{yx} &= 0. \end{aligned} \tag{3.13}$$

From the form of the stress tensor it is obvious that the twisting moments M_{xy} and M_{yx} are equal. The shearing forces T_x and T_y, perpendicular to the $x\,y$-plane may be immediately eliminated and we obtain

$$\begin{aligned} &\frac{\partial^2 M_x}{\partial x^2} + 2 \frac{\partial^2 M_{xy}}{\partial x \, \partial y} + \frac{\partial^2 M_y}{\partial y^2} + N_x \frac{\partial^2 w}{\partial x^2} + 2 N_{xy} \frac{\partial^2 w}{\partial x \, \partial y} + N_y \frac{\partial^2 w}{\partial y^2} \\ &\quad = -p_z + p_x \frac{\partial w}{\partial x} + p_y \frac{\partial w}{\partial y}, \end{aligned} \tag{3.14}$$

where we have utilized again the Eq. (3.12).

The right hand member of Eq. (3.14) obviously represents the external force in the negative direction of the normal of the deformed middle surface.

4. Non-validity of the Analogue

The Eqs. (3.12), (3.13) together with (3.11), (3.09), (3.02), (2.13) and appropriate boundary conditions form a complete system for the solution of the problems of type E, i. e. those of primary creep or non-linear elasticity for plates and membranes.

Correspondingly the Eqs. (3.12), (3.13), (3.11), (3.10), (3.03), (2.08) and appropriate boundary conditions form a complete system for the solution of the problems of type C, i. e. those of secondary creep in plates and membranes.

More or less complete solutions for problems of type E in the case of linear elasticity have existed for a long time for special loading cases [*8*, *9*]. In the case of non-linear elasticity new solutions have been recently developed [*10*, *11*].

If the non-linear terms in (3.02) and (3.03) are neglected, HOFF's elastic analogue becomes applicable. This means that a series of known solutions to problems of type E may be interpreted from the point of view of creep theory [*2*]. New solutions, utilizing digital and analogue machines have also been developed [*12*, *13*]. A complete solution of the problem of incipient creep in circular membranes, where non-linear terms are taken into account, has also been given [*14*].

A short account of the results of the papers [*10*] and [*14*] will first be given below in paragraph 5. In these cases the elastic analogue is not applicable.

In the quoted papers [*8*] to [*14*] both methods of differential equations and energy methods have been used. Most of the work is concerned with clamped or freely supported circular or rectangular plates as well as beams, loaded with uniform pressure over part of, or over the entire surface. In some cases concentrated loads are treated.

If the non-linear terms in (3.02) and (3.03) are retained it is obvious that the elastic analogue is no more applicable. In this case both the displacement vector u, v, w and the velocity vector $\dot{u}$, $\dot{v}$, $\dot{w}$ occur in the equations and the problem essentially becomes a very complex non-stationary one. In fact *the problem is non-linear for three different reasons* (1) *geometric*, represented by the non-linear terms in (3.02) or (3.03), (2) *static*, represented by the non-linear terms in (3.13) or (3.14), (3) *physical*, represented by the non-linearities contained in (3.09) or its equivalents. The elastic analogue can take care only of the physical non-linearity. In the sequel examples will be treated, where all three kinds of non-linearity will be taken into consideration.

5. Non-stationary Creep of Circular Membrane under Uniform Lateral Pressure

We shall consider, in this paragraph, the creep of a plane circular membrane, freely supported at its outer edge $r = a$ and loaded at the time $t = 0$ by a constant lateral pressure p, uniformly distributed over its surface. Bending creep will be neglected, i. e. we shall drop all terms containing z in (3.02) and (3.03). The problem will be split up in two, corresponding to the two terms in (2.03). We shall first treat the second term resulting in a problem of type E, using (3.02) treated in detail in [*10*]. The solution then gives us the initial deformation of the plate or that of primary creep if the usual terminology of creep theory be used. Subsequently we shall treat the first term in (2.03) resulting in a problem of type C, using (3.03).

Transforming the problem (3.02), (3.12) and (3.14) to polar coordinates, we obtain for the strains in the radial and tangential directions

$$\varepsilon_r = \frac{du}{dr} + \frac{1}{2}\left(\frac{dw}{dr}\right)^2, \tag{5.01}$$

$$\varepsilon_t = \frac{u}{r},$$

where u is the radial and w the transverse displacement of the middle surface. Further we have

$$\frac{d}{dr}(r\,N_r) = N_t, \tag{5.02}$$

$$\frac{1}{r}\frac{d}{dr}\left(r\,N_r\frac{dw}{dr}\right) = -p. \tag{5.03}$$

With the requirement that $N_r(dw/dr)$ should remain finite for $r = 0$, Eq. (5.03) may be immediately integrated:

$$N_r\frac{dw}{dr} = -\frac{p\,r}{2}. \tag{5.04}$$

We shall use (2.11) in the form

$$\varepsilon_r = H(N_e)\,[N_r - (N_t/2)], \tag{5.05}$$

$$\varepsilon_t = H(N_e)\,[N_t - (N_r/2)], \tag{5.06}$$

with

$$N_e^2 = N_r^2 - N_r\,N_t + N_t^2,$$

$$H(N_e) = (N_e/2\,\sigma_0\,h)^{n_0-1}/2\,\sigma_0\,h. \tag{5.07}$$

Introducing dimensionless variables R, T, U, W, ϱ instead of N_r, N_t, u, w and r, we put

$$\begin{aligned} N_r &= 2\,R\,\sigma_0\,h, & N_t &= 2\,T\,\sigma_0\,h, \\ u &= 2\,U\,\sigma_0\,h/p, & w &= 2\,W\,\sigma_0\,h/p, \quad r = 2\,\varrho\,\sigma_0\,h/p. \end{aligned} \tag{5.08}$$

Thus we obtain the system

$$\frac{d}{d\varrho}(\varrho R) = T, \tag{5.09}$$

$$R\frac{dW}{d\varrho} = -\frac{\varrho}{2}, \tag{5.10}$$

$$\frac{dU}{d\varrho} + \frac{1}{2}\left(\frac{dW}{d\varrho}\right)^2 = [R^2 - R\,T + T^2]^{\frac{n_0-1}{2}}\,[R - (T/2)], \tag{5.11}$$

$$\frac{U}{\varrho} = [R^2 - R\,T + T^2]^{\frac{n_0-1}{2}}\,[T - (R/2)] \tag{5.12}$$

with the boundary conditions $U = 0$, $W = 0$ for

$$\varrho = \varrho_1 = \frac{a\,p}{2\,\sigma_0\,h}\,. \tag{5.13}$$

The system (5.09) to (5.12) can be solved, if HENCKY's method is used and a solution in the form of a power series in ϱ is assumed for the unknown quantities R and T. Then the quantity $dU/d\varrho + [(dW/d\varrho)^2/2]$ can be formed in two different ways, once from (5.11) and once using (5.10) and (5.12). Equating these expressions yields recurrence formulae for the coeffcients of the power series. Satisfaction of the boundary condition (5.13) for U leads to an equation for the quantity X defined by

$$X = \varrho_1^2/16\,(n_0 + 3)\,S^{n_0+2}, \tag{5.14}$$

where S is the first constant term in the power series for R. The equation for X reads:

$$X = \frac{1}{2\,n_0 + 3}\,[1 - 3\,(2\,n_0^2 + 5\,n_0 + 9)\,X^2/2\,(n_0 + 3)$$
$$- (56\,n_0^4 + 582\,n_0^3 + 855\,n_0^2 + 2700\,n_0 + 1215)\,X^3/36\,(n_0 + 3)^2 - \cdots]. \tag{5.15}$$

Hence the only parameter in this equation is the exponent n_0. The quantity X thus being known, we may compute S from (5.14) and subsequently we can determine the other coefficients of the power series from the recurrence formulae. Then W may be obtained from (5.10) by integration, provided the boundary condition (5.13) is utilized. This yields

$$W = \frac{\varrho_1^2}{4\,S}\left\{1 - \left(\frac{\varrho}{\varrho_1}\right)^2 + \frac{X}{2}\left[1 - \left(\frac{\varrho}{\varrho_1}\right)^4\right]\right.$$
$$\left. + \frac{4\,n_0^2 + 27\,n_0 + 9}{18\,(n_0 + 3)}\left[1 - \left(\frac{\varrho}{\varrho_1}\right)^6\right] X^2 + \cdots\right\}. \tag{5.16}$$

The form of the meridian curve of the membrane thus will deviate but slightly from a parabola. The central deflection $w(0)$ will be

$$w(0) = 2\,\sigma_0\,h\,W(0)/p = a\left(\frac{p\,a}{2\,\sigma_0\,h}\right)^{\frac{n_0}{n_0+2}}\Phi, \tag{5.17}$$

where

$$\Phi = \left[16(n_0+3)\,X\right]^{\frac{1}{n_0+2}}$$

$$\left\{1 + X/2 + (4\,n_0^2 + 27\,n_0 + 9)\,X^2/18(n_0+3) + \cdots\right\}/4. \quad (5.18)$$

Numerical calculations with the solution (5.15) and (5.18) yield

$n_0 =$	1	2	3	5	7	10	∞
$X =$	0.16	0.1210	0.0972	0.0694	0.0541	0.0407	0
$\Phi =$	0.60	0.4737	0.4144	0.3563	0.3285	0.3063	0.25

We now proceed to the problem of type C, treated in the paper [*14*], i. e. the non-steady creep starting at $t = 0$ from the form given by (5.16) and governed by the system of equations

$$\dot{\varepsilon}_r = \frac{d\dot{u}}{dr} + \frac{dw}{dr}\frac{d\dot{w}}{dr} = F(N_e)\,[N_r - (N_t/2)], \quad (5.19)$$

$$\dot{\varepsilon}_t = \frac{\dot{u}}{r} = F(N_e)\,[N_t - (N_r/2)] \quad (5.20)$$

with

$$F(N_e) = (N_e/2\,\sigma_c\,h)^{n-1}/2\,\sigma_c\,h. \quad (5.21)$$

Introduction of dimensionless variables (5.08) and substitution of σ_c. for σ_0 leads to a system of equations similar to (5.09) to (5.12); however, the two last equations now read

$$\frac{d\dot{U}}{d\varrho} + \frac{dW}{d\varrho}\frac{d\dot{W}}{d\varrho} = [R^2 - R\,T + T^2]^{\frac{n_0-1}{2}}\,[R - (T/2)], \quad (5.22)$$

$$\frac{\dot{U}}{\varrho} = [F^2 - R\,T + T^2]^{\frac{n-1}{2}}\,[T - (R/2)]. \quad (5.23)$$

The dots as usual denote differentiation with respect to the time t. The boundary condition is the same as (5.13) with σ_c substituted for σ_0. The method of solution is similar but now the coefficients of the power series are functions of t. Equating of the two different expressions obtained for $(d\dot{U}/d\varrho) + (d\dot{W}/d\varrho)\,(dW/d\varrho)$ now yields a recurrence system of ordinary differential equations for the coefficients. Inspection shows however that the system admits of a comparatively simple solution, all coefficients being proportional to a single function $S(t)$ which may be determined from the first equation of the system

$$\frac{\dot{S}\,\varrho_1^2}{8(n+3)\,S^{n+3}} = -C, \quad (5.24)$$

where C is a constant, to be determined from a transcendental equation derived from the boundary condition $U(\varrho_1) = 0$, *viz.*

$$1 - (2n + 3) C - 3(2n^2 + 5n + 9) C^2/2(n + 3) + \cdots = 0. \tag{5.25}$$

Recurrence formulae for the other coefficients are given in the original publication.

It is seen that the functions $X = X(n_0)$ and $C = C(n)$ are identical and the table above could be used also for computing C for a given n. As the constants n_0 and n are generally given independently of each another (and for a number of materials $n_0 < n$) we see that in a given problem C will be different from X and usually C will be smaller than X.

Eq. (5.24) can now be integrated to obtain

$$S = [1/S_0^{n+2} + 8(n + 2)(n + 3) C t/\varrho_1^2]^{-1/(n+2)}, \tag{5.26}$$

where $S = S_0$ for $t = 0$. Integration of (5.10) yields

$$W = \frac{\varrho_1^2}{4S}\left\{1 - \left(\frac{\varrho}{\varrho_1}\right)^2 + \frac{C}{2}\left[1 - \left(\frac{\varrho}{\varrho_1}\right)\right]^4 + \frac{(4n^2 + 27n + 9)\,C^2}{18(n + 3)}\left[1 - \left(\frac{\varrho}{\varrho_1}\right)\right]^6 + \cdots\right\}. \tag{5.27}$$

Comparison of (5.16) and (5.27) shows that the initial condition can be only approximately satisfied.

From (5.14) we obtain

$$S_0 = [16(n_0 + 3) X/\varrho_1^2]^{1/(n_0+2)}. \tag{5.28}$$

Eq. (5.26) shows that S will decrease monotonously with t and thus W will increase correspondingly. The expressions for the dimensionless stresses are:

$$R = S\left\{1 - C\left(\frac{\varrho}{\varrho_1}\right)^2 - \frac{(4n^2 + 21n - 9)\,C^2}{6(n + 3)}\left(\frac{\varrho}{\varrho_1}\right)^4 + \cdots\right\}, \tag{5.29}$$

$$T = S\left\{1 - 3C\left(\frac{\varrho}{\varrho_1}\right)^2 - \frac{5(4n^2 + 21n - 9)\,C^2}{6(n + 3)}\left(\frac{\varrho}{\varrho_1}\right)^4 + \cdots\right\}. \tag{5.30}$$

They indicate that the stresses will increase from their initial values, which deviate but slightly from the common value S. Appreciable deviations will occur only in the neighbourhood of the boundary $r = a$.

I have taken this opportunity to correct a couple of misprints in the formulas (5.27), (5.29) och (5.30) in the previous publication [*14*].

6. Bending Creep of Circular Plates under Uniform Lateral Pressure

If, from the outset, we had also taken account of the bending creep, the problem of paragraph 5 would have been much more complicated and the series method employed would have proved useless. On the other hand, for comparatively thick plates with small deflections bending creep plays the most important part in the initial stages of the motion. Then the geometric and static non-linearities can be neglected and the elastic analogue can be used.

Consider the case of a circular plate of radius a and thickness h, subjected to uniform lateral pressure p. Alternatively we shall use the boundary conditions

(α) built-in at $r = a$,

(β) freely supported at $r = a$.

The equilibrium condition for the moments M_r in the radial direction and the moments M_t in the tangential direction is

$$d(r\, M_r)/dr - M_t = -p\, r^2/2 \tag{6.01}$$

if a concentrated force at the origine is excluded. From (3.10) we obtain for the stresses

$$\begin{aligned} \sigma_r &= 2\, f(\dot{\varepsilon}_e)\, [2\, \dot{\varepsilon}_r + \dot{\varepsilon}_t]/3, \\ \sigma_t &= 2\, f(\dot{\varepsilon}_e)\, [\dot{\varepsilon}_r + 2\, \dot{\varepsilon}_t]/3, \end{aligned} \tag{6.02}$$

where the strain rates, according to (3.03) and (3.05), are

$$\dot{\varepsilon}_r = -z\, \dot{w}'', \quad \dot{\varepsilon}_t = -z\, \dot{w}'/r,$$

$$\dot{\varepsilon}_e^2 = 4(\dot{\varepsilon}_r^2 + \dot{\varepsilon}_r\, \dot{\varepsilon}_t + \dot{\varepsilon}_t^2)/3 = z^2\, A^2$$

with

$$A^2 = 4[\dot{w}''^2 + \dot{w}'\, \dot{w}''/r + (\dot{w}'/r)^2]/3$$

and from (2.09)

$$f(\dot{\varepsilon}_e) = \sigma_c\, A^{1/n-1}\, z^{1/n-1}.$$

We may now compute the moments M_r and M_t from (6.02). Insertion in (6.01) yields the differential equation

$$\frac{d}{dr}\left[\frac{2\, r\, \dot{w}'' + \dot{w}'}{A^{1-1/n}}\right] - \frac{r\, \dot{w}'' + 2\, \dot{w}'}{r\, A^{1-1/n}} = \frac{2\, n+1}{2\, n}\left(\frac{2}{h}\right)^{2+1/n} \frac{3\, p\, r^2}{4\, \sigma_c} \tag{6.03}$$

which is a non-linear ordinary differential equation for $\dot{w}$.

We may now introduce dimensionless variables

$$w = a\, \Pi^n\, V, \tag{6.04}$$

$$r = a \cdot \varrho$$

with

$$\Pi = \frac{2\, n+1}{n} \left(\frac{3\, a^2}{h^2}\right)^{\frac{n+1}{2n}} \frac{p\, a}{h\, \sigma_c}. \tag{6.05}$$

If further $dV/d\varrho$ be denoted V', and

$$\begin{aligned} A^2 &= 4\, \Pi^{2n}\, [V''^2 + V'\, V''/\varrho + (V'/\varrho)^2]/3\, a^2 \\ &= 4\, \Pi^{2n}\, B^2/3\, A^2 \end{aligned} \tag{6.06}$$

we finally obtain

$$d[(2\varrho V'' + V').B^{1/n-1}]/d\varrho - (V'' + 2V'/\varrho)B^{1/n-1} = \varrho^2 \quad (6.07)$$

with B defined in (6.06) and the boundary conditions in the form

$$(\alpha) \quad V(1) = 0, \quad V'(1) = 0,$$

$$(\beta) \quad V(1) = 0, \quad 2V''(1) + V'(1) = 0.$$

In (6.07) only the exponent n remains as a parameter. For the numerical calculations it is advantageous to solve this boundary value problem starting the integration from $\varrho = 0$. It is easy to prove that, at $\varrho = 0$, we have

$$V'(0) = 0, \quad V'''(0) = 0$$

leaving but one parameter open, say $V''(0)$. The second boundary condition of (α) and (β), respectively, will be fulfilled by interpolation at $\varrho = 1$. The computations were carried out independently with an electromagnetic analogue machine FIDA and with the aid of the KUTTA-RUNGE method on the digital computer BESK. The details are given in the original paper [*13*]. Solution method was suggestad by H. O. KREISS.

It is of particular interest to compute the central deflection rate $\dot{w}(0)$ of the plate, which may be expressed with $V(0)$ according to (6.04) and (6.05). The result is seen in Fig. 1. It should be noted that the reducing

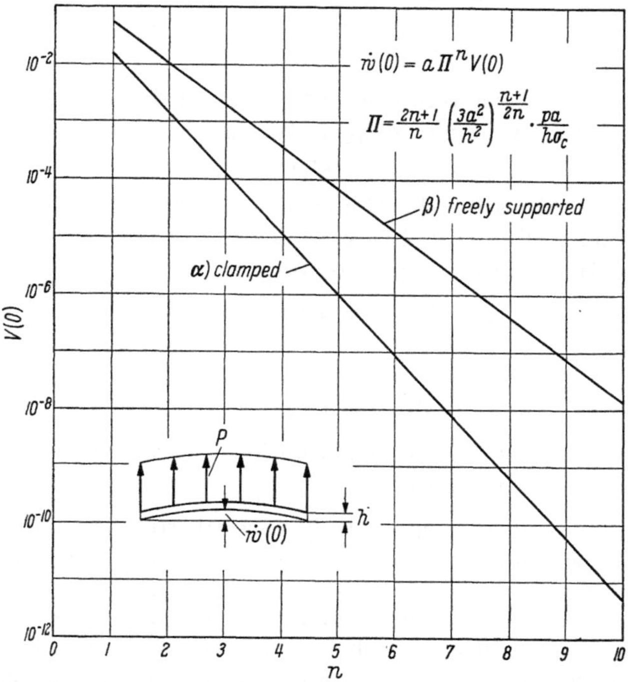

Fig. 1. Central deflection rate $\dot{w}(0)$ of circular plate, according to ODQVIST [*13*]

influence on $\dot{w}(0)$ by clamping increases considerably with the exponent n. Similar results have been obtained by N. N. MALININ as quoted in [2] and reproduced in [16]. This author also treats the case of a concentrated load at the centre of the plate. V. V. SOKOLOVSKY [2] treats the special case of Eq. (6.03) for $n = 3$ and the boundary condition (β) with an entirely different method. His numerical result does not check with ours.

Further solutions of the Eqs. (3.03) and (3.04) may be looked for. The problem of type E for a rectangular plate has been solved by L. WALLIN [11] with the aid of the energy method. On the other hand, if bending creep should also be taken into account the problem becomes much more complicated.

Consider e. g. *the simple case of plane motion parallel to to the x z-plane.* We have

$$\dot{\varepsilon}_y = 0, \quad \sigma_y = \sigma_x/2 \quad \text{and} \quad \sigma_e = \sigma_x \sqrt{3}/2$$

as well as $\dot{\varepsilon}_x = -\dot{\varepsilon}_z$ and thus $\dot{\varepsilon}_e = 2\,\dot{\varepsilon}_x/\sqrt{3}$. Also from (2.09) we take

$$f(\dot{\varepsilon}_e) = \sigma_c \left(\sqrt{3}/2\right)^{\frac{n-1}{n}} \dot{\varepsilon}_x^{\,-\frac{n-1}{n}} \tag{6.08}$$

and from (3.03) we obtain

$$\dot{\varepsilon}_x = \frac{d\dot{u}}{dx} + \frac{dw}{dx}\frac{d\dot{w}}{dx} - z\,\frac{d^2\dot{w}}{dx^2} \tag{6.09}$$

because all derivatives with respect to y vanish. In this case the plate degenerates into an infinite strip or slab parallel to the y-axis and all quantities become functions of x and t only.

Similar to this case is *that of a beam, extending in the x-direction* and with the center of gravity of its cross sections (symmetric with respect to the $x\,z$-plane) in the undeformed state on the x-axis. We then have $\sigma_y = 0$ and $\dot{\varepsilon}_y = \dot{\varepsilon}_z = -\dot{\varepsilon}_x/2$. Thus we obtain $\sigma_e = \sigma_x$ and $\dot{\varepsilon}_e = \dot{\varepsilon}_x$ and Eq. (2.09) reduces to

$$f(\dot{\varepsilon}_e) = \sigma_c\,\dot{\varepsilon}_x^{\,-\frac{n-1}{n}} \tag{6.10}$$

whereas all other equations remain unchanged from the case of plane motion parallel to the $x\,z$-plane.

The latter problem is one of truly uniaxial stress. A solution of the former may be obtained merely by changing the value of σ_c of Eq. (6.10) to $\sigma_c' = \sigma_c \left(\sqrt{3}/2\right)^{\frac{n-1}{n}}$ according to Eq. (6.08). In spite of the far-reaching simplification made, it is difficult to obtain any further progress along the lines of this paragraph, as M_x and N_x take a rather complex form.

7. Beams with Idealized H-Section

For further progress we shall simplify our problem by considering in the following paragraph *a beam with idealized H-section* or a slab in the form of a sandwich plate or a double deck. The treatment of these two cases is identical but for a constant factor to be applied to the constant σ_c; the difference between the two cases is similar to that found between Eqs. (6.08) and (6.10) above. It is thus sufficient to consider the beam problem only in the sequel.

The upper and lower flanges of the beam may have a cross section A each and their distance may be $2h$. Using subscripts u and l to indicate the quantities pertaining to the upper and lower flange we then have

$$\dot{\varepsilon}_{xu} = \dot{u}' + \dot{w}' w' - h \dot{w}'' = \left(\frac{\sigma_{xu}}{\sigma_c}\right)^n, \tag{7.01}$$

$$\dot{\varepsilon}_{xl} = \dot{u}' + \dot{w}' w' + h \dot{w}'' = \left(\frac{\sigma_{xl}}{\sigma_c}\right)^n \tag{7.02}$$

where the dashes denote differentiation with respect to x and the dots with respect to time t as before. From (7.01) and (7.02) we immediately obtain

$$2 h \dot{w}'' = \left(\frac{\sigma_{xl}}{\sigma_c}\right)^n - \left(\frac{\sigma_{xu}}{\sigma_c}\right)^n, \tag{7.03}$$

$$2(\dot{u}' + \dot{w}' w') = \left(\frac{\sigma_{xl}}{\sigma_c}\right)^n + \left(\frac{\sigma_{xu}}{\sigma_c}\right)^n. \tag{7.04}$$

From now on we shall assume *n to be a positive odd integer*. This assumption simplifies somewhat the formulae without restricting the generality of the treatment.

Equivalence of external and internal forces requires

$$N_x = A(\sigma_{xu} + \sigma_{xl}),$$

$$M_x = A h(\sigma_{xu} - \sigma_{xl})$$

or

$$2 A h \sigma_{xu} = N_x h + M_x,$$

$$2 A h \sigma_{xl} = N_x h - M_x$$

or, upon substitution in (7.03) and (7.04)

$$2 h \dot{w}'' = \left[\frac{N_x h - M_x}{2 A h \sigma_c}\right]^n - \left[\frac{N_x h + M_x}{2 A h \sigma_c}\right]^n \tag{7.05}$$

$$2(\dot{u}' + \dot{w}' w') = \left[\frac{N_x h - M_x}{2 A h \sigma_c}\right]^n + \left[\frac{N_x h + M_x}{2 A h \sigma_c}\right]^n. \tag{7.06}$$

If the external forces per unit length of the beam are p_x, p_z and if T is the shearing force, we further have the equilibrium conditions for the deformed beam element

$$\frac{\partial N_x}{\partial x} = - p_x, \tag{7.07}$$

$$\frac{\partial T}{\partial x} = - p_z, \tag{7.08}$$

$$\frac{\partial M_x}{\partial x} = T - N_x \frac{\partial w}{\partial x}. \tag{7.09}$$

Partial differentiation with respect to x indicates that in general all functions depend explicitely on time t, also.

Elimination of T yields

$$\frac{\partial^2 M_x}{\partial x^2} + N_x \frac{\partial^2 w}{\partial x^2} - p_x \frac{\partial w}{\partial x} + p_z = 0 \tag{7.10}$$

which is of course also a special case of Eq. (3.14). In the most general case M_x and N_x are statically indeterminate and depend on the boundary conditions. For simplicity we shall assume in the sequel $p_x = 0, p_z = p =$ constant, independent of time.

8. Beams with no Normal Force

In this section we shall treat the elementary case when $N_x \equiv 0$, *statically determinate*. Then from (7.05) we obtain

$$h \dot{w}'' = - (M_x/2A\, h\, \sigma_c)^n, \tag{8.01}$$

$$\dot{u}' = - \dot{w}' w'. \tag{8.02}$$

This problem is still within the domain of applicability of the elastic analogue, cf. N. J. HOFF [4], particularly p. 261. We shall treat two special cases.

(a) Simply Supported Beam

Consider a beam of length $2b$, simply supported at $x = \pm b$, so that we have

$$M_x = p(b^2 - x^2)/2.$$

Then (8.01) yields

$$h \dot{w}'' = - (p\, b^2/4\, A\, h\, \sigma_c)^n (1 - x^2/b^2)^n.$$

Take for simplicity $n = 3$, then integration is easily achieved and the boundary condition $w(\pm b, t) = 0$ as well as the initial condition $w(x, 0) \equiv 0$ are satisfied by

$$w(x, t) = \frac{b^2}{h} \left[\frac{p\, b^2}{4\, A\, h\, \sigma_c}\right]^3 \left[\frac{93}{280} - \frac{x^2}{2\, b^2} + \frac{x^4}{4\, b^4} - \frac{x^6}{10\, b^6} + \frac{x^8}{56\, b^8}\right] t. \tag{8.03}$$

Subsequently u may be obtained from (8.02) by integration, if the boundary condition $u(0, t) = 0$ is taken into account. This is an exact solution in the case $n = 3$. We shall give later in section 10 an approximate solution for arbitrary n.

(b) Beam, Clamped at both Ends

In this case we have to introduce a statically indeterminate moment M_1:

$$M_x = p\,(b^2 - x^2)/2 + M_1.$$

In the special case $n = 3$, putting $1 + 2\,M_1/p\,b^2 = m_1$, we obtain

$$h\,\dot{w}'' = -\left[\frac{p\,b^2}{4\,A\,h\,\sigma_c}\right]^3 [m_1 - x^2/b^2]^3$$

$$= -\left[\frac{p\,b^2}{4\,A\,h\,\sigma_c}\right]^3 [m_1^3 - 3\,m_1^2\,x^2/b^2 + 3\,m_1\,x^4/b^4 - x^6/b^6]$$

which becomes upon integration

$$h\,\dot{w}' = -\,b\left[\frac{p\,b^2}{4\,A\,h\,\sigma_c}\right]^3 \left[m_1^3\,\frac{x}{b} - m_1^2\,\frac{x^3}{b^3} + \frac{3\,m_1}{5}\,\frac{x^5}{b^5} - \frac{x^7}{7\,b^7}\right]$$

provided the symmetry condition $\dot{w}'(0, t) = 0$ is duly taken into account. The boundary condition $w'(\pm b, t) = 0$ yields an equation for the determinations of m_1

$$m_1^3 - m_1^2 + 3\,m_1/5 - 1/7 = 0$$

from which we obtain

$$m_1 = 0{,}3959.$$

In the case of $n = 1$ we would have $m_1 = 1/3$. Further integration and satisfaction of the boundary condition $\dot{w}(\pm b, t) = 0$ yield

$$w(x, t) = \frac{b^2}{h}\left[\frac{p\,b^2}{4\,A\,h\,\sigma_c}\right]^3 \left[\frac{m_1^3}{2}\left(1 - \frac{x^2}{b^2}\right) - \frac{m_1^2}{4}\left(1 - \frac{x^4}{b^4}\right) + \frac{m_1}{10}\left(1 - \frac{x^6}{b^6}\right)\right.$$
$$\left. - \frac{1}{56}\left(1 - \frac{x^8}{b^8}\right)\right] t = \frac{b^2}{h}\left[\frac{p\,b^2}{4\,A\,h\,\sigma_c}\right]^3 \cdot [0{,}01358 - 0{,}03105\,x^2/b^2 + \cdots]\,t. \tag{8.04}$$

Comparison with (8.03) shows that in this case clamping of the beam ends reduces the deflection at the origin in the proportion

$$\frac{93}{280} : 0{,}01358 \simeq 24{,}47 : 1$$

as compared with the proportion 5 : 1 for elastically deflected beams. This tendency of clamping to reduce deflection increases rapidly with n. It is also noticable in the case of bending creep in plates, though to a slightly minor degree, cf. paragraph 6 and the figure, from which the proportions 17.6 : 1 and 3.70 : 1 can be taken for $n = 3$ and 1, respectively.

9. Doubly Hinged Beams. Fundamental equations

We now proceed to the case when N_x is statically indeterminate and we have the additional boundary condition $u(\pm b, t) = 0$ for *the doubly hinged beam* as distinguished from the simply supported beam treated in the preceding paragraph. In this case the elastic analogue does no longer help us. From (7.07) we infer that $N_x = N_x(t)$ is independent of x, and from (7.10) we obtain

$$M_x = p(b^2 - x^2)/2 - N_x w \tag{9.01}$$

if we take into account the boundary conditions $M_x(\pm b, t) = 0$, $w(\pm b, t) = 0$. Then (7.05) takes the form

$$2\,h\,\dot{w}'' = \left[\frac{N_x(h+w) - p(b^2 - x^2)/2}{2\,A\,h\,\sigma_c}\right]^n - \left[\frac{N_x(h-w) + p(b^2 - x^2)/2}{2\,A\,h\,\sigma_c}\right]^n \tag{9.02}$$

and correspondingly (7.06) becomes

$$2(\dot{u}' + \dot{w}'\,w') = \left[\frac{N_x(h+w) - p(b^2 - x^2)/2}{2\,A\,h\,\sigma_c}\right]^n + \left[\frac{N_x(h-w) + p(b^2 - x^2)/2}{2\,A\,h\,\sigma_c}\right]^n. \tag{9.03}$$

Eq. (9.02) comprises as a special case for $N_x =$ negative constant and $p = 0$ the equation for creep buckling of columns, derived by HOFF [*15*]. Our method of solution will be an adaptation of HOFF's method. We shall assume the initial condition $w(x, 0) = 0$, $N_x(0) = 0$. Then it is natural to linearize Eq. (9.02) and (9.03) with respect to N_x. Thus we obtain

$$h\,\dot{w}'' = -\left[\frac{p(b^2 - x^2)}{4\,A\,h\,\sigma_c}\right]^n + n\,\frac{N_x\,w}{A\,h\,\sigma_c}\left[\frac{p(b^2 - x^2)}{4\,A\,h\,\sigma_c}\right]^{n-1}, \tag{9.04}$$

$$\dot{u}' + \dot{w}'\,w' = n\,\frac{N_x}{A\,\sigma_c}\left[\frac{p(b^2 - x^2)}{4\,A\,h\,\sigma_c}\right]^{n-1}. \tag{9.05}$$

We now introduce dimensionless variables U, W, N and ξ, setting

$$\frac{w}{h} = W, \qquad \frac{N_x}{A\,\sigma_c} = N, \qquad \frac{u}{h} = U, \qquad \frac{x}{b} = \frac{2}{\pi}\,\xi. \tag{9.06}$$

We further put

$$\frac{p\,b^2}{4\,A\,h\,\sigma_c} = \Pi. \tag{9.07}$$

Then we obtain the system

$$\frac{\pi^2 h^2}{4\,b^2}\,\frac{\partial^3 W}{\partial \xi^2\,\partial t} = -\Pi^n\,(1 - 4\,\xi^2/\pi^2)^n + n\,\Pi^{n-1}\,W\,N(1 - 4\,\xi^2/\pi^2)^{n-1}, \tag{9.08}$$

$$\frac{\pi\,h}{2\,b}\,\frac{\partial^2 U}{\partial \xi\,\partial t} = -\frac{\pi^2 h^2}{8\,b^2}\,\frac{\partial}{\partial t}\left(\frac{\partial W}{\partial \xi}\right)^2 + n\,\Pi^{n-1}\,N(1 - 4\,\xi^2/\pi^2)^{n-1} \tag{9.09}$$

for the two unknown functions $U(\xi, t)$, $W(\xi, t)$, which most fulfill the boundary conditions

$$U(\pm\,\pi/2, t) = 0, \quad W(\pm\,\pi/2, t) = 0. \tag{9.10}$$

The initial conditions are

$$U(\xi, 0) = 0, \quad W(\xi, 0) = 0, \quad N(0) = 0. \tag{9.11}$$

These are sufficient to determine the functions $W(\xi, t)$, $U(\xi, t)$ as well as $N(t)$.

The method of solution will be to assume $W(\xi, t)$ in the form of a FOURIER series

$$W = W_1(t)\cos\xi + W_3(t)\cos 3\xi + \cdots \tag{9.12}$$

which obviously satisisies the boundary condition (9.10). Assuming further that we may write

$$(1 - 4\xi^2/\pi^2)^n = C_1^{(n)}\cos\xi + C_3^{(n)}\cos 3\xi + \cdots \tag{9.13}$$

which will be considered in more detail later, we may now develop both sides of Eq. (9.08) in FOURIER series and equate coefficients on the two sides. We may further develop the right side of (9.09) in FOURIER series, integrate that equation with respect to ξ and satisfy the first boundary condition (9.10). In doing so we obtain the system of ordinary differential equations for the functions $W_1(t)$, $W_3(t)$, ...

$$-\frac{\pi^2 h^2}{4 b^2}\frac{dW_1}{dt} = -\Pi^n C_1^{(n)} + n\,\Pi^{n-1} N\frac{4}{\pi}$$
$$\times\left[\frac{2}{3}W_1 C_1^{(n-1)} + \frac{2}{15}(W_1 C_3^{(n-1)} + W_3 C_1^{(n-1)}) + \frac{18}{35}W_3 C_3^{(n-1)}\right]$$
$$-9\frac{\pi^2 h^2}{4 b^2}\frac{dW_3}{dt} = -\Pi^n C_3^{(n)} + n\,\Pi^{n-1} N\frac{4}{\pi}$$
$$\times\left[\frac{2}{15}W_1 C_1^{(n-1)} + \frac{18}{35}(W_1 C_3^{(n-1)} + W_3 C_1^{(n-1)}) - \frac{2}{9}W_3 C_3^{(n-1)}\right]$$

. . . .

$$-\frac{\pi^2 h^2}{8 b^2}\frac{4}{\pi}\frac{d}{dt}$$
$$\left[\frac{1}{3}W_1^2 + \frac{6}{5}W_1 W_3 + \frac{153}{35}W_3^2 - \frac{1}{3}\left(-\frac{7}{15}W_1^2 + \frac{18}{35}W_1 W_3 - W_3^2\right) + \cdots\right]$$
$$+ n\,\Pi^{n-1} N\,(C_1^{(n-1)} - C_3^{(n-1)}/3 + \cdots) = 0. \tag{9.14}$$

In the system (9.14) the terms with subscript 5 and higher have been neglected. Together these equations form a complete system for the determination of the quantitites W_1, W_3 and N as functions of t and the initial conditions

$$W_1(0) = 0,\ W_3(0) = 0, \ldots, N(0) = 0 \tag{9.15}$$

are duly taken into account. Still the system (9.14) is rather complicated and only an approximate solution will be attempted here. Before doing so we shall study the coefficients $C_{2m+1}^{(n)}$ of the series (9.13).

We have the general representation

$$C_{2m+1}^{(n)} = \frac{4}{\pi}\int_0^{\pi/2}(1 - 4\xi^2/\pi^2)^n\cos(2m+1)\xi\,d\xi$$
$$= 2\int_0^1 (1-\zeta^2)^n\cos(2m+1)\pi\zeta/2\,d\zeta. \tag{9.16}$$

Double integration by parts yields the relation

$$C^{(n)}_{2m+1} = \frac{8\,n}{(2\,m+1)^2\,\pi^2}\left[(2\,n-1)\,C^{(n-1)}_{2m+1} - (2\,n-2)\,C^{(n-2)}_{2m+1}\right] \tag{9.17}$$

provided n is greater than 1. We can compute immediately

$$C^{(0)}_{2m+1} = \frac{4}{\pi}\,\frac{\sin(2\,m+1)\,\pi/2}{2\,m+1}, \tag{9.18}$$

$$C^{(1)}_{2m+1} = \frac{32}{\pi^3}\,\frac{\sin(2\,m+1)\,\pi/2}{(2\,m+1)^3}. \tag{9.19}$$

Hence (9.17) forms a recurrence formula for higher values of n.

Table 1 contains numerical values of several coefficients $C^{(n)}_{2m+1}$. When the formula (9.17) is used it is necessary to start with a large number of decimals in the expressions for the lowest values of n as given by (9.18) and (9.19). Accuracy is rapidly lost in each step of the calculations due to the formation of differences between almost equal numbers. It can be shown that $C^{(n)}_{2m+1}$ is proportional to $(2\,m+1)^{-(n+2)}$, so that the series (9.13) converges very rapidly for large values of n.

Table 1. *The Quantity* $C^{(n)}_{2m+1}$

	$2m+1=1$	3	5	7
$n=0$	1.273239546	−0.424413182	0.254647909	−0.1818913637
1	1.032049104	−0.03822404089	0.00825639283	−0.003008889516
2	0.891088551	0.13224067	−0.031419400	0.00173691555
3	0.79576768	0.2199614	−0.01849291	0.0010282720
4	0.7257593	0.2688518	0.007660343	−0.000021330147
5	0.671527	0.297197	0.03516029	−0.0008391783
6	0.62836	0.313774	0.0603374	−0.0008950335
7	0.6258	0.32322	0.082265	−0.000181254

10. Doubly Hinged Beams. Approximate Solution

The properties of the coefficients $C^{(n)}_{2m+1}$ which we have just established indicate how to obtain an approximate solution of the system (9.14). Since $|C^{(n)}_{2m+1}|$ and $|C^{(n-1)}_{2m+1}|$ decrease rapidly with m, this may also be expected of $|W_{2m+1}|$. Hence an approximate solution can be obtained if we retain only the first and last equations of the system (9.14) and neglect all FOURIER coefficients with subscripts greater than 1. From the last equation we then have

$$N = \frac{22\,\pi}{45\,C_1^{(n-1)}\,n\,\Pi^{n-1}}\cdot\frac{h^2}{b^2}\cdot W_1\,\frac{dW_1}{dt}. \tag{10.01}$$

Elimination of N from the first yields

$$\frac{dW_1}{dt} + \frac{704}{135\,\pi^2}\,W_1^2\,\frac{dW_1}{dt} = \frac{4\,b^2}{\pi^2\,h^2}\,\Pi^n\,C_1^{(n)}.$$

This equation can be integrated without difficulty and the initial condition $W_1(0) = 0$ can be satisfied. Since at the same time from (10.01) $N(0) = 0$, we have

$$W_1 + \frac{704}{405\,\pi^2}\,W_1^3 = \frac{4\,b^2}{\pi^2\,h^2}\,\Pi^n\,C_1^{(n)}\,t. \tag{10.02}$$

Incidentally, if the cubic term on the left-hand side is neglected, Eq. (10.02) gives an approximate solution, for arbitrary n, of problem (a) treated above under 8. In the special case $n = 3$, the central deflection becomes

$$w(0) = \frac{4}{\pi^2}\cdot\frac{b^2}{h}\left[\frac{p\,b^2}{4\,A\,h\,\sigma_c}\right]^3\cdot 0{,}7958\cdot t = 0{,}323\,\frac{b^2}{h^2}\left[\frac{p\,b^2}{4\,A\,h\,\sigma_c}\right]^3 t$$

as compared with the exact value $(93/280)\,(b^2/h^2)\,(p\,b^2/4\,A\,h\,\sigma_c)^3\,t = 0.332\,(b^2/h^2)\,(p\,b^2/4\,A\,h\,\sigma_c)^3\,t$ of Eq. (8.03). In this case the error of Eq. (10.02) stays within 3 per cent. Eq. (10.02) may be solved for W_1 during the initial stages of the motion

$$W_1 = \frac{4\,b^2}{\pi^2\,h^2}\,\Pi^n\,C_1^{(n)}\,t - \frac{704}{405\,\pi^2}\left[\frac{4\,b^2}{\pi^2\,h^2}\,\Pi^n\,C_1^{(n)}\,t\right]^3 + \cdots, \tag{10.03}$$

$$N = \frac{88\,C_1^{(n)}}{45\,\pi\,n\,C_1^{(n-1)}}\left\{\frac{4\,b^2}{\pi^2\,h^2}\,\Pi^n\,C_1^{(n)}\,t - \frac{2816}{405\,\pi^2}\left[\frac{4\,b^2}{\pi^2\,h^2}\,\Pi^n\,C_1^{(n)}\,t\right]^3 + \cdots\right\}. \tag{10.04}$$

The solution (10.03), (10.04) shows creep in a beam or plate strip under the combined action of bending and stretching. The rate of deflection is diminished as a consequence of the built-up normal force. The normal force itself is reduced due to bending action as compared with that of a membrane.

In order to investigate more closely the statement above that $W_1 \cos\xi$ may be considered as a good approximation to the function $W(\xi, t)$ one may proceed to an approximate determination of W_3 from the second Eq. (9.14) leaving out the terms containing W_3 on the right side. It will thus be seen that $|W_3|$ will be but a small fraction of $|W_1|$.

11. Strings

The limiting case $h = 0$, i. e. the case of a string or of a beam or plate with vanishing flexural rigidity — in the terminology of H. Hencky [*8*] — the solution (10.03), (10.03) will fail. This case corresponds to the membrane problem treated in paragraph 5 and admits of a similar solution.

In this case we may start directly from (2.03) to (2.05) which for uniaxial stress σ yield

$$\frac{d\varepsilon}{dt} = \dot{u}' + \dot{w}'\,w' = \frac{d}{dt}\left[\frac{\sigma}{\sigma_0}\right]^{n_0} + \left[\frac{\sigma}{\sigma_c}\right]^n. \tag{11.01}$$

Remembering that $N_x = 2\,A\,\sigma$ we also have from a statically determinate solution of (3.14)

$$w = p\,(b^2 - x^2)/4\,A\,\sigma\,. \tag{11.02}$$

If we neglect the second term in the right-hand member of (11.01) and integrate with respect to the time, we obtain a problem of type E for the determination of the initial deformation u_1, w_1 in primary creep as well as the corresponding stress σ_1, *viz.*

$$u_1' + \frac{w_1'^2}{2} = \left[\frac{\sigma_1}{\sigma_0}\right]^{n_0}, \qquad w_1 = p\,(b^2 - x^2)/4\,A\,\sigma_1\,.$$

Upon elimination of w_1 we obtain

$$u_1' = \left[\frac{\sigma_1}{\sigma_0}\right]^{n_0} - \frac{p^2\,x^2}{8\,A^2\,\sigma_1^2}\,.$$

Integration and satisfaction of the boundary condition $u_1(\pm b) = 0$ yield

$$\sigma_1 = \left[\frac{p\,b}{2\,A\,\sigma_0}\right]^{\frac{2}{n_0+2}} \cdot \frac{\sigma_0}{6^{1/(n_0+2)}}\,. \tag{11.03}$$

Hence

$$w_1 = \frac{b}{2}\left[1 - \frac{x^2}{b^2}\right]\left[\frac{p\,b}{2\,A\,\sigma_0}\right]^{\frac{n_0}{n_0+2}} 6^{\frac{1}{n_0+2}}\,. \tag{11.04}$$

Upon application of the load p the beam will carry the stress σ_1 and reach the deflection w_1 according to (11.03) and (11.04). From now on the second term in the right-hand member of (11.01) will come into play and this will cause a decrease in the stress σ with time. Hence the first term must now be dropped and we arrive at a problem of type C. Eq. (11.02) still holds true and we have

$$\dot{w} = -\,p\,(b^2 - x^2)\,\dot{\sigma}/4\,A\,\sigma^2\,. \tag{11.05}$$

Consequently from (11.01)

$$\dot{u}' = \left[\frac{\sigma}{\sigma_c}\right]^n + \frac{p^2\,\dot{\sigma}\,x^2}{4\,A^2\,\sigma^3}\,. \tag{11.06}$$

Satisfying the boundary condition $u(\pm b) = 0$, we then obtain

$$\left[\frac{\sigma}{\sigma_c}\right]^n + \left[\frac{p\,b}{2\,A\,\sigma_c}\right]^2 \frac{\sigma_c^2\,\dot{\sigma}}{3\,\sigma^3} = 0 \tag{11.07}$$

which is a differential equation for the determination of σ as a function of t, to be integrated with the initial condition $\sigma = \sigma_1$ for $t = 0$. This yields

$$\left[\frac{\sigma_c}{\sigma}\right]^{n+2} - \left[\frac{\sigma_c}{\sigma_1}\right]^{n+2} = 3\,(n+2)\left[\frac{2\,A\,\sigma_c}{p\,b}\right]^2 t \tag{11.08}$$

giving a σ decreasing to zero as t approaches infinity. The corresponding deflection is given by (11.02). The value of σ_1 should be taken from (11.03).

12. Conclusions

In general, the problems of creep are non-linear for three different reasons: geometric, static and physical. The elastic analogue is applicable only as long as the first two types of non-linearity are neglected. Thus in the present paper a series of problems have been treated, by way of example, where geometric or static non-linearity prevents an application of the elastic analogue. These problems are those of membranes, beams of vanishing flexural rigidity and sandwich or double deck plates (or beams with idealized H-section) subjected to large deformations.

References

[1] ILYUSHIN, A. A.: Prikl. Math. Mech., **9**, (3), (1945).
[2] SOKOLOVSKY, V. V.: Theorie der Plastizität, Berlin 1955.
[3] HOFF, N. J.: Quarterly of Applied Mathematics, Vol. XII, (1954) p. 49—55.
[4] HOFF, N. J.: High Temperature Effects in Aircraft Structures, N. Y. (1958) p. 255—266.
[5] SOKOLNIKOFF, I. S.: Math. Th. of Elast., Sec. ed., N. Y. (1956) p. 29—31.
[6] ODQVIST, F. K. G.: Symposium su la plasticità nella scienza delle costruzioni, Varenna, Sett. 1956, in onore di Arturo Danusso, Bologna (1956) p. 204—215.
[7] VON KÁRMÁN, TH.: Enc. d. math. Wiss., Vol. IV, 2, 2, Leipzig (1910) p. 349.
[8] HENCKY, H.: Zeitschr. f. Math. u. Phys., **63**, 311 (1915).
[9] TIMOSHENKO, S.: Th. of Plates and Shells, N. Y. (1940) p. 329—350.
[10] ODQVIST, F. K. G.: Arkiv f. fysik, K. V. A., Stockholm, **16**, 113 (1959).
[11] WALLIN, L.: Arkiv f. fysik, K. V. A., Stockholm, **17**, 89 (1960).
[12] PATEL, SHARAD A., B. VENKATRAMAN and P. G. HODGE: PIBAL Rep. Nr. 351, Dec. 1956.
[13] ODQVIST, F. K. G.: Symposium on "Non-linear Physical Problems", Roorkee, Dec. 1959, (with Appendix by H. O. Kreiss), Proceedings to be published.
[14] ODQVIST, F. K. G.: Arkiv f. fysik, K. V. A., Stockholm, **16**, 527 (1960).
[15] HOFF, N. J.: Wilbur Wright Memorial Lecture, Journal of the Royal Aeron. Soc., **58**, 31 (1954).
[16] FINNIE, L., and W. R. HELLER: Creep of Engineering Materials, N. Y. (1959) p. 191.

Discussion

E. T. ONAT, Brown University: In the constitutive equations (2.03) Professor ODQVIST employs the term $\frac{3}{2} G(\sigma_e)\, \dot{s}_{ij}$ to describe approximately the rate of strain due to primary creep. Since the creep phenomenon is closely related to plastic action, one wonders whether it would not be better to use an expression of the type

$H(\sigma_e)\, s_{ij} \frac{d\sigma_e}{dt}$ together with the loading condition introduced by Professor ODQVIST to represent the strain rate due to primary creep. It should be possible to devise experiments by which one may be able to decide which of the two expressions mentioned above represents more closely the actual material behavior.

Another point which may be of interest is that when equations (2.03) are modified on the basis of TRESCA's yield criterion of plasticity, the analysis of the finite deformations of a circular membrane under hydrostatic pressure becomes quite simple and can be carried out in closed form. It would be interesting to compare the results of Professor ODQVIST with the results of ONAT and YÜKSEL ("On the steady creep of shells" Proc. Third US Nat. Cong. of Appl. Mech., ASME, pp. 625—630) based on the modified constitutive equations.

Editor's note: On the basis of the suggestion contained in the first paragraph of Professor ONAT's discussion, Professor ODQVIST revised Eqs. (2.03), (2.04), (2.06) and (2.10) of his manuscript. In the paper printed in this volume the equations appear in the revised form.

Oil Canning Problems in Creep

By

Jan Hult

Royal Institute of Technology, Stockholm, Sweden

Abstract

The stability of some transversely loaded nearly flat structures on hinged supports is studied, assuming the material to be subject to creep. It is shown that a sudden jump in the equilibrium configuration will occur after a certain time, due to simultaneous compression and bending of the structure.

Nomenclature

A	cross sectional area
E	modulus of elasticity
I	moment of inertia of cross section
k	creep rate parameter
K	dimensionless load parameter $= Q\,L/4\,\pi\,\varrho\,S_2$ (two-bar system), $= Q\,L/4\,\pi\,\varrho\,P_2$ (curved beam)
L	length of beam
M	bending moment
P, S	compressive axial force
P_2, S_2	elastic buckling force
q	distributed lateral load
Q	concentrated lateral load
s	dimensionless force parameter $= P/P_2,\ S/S_2$
t	time
w	deflection of beam
x	axial coordinate
z	dimensionless deflection parameter $= (L\,\vartheta/2\,\pi\,\varrho)^2$ (two-bar system), $= (\delta_1/\varrho)^2$ (curved beam)
δ	midspan deflection of beam
δ_1, δ_2	amplitudes of first and second harmonics in beam deflection
ε	compressive strain
ϑ	angle
ϱ	radius of inertia of cross section $= \sqrt{I/A}$
σ	compressive stress
τ	dimensionless time parameter $= E\,k\,t$
ζ	dimensionless deflection parameter $= (\delta/\varrho)^2/2$ (two-bar system), $= (2\,\delta_2/\varrho)^2$ (curved beam)

Subscript j refers to jump condition

Superscript 00 refers to the initial state prior to application of load

Superscript 0 refers to the initial state immediately after application of load

1. Introduction

This paper deals with the stability of some transversely loaded nearly flat structures subject to creep. PIAN [1] (1958) has shown that in a slightly curved beam mounted on fixed hinged supports, and acted upon by a constant uniform transverse load, the rate of deflection becomes infinite after a certain time, which causes the beam to snap over from one equilibrium position to another. This kind of instability has been extensively studied in elastic structures, cf. MARGUERRE [2] (1950), HOFF and BRUCE [3] (1952), and is sometimes termed oil canning.

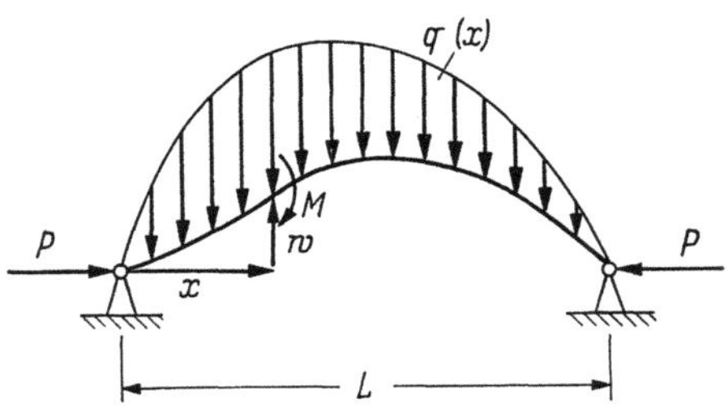

Fig. 1.

Consider the beam shown in Fig. 1. If the curvature of the beam is small, the transverse load q will give rise to a large axial compressive force P. This force gives rise to a certain rate of shortening of the beam, and as a result its deflection w will decrease, which in turn causes an increasing rate of compression, leading eventually to a snap.

The overall deflection will decrease further, if the second or higher harmonics in the deflections w increase during the deformation process. Hence the occurrene of such higher harmonics will accelerate the rate of lowering of the beam.

If the material creeps according to a non-linear creep law, increasing higher harmonics in the deflection will always develop, even if they are not present at the outset, cf. HOFF [4] (1954). The true lifetime is therefore always shorter than the lifetime determined only from the rate of compression of the beam.

This paper is limited to linearly viscoelastic materials. The creep jump, first studied by PIAN, results from simultaneous elastic and creep deformation and is not *per se* a consequence of the creep law having some particular form. However, it will be seen that if the elastic part of the total deformation is neglected, the resulting equations will not predict such a sudden finite jump.

Oil canning effects in non-linear viscous creep will be discussed in a subsequent paper.

2. Two-bar System, Straight Bars

Some of the characteristic features of the curved beam, which will be studied in a later section, are displayed also by the two-bar system shown in Fig. 2. This system, however, is statically determinate, and therefore the analysis is much simplified.

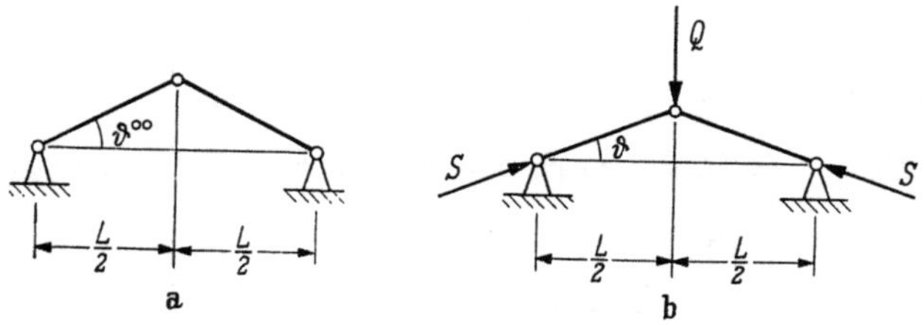

Fig. 2.

An equilibrium equation yields the compressive bar force

$$S = Q/2\,\vartheta\,, \tag{1}$$

and hence the compressive stress is

$$\sigma = Q/2\,A\,\vartheta\,. \tag{2}$$

If ϑ^{00} and ϑ are small, the compressive strain is

$$\varepsilon \simeq (\vartheta^{00^2} - \vartheta^2)/2\,. \tag{3}$$

The linearly visco-elastic creep law

$$\dot{\varepsilon} = \dot{\sigma}/E + k\,\sigma\,, \tag{4}$$

where a dot denotes differentiation with respect to the time t, then yields with (2) and (3) the differential equation

$$-\vartheta\,\dot{\vartheta} = -\frac{Q}{2\,A\,E}\cdot\frac{\dot{\vartheta}}{\vartheta^2} + \frac{k\,Q}{2\,A}\cdot\frac{1}{\vartheta}\,, \tag{5}$$

which should be solved subject to the initial condition $\vartheta(0) = \vartheta^0$, where the superscript ^{0} denotes the initial state existing immediately after the load application. Inertia effects will be disregarded throughout the analysis.

(a) Initial State

If the load is applied instantaneously at $t = 0$, the material behaves elastically, and the creep law (4) is replaced by HOOKE's law

$$\varepsilon = \sigma/E\,. \tag{6}$$

From (2), (3), and (6) then follows

$$\vartheta^0\,(\vartheta^{00^2} - \vartheta^{0^2}) = Q/A\,E\,, \tag{7}$$

and from (1) the compressive bar force is

$$S^0 = Q/2\,\vartheta^0\,. \tag{8}$$

Introduction of the elastic buckling force in the second mode of the entire system (the first mode buckling force being $S_1 = 0$)

$$S_2 = 4\,\pi^2\,E\,I/L^2, \tag{9}$$

and of the notation

$$s = S/S_2,\; z = (L\,\vartheta/2\,\pi\,\varrho)^2,\; \varrho = \sqrt{I/A},\; K = Q\,L/4\,\pi\,\varrho\,S_2, \tag{10}$$

yields with (7) and (8)

$$\sqrt{z^0}\,(z^{00} - z^0) = 2\,K, \tag{11}$$

$$z^{00} - z^0 = 2\,s^0, \tag{12}$$

$$s^0\sqrt{z^0} = K, \tag{13}$$

where the superscript ^{00} denotes the state prior to loading.

From (11) it follows that a jump due to compression of the bars will occur, i. e. $dK/dz^0 = 0$, when

$$z^0 = z_j^0 = z^{00}/3,$$
$$s^0 = s_j^0 = z^{00}/3,$$

and hence

$$z_j^0 = s_j^0. \tag{14}$$

Another kind of instability, viz. a jump due to an elastic buckling of the bars will occur when $S^0 = S_2$, i. e. when

$$s^0 = s_j^0 = 1, \tag{15}$$

irrespective of z_j^0.

The state of the system is uniquely defined by the two quantities s^0 and z^0, and it can therefore be represented by the location of the point (s^0, z^0) in an $s - z$-plane.

In this plane there exists a region of stable equilibrium, bounded by the straight lines (14) and (15), shown as OA and AB, respectively, in Fig. 3. The shape of this elastic instability boundary is a property of the system itself, and does not depend on its initial state.

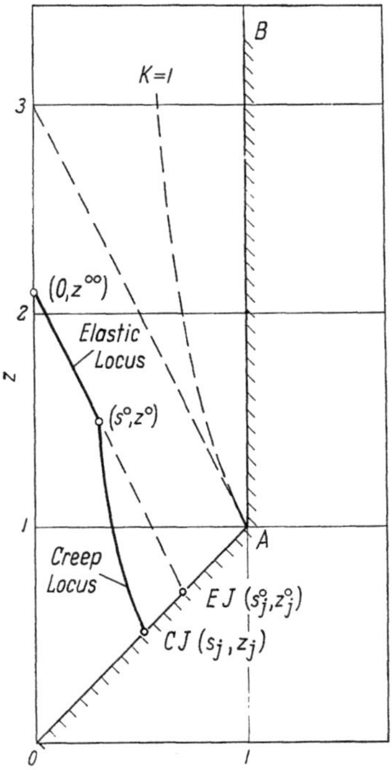

Fig. 3.

The initial unloaded state ($K = 0$), $s = 0$, $z = z^{00}$ corresponds to a point on the ordinate axis. When the load K is increased, the state point moves along the straight line (12), which will therefore be called the elastic locus. The intersection between the elastic locus and the elastic instability boundary OAB is called the elastic jump point E. J. (s_j^0, z_j^0). The jump point load is given by (13)

$$K_j = s_j^0 \sqrt{z_j^0}\,.$$

For any load $K < K_j$ the system assumes a stable equilibrium position (s^0, z^0) defined by (12), (13).

It follows that if $z^{00} < 3$, a jump due to compression of the bars will occur at a load $K_j < 1$, while if $z^{00} > 3$, a jump due to buckling of the bars will occur at a load $K_j > 1$ (cf. Fig. 3).

(b) Creep Deformation

With the notation (10) and

$$\tau = E\,k\,t, \tag{16}$$

(5) takes the form

$$\dot{z}\left(1 - \frac{s}{z}\right) = -2\,s, \tag{17}$$

where now a dot denotes differentiation with respect to the dimensionless time τ.

The deformation velocity becomes infinite, when

$$z = s\,.$$

Further, elastic buckling will occur, when

$$s = 1,$$

and hence there exists a creep instability boundary in the $s - z$-plane, which coincides with the elastic instability boundary OAB.

The relation (13), which is an equilibrium equation, holds also in this case

$$s\sqrt{z} = K, \tag{18}$$

and since K is constant during the creep process, the corresponding curve will be called the creep locus. The creep deformation is then characterized by the state point (s, z) moving along the creep locus, starting at (s^0, z^0). A sudden jump will occur when the state point reaches the instability boundary; the intersection is called the creep jump point C. J. (s_j, z_j). If $K < 1$, this point lies on OA $(s_j = z_j = K^{2/3})$, and the jump is caused by pure compression of the bars. If $K > 1$, the jump point lies on OB $(s_j = 1, z_j = K^2)$, and the jump is caused by buckling of the bars.

The slope of the creep locus is larger than the slope of the elastic locus at all points within the stable region, and it then follows that a creep jump will always occur at a smaller deflection and a smaller axial force than those of the corresponding elastic jump.

If the load is removed before the creep jump point is reached, the state point will move back to the ordinate axis from its instantaneous location C along that elastic locus which passes through C. If the same load is applied again, the state point will move down along the same elastic locus until it reaches C, and then it will continue again along the old creep locus. The point C can be chosen arbitrarily close to the creep instability boundary, and it is then evident that the elastic instability boundary and the creep instability boundary must coincide.

The existence of a fixed creep instability boundary is itself a consequence of the assumption of the creep law (4) as an equation of state.

By inserting (18) into (17) a differential equation is obtained, which permits calculation of the time τ_j at which the jump occurs; this time will be termed the lifetime of the system studied. There results

$$\begin{aligned} K \leq 1 : \tau_j &= \frac{1}{3}\left(\frac{K^2}{s_0^3} - 1 - \ln \frac{K^2}{s_0^3}\right), \\ K \geq 1 : \tau_j &= \frac{1}{3}\left(\frac{K^2}{s_0^3} - K^2 - \ln \frac{1}{s_0^3}\right). \end{aligned} \tag{19}$$

If the elastic part of the strain rate (4) were neglected, (17) would read $= -2\,s = -2\,K/\sqrt{z}$, and hence the deformation velocity would not become infinite until the time when $z = 0$. A sudden finite jump is therefore the result of the simultaneous occurrence of elastic and creep deformation.

During the whole deformation process the shape of the two-bar system is identical with the funicular shape that corresponds to the concentrated load Q. It is characteristic of such a case that different analytical expressions have to be used, depending on whether Q is smaller or larger than a certain value. If the bars were initially curved, no such distinction would have to be made, since both compression and bending would always be present. Such a system is studied in the next section.

3. Two-bar System, Curved Bars

Consider the two-bar system shown in Fig. 4. The bars are assumed to be slightly curved with a sinusoidal shape in the unloaded state.

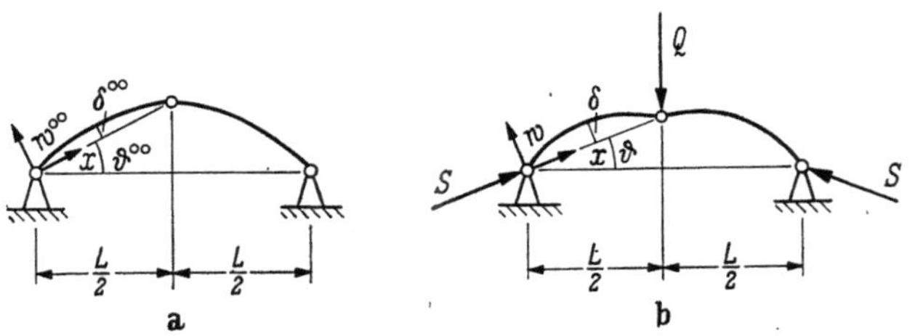

Fig. 4.

The compressive force S gives rise to a bending moment, which increases the deflections of the bars. In the derivation of the relation between this force and the deflection, the distance between the two end points of the bar many temporarily be assumed to be constant and equal to $L/2$.

The initial deflection of the bars can then be written

$$w^{00} = \delta^{00} \sin 2\pi x/L. \tag{20}$$

The linear viscoelastic creep law (4) yields the following general relation between the deflection w and the bending moment M of the bars (cf. Hult [5] (1960)):

$$-E I \frac{\partial^3 w}{\partial x^2 \partial t} = \frac{\partial M}{\partial t} + E k M. \tag{21}$$

Here the bending moment is

$$M = S w, \tag{22}$$

and hence there results the following differential equation for the deflection w:

$$E I \frac{\partial^3 w}{\partial x^2 \partial t} + \frac{\partial}{\partial t}(S w) + E k S w = 0. \tag{23}$$

With the boundary conditions

$$w(0) = 0, \ w(L/2) = 0, \tag{24}$$

the solution of (23) takes the form

$$w = \delta \sin 2\pi x/L, \tag{25}$$

where δ is a solution of

$$-\frac{4\pi^2 E I}{L^2}\frac{d\delta}{dt} + \frac{d}{dt}(S\delta) + E k S \delta = 0. \tag{26}$$

Moreover, Eqs. (1) and (2) above are still valid, σ now denoting the mean stress; whereas the compressive mean strain takes the form, if (20) and (25) are considered,

$$\varepsilon = (\vartheta^{00^2} - \vartheta^2)/2 - \pi^2(\delta^2 - \delta^{00^2})/L^2. \tag{27}$$

With the notation (9), (16), and with

$$2\zeta = (\delta/\varrho)^2, \tag{28}$$

and if (27) and (2) are substituted in (4), there results

$$\dot{z} + \zeta + 2K\left(-\frac{\dot{z}}{2z\sqrt{z}} + \frac{1}{\sqrt{z}}\right) = 0, \tag{29}$$

where a dot now denotes differentiation with respect to τ.

Eqs. (1) and (26) take the forms

$$s\sqrt{z} = K, \tag{30}$$

$$-\dot{\zeta} + s\,\dot{\zeta} + 2\,\dot{s}\,\zeta + 2\,s\,\zeta = 0. \tag{31}$$

The system (29), (30), (31) should be solved subject to the initial condition $z(0) = z^0$, $\zeta(0) = \zeta^0$, $s(0) = s^0$.

(a) Initial State

If the creep law (4) is replaced by HOOKE's law (6), Eqs. (29), (30), (31) change into

$$\sqrt{z^0}\,(z^{00} - z^0) = 2\,K + \zeta^{00}\sqrt{z^0}\left[\frac{1}{(1 - K/\sqrt{z^0})^2} - 1\right], \tag{32}$$

$$z^{00} - z^0 = 2\,s^0 + \zeta^{00}\left[\frac{1}{(1 - s^0)^2} - 1\right], \tag{33}$$

$$s^0\sqrt{z^0} = K, \tag{34}$$

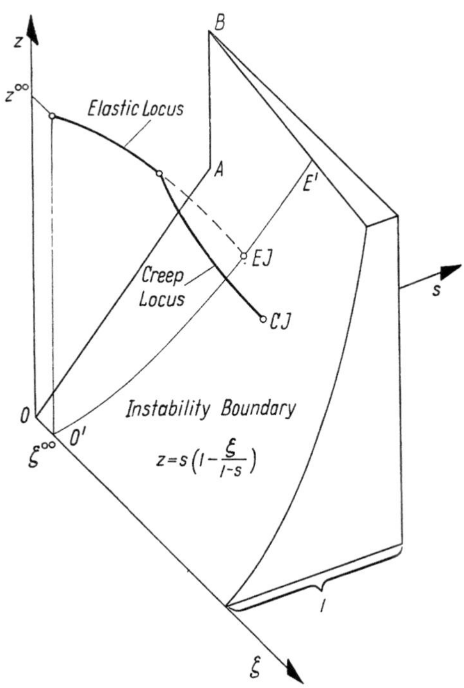

Fig. 5.

which correspond to (11), (12), (13) above. The midspan deflection of the bars is obtained from

$$\zeta^0 = \frac{\zeta^{00}}{(1 - s^0)^2}. \tag{35}$$

A jump occurs, i. e. $dK/dz^0 = 0$, when z^0, ζ^0 and s^0 satisfy the condition

$$z^0 = s^0\left(1 + \frac{\zeta^0}{1 - s^0}\right). \tag{36}$$

This is the equation of the elastic instability boundary, which is a surface in the $z - \zeta - s$-space, cf. Fig. 5. The elastic locus is a space curve defined by (33), (35), and the intersection between the elastic locus and the elastic instability boundary is the elastic jump point E.J. $(z_j^0, \zeta_j^0, s_j^0)$.

If ζ^{00} is kept constant while z^{00} is varied, the elastic jump point E. J. describes a space curve $O'E'$ located on the elastic instability boundary, and hence E. J. can be found as the intersection between the elastic locus and $O'E'$. From (35), (36) follows

$$z_j^0 = s_j^0 \left[1 + \frac{\zeta^{00}}{(1 - s_j^0)^3}\right], \tag{37}$$

which is the projection of $O'E'$ on the plane $\zeta = 0$. Further, (33) is the projection of the elastic locus on the plane $\zeta = 0$, and hence z_j^0 and s_j^0 can be found as the intersection of the plane curves (33) and (37). Since the projected elastic locus (33) is steeper than that found in the case of straight bars, and since the projected instability boundary (37) lies above OA, it follows that a jump always occurs at a smaller axial force s_j^0 when the bars are initially curved than when they are initially straight. The jump point load K_j is given by (34)

$$K_j = s_j^0 \sqrt{z_j^0}.$$

For any load $K < K_j$ the system assumes a stable equilibrium position (z^0, ζ^0, s^0), defined by (32), (34), and (35).

(b) Creep Deformation

Elimination between (29), (30), (31) yields

$$\dot{z}\left[z - s\left(1 + \frac{\zeta}{1 - s}\right)\right] = -2\, s\, z \left(1 + \frac{\zeta}{1 - s}\right), \tag{38}$$

and hence the deformation velocity becomes infinite, when (z, ζ, s) is located on the surface

$$z = s\left(1 + \frac{\zeta}{1 - s}\right), \tag{39}$$

which implies that again the creep instability boundary coincides with the elastic instability boundary (36). The creep locus is a space curve, defined by (29), (30), (31), and its intersection with the instability boundary is the creep jump point C. J. (z_j, ζ_j, s_j). Since (29) and (31) are differential relations, the shape of the creep locus and the location of the creep jump point depend implicitly on the deformation history, and therefore they cannot be determined beforehand, as was possible in the case of straight bars. This is clearly due to fact that the system studied has two degrees of freedom (z, ζ), whereas the straight bar system has only one (z).

Since the location of C. J. is not explicitly known, it also follows that the creep deformation process cannot be followed in the projected plane $\zeta = 0$ as was possible in the case of elastic deformation. A comparison between (35) and the following equation (40) actually shows that the projection of C. J. on the plane $\zeta = 0$ is located above the projected elastic instability boundary (37).

The relation (31) can be written

$$\frac{d\zeta}{\zeta} = \frac{2\,ds}{1-s} + \frac{2\,s\,d\tau}{1-s}\,.$$

If (35) is duly considered, integration gives

$$\zeta = \frac{\zeta^{00}}{(1-s)^2} \cdot g(\tau)\,, \tag{40}$$

where

$$g(\tau) = \exp \int_0^\tau \frac{2\,s(v)}{1-s(v)}\,dv\,. \tag{41}$$

Because of (30) and (40), (38) can be written as

$$\frac{ds}{d\tau} = -s + \frac{K^2 s}{K^2 - s^3 \left[1 + \frac{\zeta^{00}}{(1-s)^3} \cdot g(\tau)\right]}\,. \tag{42}$$

The two relations (41), (42) permit calculation of the lifetime τ_j by means of step by step integration. When the creep deformation starts, $\tau = 0$, $s = s^0$, $g = 1$, and hence $(ds/d\tau)^0$ is found from (42). After a time interval $\Delta\tau$

$$g \simeq \exp\left(\frac{2\,s^0}{1-s^0} \cdot \Delta\tau\right),$$

$$s \simeq s^0 + (ds/d\tau)^0 \cdot \Delta\tau\,.$$

This procedure is repeated, and the integration terminates when the denominator of (42) vanishes. From (30) it follows that $ds/d\tau$ becomes infinite at the same instant as does $dz/d\tau$, viz. when $\tau = \tau_j$.

It is found that an initial out-of-straightness of the bars may have a strong influence on the lifetime of the system. For example with $z^{00} = 2$, $K = 0.5$ the lifetime τ_j equals 0.1 if the bars are straight, and it becomes zero if the initial midspan deflection of the bars is around $\varrho/5$, where ϱ is the radius of inertia of the cross section.

4. Curved Beam

From the preceding analysis of the two-bar system it is concluded that in a laterally loaded curved beam subject to creep the second and higher harmonics in the deflection curve play a particularly important role, at least if the beam is rather slender. This is in contrast with the creep buckling behavior of a column, where such higher harmonics have been shown to be of relatively minor importance.

Consider the beam shown in Fig. 1. The beam material is assumed to be linearly viscoelastic, which implies the relation (21) between the deflection w and the bending moment M. If this is combined with the equilibrium relation

$$q = \frac{\partial^2 M}{\partial x^2} - P \frac{\partial^2 w}{\partial x^2},$$

where q is the applied load per unit length and P is the compressive axial force, there results the differential equation

$$E\, I\, w_{xxxxt} + P\, w_{xxt} + P_t\, w_{xx} + E\, k\, P\, w_{xx} = - q_t - E\, k\, q, \quad (43)$$

where the index notation is used to indicate differentiation. If the load is assumed as

$$q(x, t) = \frac{\pi}{2\,L}\, Q(t) \sin \pi\, x/L, \quad (44)$$

Q being the total lateral force, and it the initial deflection is

$$w^{00}(x) = \delta_1^{00} \sin \pi\, x/L + \delta_2^{00} \sin 2\, \pi\, x/L, \quad (45)$$

it follows that the deflection takes the form

$$w(x, t) = \delta_1(t) \sin \pi\, x/L + \delta_2(t) \sin 2\, \pi\, x/L. \quad (46)$$

The mean compressive strain is obtained from (45) and (46),

$$\varepsilon = \frac{\pi^2}{4\, L^2} [\delta_1^{00^2} - \delta_1^2 + 4(\delta_2^{00^2} - \delta_2^2)], \quad (47)$$

and the mean compressive stress is

$$\sigma = P/A. \quad (48)$$

If (46) and (44) are substituted in (43), and (47) and (48) are substituted in (4), there result the following three relations

$$\dot{z}(4\, s - 1) + 8\, \dot{s}\, z + 8\, s\, z = 16\, K \sqrt{z}, \quad (49)$$

$$\dot{\zeta}(1 - s) - 2\, \dot{s}\, \zeta - 2\, s\, \zeta = 0, \quad (50)$$

$$16\, \dot{s} + 16\, s + \dot{z} + \dot{\zeta} = 0, \quad (51)$$

where the new dimensionless variables are

$$z = (\delta_1/\varrho)^2,\ \zeta = (2\, \delta_2/\varrho)^2,\ s = P/P_2,\ K = Q\, L/4\, \pi\, \varrho\, P_2, \quad (52)$$

and where a dot indicates differentiation with respect to $\tau = E\, k\, t$. P_2 is the elastic buckling force in the second mode

$$P_2 = 4\, \pi^2\, E\, I/L^2. \quad (53)$$

The system (49), (50), (51) should be solved subject to the initial condition $z(0) = z^0$, $\zeta(0) = \zeta^0$, $s(0) = s^0$.

(a) Initial State

The initial state, existing immediately after load application, is obtained if the creep law (4) is replaced by HOOKE's law (6) in the derivations above. Eqs. (49), (50), (51) are then replaced by

$$\sqrt{z^{00}} - \sqrt{z^0} + 4\, s^0 \sqrt{z^0} = 8\, K, \tag{54}$$

$$\zeta^0 = \zeta^{00}/(1 - s^0)^2, \tag{55}$$

$$16\, s^0 + z^0 - z^{00} + \zeta^0 - \zeta^{00} = 0. \tag{56}$$

A jump occurs, i. e. $dK/dz^0 = 0$, when z^0, ζ^0, s^0 satisfy the condition

$$z^0 = 2(4\, s^0 - 1)\left[1 + \frac{\zeta^0}{8(1 - s^0)}\right]. \tag{57}$$

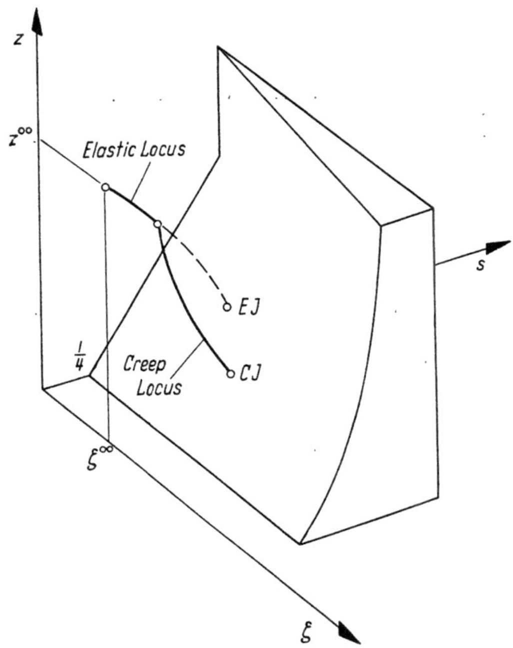

Fig. 6.

This elastic instability boundary is a surface in the $z - \zeta - s$-space, cf. Fig. 6. The elastic locus is a space curve defined by (55), (56). Its intersection with the elastic instability boundary is the elastic jump point E. J. $(z_j^0, \zeta_j^0, s_j^0)$.

It is found that E. J. falls on the line $z = 0$, $s = 1/4$ if

$$z^{00} - \frac{7}{9}\zeta^{00} = 4. \tag{58}$$

If the starting point (z^{00}, ζ^{00}) is located below the line (58), i. e. if

$$z^{00} - \frac{7}{9}\zeta^{00} < 4,$$

then the equilibrium will always be stable. Hence if the beam is flat enough, the jump phenomenon will not occur, cf. HOFF and BRUCE [*3*] (1952).

The jump point load K_j is obtained from (54) by inserting z_j^0, s_j^0. For any load $K < K_j$ the beam assumes a stable equilibrium position defined by (54), (55), (56).

(b) Creep Deformation

Elimination within (49), (50), (51) yields

$$\frac{ds}{d\tau} = -s + \frac{8\, K \sqrt{z}\,(1 - s)}{4\, z(1 - s) - \zeta(4\, s - 1) - 8(1 - s)\,(4\, s - 1)}, \tag{59}$$

and it follows that the creep instability boundary is identical with the elastic instability boundary (57).

The relations (49) and (50) can each be integrated as was done earlier in the case of (31). The result is

$$z = \left(\frac{8K - \sqrt{z^{00}}}{4s - 1}\right)^2 \cdot f(\tau), \; f(\tau) = \exp \int_0^\tau \frac{8\,[2K - s(v)\sqrt{z(v)}]}{[4s(v) - 1]\sqrt{z(v)}}\,dv, \tag{60}$$

$$\zeta = \frac{\zeta^{00}}{(1-s)^2} \cdot g(\tau), \; g(\tau) = \exp \int_0^\tau \frac{2s(v)}{1 - s(v)}\,dv. \tag{61}$$

From (61) it follows that if the beam deflection is purely sinusoidal before the load is applied ($\zeta^{00} = 0$), it will remain purely sinusoidal throughout the creep deformation. A jump will then eventually occur either because the rate of compression of the beam becomes infinite or in consequence of an elastic buckling of the beam in its second mode. If $\zeta^{00} \neq 0$ the jump will be caused by the simultaneous occurrence of compression and bending in the beam.

The three relations (59), (60), (61) permit the calculation of the lifetime τ_j by means of step by step integration, in the same way as was done with the curved bar system.

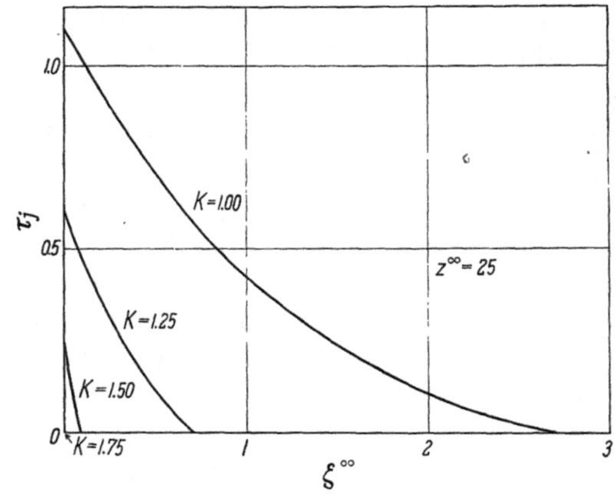

Fig. 7.

Results from such an integration are shown in Fig. 7. It is evident that the lifetime is strongly influenced by the second harmonic of the deflection curve. If the deflection curve were assumed to be purely sinusoidal, the calculated lifetime would always be larger than the true lifetime.

References

[1] Pian, T. H. H.: Creep Buckling of Curved Beam Under Lateral Loading. 3rd US Congr. Appl. Mech. Proc. (1958) pp. 649—654.

[2] Marguerre, K.: Neuere Festigkeitsprobleme des Ingenieurs. Berlin/Göttingen/Heidelberg: Springer 1950, Ch. V.

[3] Hoff, N. J., and V. G. Bruce: Dynamic Analysis of the Buckling of Laterally Loaded Flat Arches. VIII Int. Congr. Theor. Appl. Mech. Istanbul. J. Math. Phys. XXXII, 276—288 (1952).

[4] Hoff, N. J.: Buckling and Stability. J. Roy. Aer. Soc. 58, 1—52 (1954).

[5] Hult, J.: Creep Buckling of Plane Frameworks. Durand Centennial Conference, Stanford University, Proc. (1960) pp. 227—246.

Investigation of Transient Creep in Thick-walled Tubes under Axially Symmetric Loading

By

J. F. Besseling

Technical University, Delft, Holland

Summary

The feasibility of a numerical analysis of nonstationary creep problems is investigated for thick-walled tubes under axially symmetric loading. It is shown how approximate solutions may be obtained with the aid of automatic digital computers by means of a numerical method based upon the application of an extremum principle for the rate of deformation. The primary creep phase can be included in the analysis if creep equations based upon the concept of microscopic inhomogeneity of the material give an adequate description of this phenomenon.

As an illustration of the method here presented the exact integro-differential equation for the transition from the elastic to the steady state creep solution in the case of an incompressible material, showing only secondary creep governed by a power law in the stress, has been solved numerically for four combinations of the parameters.

Nomenclature

The following nomenclature is used in this paper:

A	matrix of coefficients of unknowns
C	bulk modulus of elasticity
G	modulus of rigidity
I_2	second invariant of elastic strain tensor $I_2 = \frac{1}{2}(e_{\alpha\beta} - \bar{e}_{\alpha\beta})(e_{\beta\alpha} - \bar{e}_{\beta\alpha})$
J_2	second invariant of strain rate tensor $J_2 = \frac{1}{2}\dot{e}_{\alpha\beta}\dot{e}_{\beta\alpha}$
N	number of subelements of element of volume
T	absolute temperature
T_c	characteristic temperature of creepprocess
V	volume
a_i	deformation parameter
b, c, d	column vectors
e_{ij}	strain-deviation tensor, $e_{ij} = \varepsilon_{ij} - \varepsilon\,\delta_{ij}$, $e_{\alpha\alpha} = 0$
$\bar{e}_{ij}$	inelastic strain tensor
$f(I_2)$	function determining stress dependence of creep rate

$g(T)$	function determining temperature dependence of creep rate
$h(J_2, T)$	function determining dependence of I_2 on J_2 and T
p	constant appearing in $f(I_2)$
p_1	internal pressure on tube
p_2	external pressure on tube
p_a	pressure determining total axial force in tube by $p_a r_1^2/2$
q	exponent of power law in the stress for creep rate
q'	temperature corrected exponent
r	radial coordinate
r_1	inner radius of tube
r_2	outer radius of tube
s_{ij}	stress-deviation tensor, $s_{ij} = \sigma_{ij} - \sigma \delta_{ij}$, $s_{\alpha\alpha} = 0$
t	time
u_i	displacement field
u	radial displacement
v', w'	row vectors
x_i	rectangular cartesian coordinates
x	radial component of elastic strain deviator, $x = e_r - \bar{e}_r$
y	tangential component of elastic strain deviator, $y = e_t - \bar{e}_t$
α	coefficient of cubic thermal expansion
δ_{ij}	Kronecker delta, $\delta_{ij} = 1$ if $i = j$, $\delta_{ij} = 0$ if $i \neq j$
ε_{ij}	strain tensor
ε	isotropic strain, $3\varepsilon = \varepsilon_{\alpha\alpha}$
$\varkappa$	ratio of bulk modulus of elasticity and modulus of rigidity, $\varkappa = C/G$
λ	ratio of outer and inner radio of tube, $\lambda = r_2/r_1$
ϱ	non-dimensional radial coordinate, $\varrho = r/r_1$
σ_{ij}	stress tensor
τ	isotropic stress, $3\sigma = \sigma_{\alpha\alpha}$
σ	temperature corrected non-dimensional time parameter
ψ^k	portion of the volume occupied by subelements of class k

1. Introduction

The solution of the stress and deformation problem of thick-walled tubes under axially symmetric loading in the elastic range is given by the well known LAMÉ equations. When the load is applied and held constant under creep conditions the rate of deformation and the stresses tend asymptotically to constant values if the material is characterized by a creep curve with a significant steady creep range. These asymptotic values have been calculated for a number of cases, such as the thick-walled tube under internal pressure (Ref. [1]). However the question how fast the transition occurs from the elastic stress distribution to the distribution of steady creep has been dealt with less satisfactory. A rigorous but cumbersome numerical method was given originally in Ref. [2]. In Ref. [3] a method of calculation is presented that is based upon a creep relation among strain, time and stress which physically can only be considered as a crude approximation of the creep behavior. Finally Ref. [4] simplifies the numerical integration procedure of Ref. [2], however with considerable loss of rigor.

Especially if the total deformations are of prime importance an analysis of the transition from the elastic state to the steady creep-state should be carried out. In addition, the more general problem of creep under varying load and temperature conditions must be considered of practical importance. The object of this paper is to investigate the feasibility of a solution of the nonstationary problem, with particular emphasis on the application of a variational principle. It will be shown how approximate solutions may be obtained with the aid of digital computers.

The primary creep phase of the material can be included in the calculations without difficulty if the creep-equations developed in Ref. [5] give an adequate description of this phenomenon.

2. Creep Equations

In the theory of Ref. [5] it was taken into account that engineering materials are macroscopically homogeneous but inhomogeneous on a microscopic scale. An element of volume is considered to be composed of various portions which in turn can be represented by subelements, all showing secondary creep and isotropic workhardening in plastic deformation but each with different creep constants and elastic limit. If it is postulated that all subelements of an element of volume have a unique measure of deformation in the total macroscopic strain then the inelastic stress-strain-time relations that follow show anisotropic strain hardening in plastic deformation and secondary creep as well as primary creep and creep recovery. The inelastic strains of the subelements play in these relations the role of deformation parameters describing the history of the deformation process. They define internal stresses which in this theory are solely responsible for the anisotropic phenomena and the after effects.

The relations of Ref. [5] give qualitatively a good description of the phenomena that have been observed in the deformation of metals and other materials. However the experimental program that is being carried out to check the accuracy of the predictions of the theory of Ref. [5] has only just started. It should be emphasized that if a material in the unstressed state is unstable, deformation parameters alone will be insufficient for an adequate description of the deformation behavior. In these instances expressions for the strain rates that depend explicitely on time as sometimes encountered in the literature may be justified, though at present the theory for these cases still does not seem to go beyond empty speculation. Therefore the analysis of this paper will be restricted to the creep equations of Ref. [5] which permit at least a physically consistent treatment of creep under varying stress and temperature conditions.

In the theory of small deformations to which the theory of Ref. [5] is restricted, it proves to be enlightening to conceive the strain tensor ε_{ij} as a superposition of an isotropic strain ε and a strain tensor that affects only the shape of an element of volume. The latter is the strain deviation tensor e_{ij}. Similarly the stress tensor σ_{ij} is decomposed into an isotropic part σ and a stress deviator s_{ij}. The deviator components of the strain and stress tensor are defined by

$$e_{ij} = \varepsilon_{ij} - \varepsilon\,\delta_{ij},\ \varepsilon = \frac{1}{3}\varepsilon_{\alpha\alpha},$$
$$s_{ij} = \sigma_{ij} - \sigma\,\delta_{ij},\ \sigma = \frac{1}{3}\sigma_{\alpha\alpha}. \tag{1}$$

Let the material initially be isotropic and let the class of subelements k represent a portion ψ^k of N portions ψ of the volume of the macroscopically homogeneous body. The stress-strain relations read then as follows (Ref. [5] and Ref. [6]).

$$\sigma = C[3\,\varepsilon - \alpha(T - T_0)],$$
$$s_{ij} = 2\,G\sum_1^N \psi^k (e_{ij} - \bar{e}_{ij}^k),\ \sum_1^N \psi^k = 1. \tag{2}$$

The quantities $\bar{e}_{ij}^k$ represent the inelastic strains of the subelements, C is the isothermal bulk modulus of elasticity and G is the modulus of rigidity or shear modulus. The initial temperature of the material is given by T_0 and α is the coefficient of cubic thermal expansion.

With respect to the creep properties of a subelement the assumption is made in Ref. [4] that the rate of energy dissipation in the creep process depends only on the value of the distortional elastic potential and the temperature T. This assumption leads to the following expression for the creep rate.

$$\dot{\bar{e}}_{ij}^k = (e_{ij} - \bar{e}_{ij}^k)\,g(T)\,f^k(I_2^k), \tag{3}$$

where

$$I_2^k = \frac{1}{2}(e_{\alpha\beta} - \bar{e}_{\alpha\beta}^k)(e_{\beta\alpha} - \bar{e}_{\beta\alpha}^k). \tag{4}$$

For $N = 1$ relations (3) reduce to the creep equations based upon the von Mises stress invariant that were first proposed by Odqvist in Ref. [7]. For any finite number N different from one relations (2) and (3) not only give a description of secondary creep, but also of primary creep as well as of creep recovery.

With respect to the temperature dependence of creep deformation it can be observed that in many physical and chemical changes the rate of change is proportional to $\exp\{-Q/(RT)\}$, where Q represents the so-called activation energy and where R is the gas constant. That the same temperature dependence may hold true for the rate of energy dissipation in creep has been confirmed in Ref. [8], though later investigations seem to indicate that more than one activation energy may be involved (Ref. [9]).

Here it will be assumed that the temperature dependence of the creep rate is given by

$$g(T) = \frac{a}{2G} e^{-\frac{T_c}{T}}, \quad T_c = \frac{Q}{R}. \tag{5}$$

The stress dependence of the secondary creep rate is determined by the function $f^k(I_2^k)$ which is assumed to be the same for all subelements except for the values of the numerical constants. If the secondary creep rate at any given temperature depends on the stress by a simple power law then function $f^k(I_2^k)$ may be written as follows

$$f^k = \frac{1}{p_k^2}\left\{\frac{(e_{\alpha\beta} - \bar{e}^k_{\alpha\beta})(e_{\beta\alpha} - e^k_{\beta\alpha})}{p_k^2}\right\}^{\frac{q-1}{2}}, \quad q = 1, 3, 5, \ldots, 2n+1. \tag{6}$$

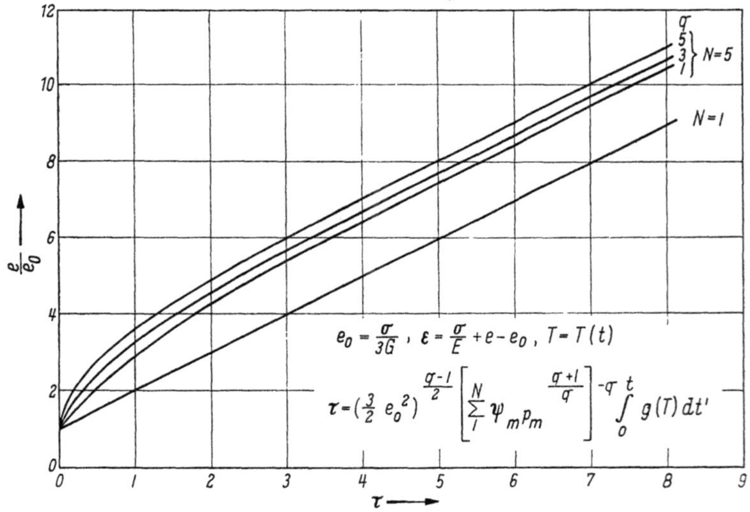

	1	2	3	4	5
ψ	0.4110	0.3636	0.1378	0.0616	0.0260
$\frac{p}{p_1}$	1	1.8334	2.9168	5.0001	9.1670

Fig. 1. Creep curves for constant tensile stress

Fig. 1 shows some non-dimensional creep curves for constant stress, determined from Eqs. (2), (3) and (6), while the numerical values for p_k and ψ^k were taken from Ref. [5]. It should be mentioned that these values have no other meaning than to provide an illustration of the consequences of the concept of subelements in creep theory.

3. Application of the Variational Principle

(a) Differential Equation and Boundary Conditions

The analysis of the thick-walled tube under axially symmetric loading will be based upon an extremum principle first given in Ref. [10]. In Ref. [11] it was shown that this extremum principle can be obtained from a more general variational principle if the effect upon the temperature of the deformation process is neglected.

The extremum principle of Ref. [10] states that the static conditions for the stress rates will be satisfied if the value of the integral expression

$$\int\limits_A \dot{p}_i^* \, \dot{u}_i \, dA - \frac{1}{2} \int\limits_V \dot{\sigma}_{ij} \left(\dot{\varepsilon}_{ij} - \sum_1^N \psi^k \, \dot{\bar{e}}_{ij}^k - \frac{\alpha}{3} \dot{T} \, \delta_{ij} \right) dV \tag{7}$$

represents a maximum with respect to all kinematically admissible displacement rates $\dot{u}_i$ provided the strain rates are linearly related to the displacement rates by

$$\dot{\varepsilon}_{ij} = \frac{1}{2} \left(\frac{\partial \dot{u}_i}{\partial x_j} + \frac{\partial \dot{u}_j}{\partial x_i} \right). \tag{8}$$

The surface integral is to be extended over the surface A where the surface tractions p_i^* are prescribed as a function of time.

The extremum principle will be applied to a long thick-walled tube subjected to an internal pressure p_1, an external pressure p_2, and a total axial force $p_a \pi r_1^2$, while the temperature distribution as function of time and radius r is assumed to be given as $T = T(r, t)$.

The inner radius of the tube will be denoted by r_1 and the outer radius by r_2.

For a long tube the assumption of a state of plane deformation independent of the axial coordinate is justified. Because of the axial symmetry of the loads and the temperature distribution the deformation is then defined by the radial displacement u as a function of the radial coordinate r and time, and by the axial strain ε_a as a function of time only. If the tangential component of the strain is indicated by a subscript t and the radial component by a subscript r the principal components of the strain tensor are given by

$$\varepsilon_a, \ \varepsilon_t = \frac{u}{r}, \ \varepsilon_r = \frac{du}{dr}.$$

The EULER-LAGRANGE equation of the variational problem formulated above reads then as follows

$$\left(C + \frac{4}{3} G \right) \left(r \frac{d^2 \dot{u}}{dr^2} + \frac{d\dot{u}}{dr} - \frac{\dot{u}}{r} \right) - C \alpha r \frac{d\dot{T}}{dr} - 2G \sum_1^N \psi^k \left(r \frac{d\dot{\bar{e}}_r^k}{dr} + \dot{\bar{e}}_r^k - \dot{\bar{e}}_t^k \right) = 0. \tag{9}$$

The natural boundary conditions are given by

$$r = r_1 : C\left(\frac{d\dot{u}}{dr} + \frac{\dot{u}}{r} + \dot{\varepsilon}_a - \alpha \dot{T}\right) + \frac{2}{3} G\left(2\frac{d\dot{u}}{dr} - \frac{\dot{u}}{r} - \dot{\varepsilon}_a\right) - 2G \sum_1^N \psi^k \dot{\bar{e}}_r^k = -\dot{p}_1, \tag{10a}$$

$$r = r_2 : C\left(\frac{d\dot{u}}{dr} + \frac{\dot{u}}{r} + \dot{\varepsilon}_a - \alpha \dot{T}\right) + \frac{2}{3} G\left(2\frac{d\dot{u}}{dr} - \frac{\dot{u}}{r} - \dot{\varepsilon}_a\right) - 2G \sum_1^N \psi^k \dot{\bar{e}}_r^k = -\dot{p}_2, \tag{10b}$$

$$\int_{r_1}^{r_2} \left\{C\left(\frac{d\dot{u}}{dr} + \frac{\dot{u}}{r} + \dot{\varepsilon}_a - \alpha \dot{T}\right) + \frac{2}{3} G\left(-\frac{d\dot{u}}{dr} - \frac{\dot{u}}{r} + 2\dot{\varepsilon}_a\right) + 2G \sum_1^N \psi^k (\dot{\bar{e}}_r^k + \dot{\bar{e}}_t^k)\right\} r\, dr = \frac{\dot{p}_a r_1^2}{2}. \tag{10c}$$

With the additional creep equations

$$\dot{\bar{e}}_r^k = (e_r - \bar{e}_r^k)\, g(T)\, f^k(I_2^k), \qquad \dot{\bar{e}}_t^k = (e_t - \bar{e}_t^k)\, g(T)\, f^k(I_2^k), \tag{11}$$

where

$$e_r = \frac{1}{3}\left(2\frac{du}{dr} - \frac{u}{r} - \varepsilon_a\right), \quad e_t = \frac{1}{3}\left(-\frac{du}{dr} + 2\frac{u}{r} - \varepsilon_a\right), \quad I_2^k = (e_r - \bar{e}_r^k)^2 + (e_r - \bar{e}_r^k)(e_t - \bar{e}_t^k) + (e_t - \bar{e}_t^k)^2. \tag{12}$$

the deformation problem is defined for given initial values of u, $\bar{e}_r^k$, $\bar{e}_t^k$, and ε_a once p_1, p_2, p_a, and $T(r, t)$ are prescribed. A solution of Eqs. (9) through (12) in closed form is not possible if the creep law is nonlinear. However before the numerical integration of the equations will be discussed, first the solution in two extreme cases will be considered.

(b) Solution for Elastic Deformation

In many cases a certain loading condition and a steady temperature distribution will be reached in a very short time, insufficiently long to allow creep strains of any significance to develop. Then the initial condition for the creep process that follows is given by the state of elastic deformation corresponding to given values of p_1, p_2, p_a, and the temperature distribution for steady heat flow through the wall of a tube which reads as follows

$$T = T_1 - \Delta T \frac{\ln \frac{r}{r_1}}{\ln \lambda}, \quad \lambda = \frac{r_2}{r_1}. \tag{13}$$

Since for $\bar{e}_r^k = \bar{e}_t^k \equiv 0$ Eqs. (9) and (10) can be integrated with respect to time, the general solution for u can be determined from the differential equation

$$r \frac{d^2u}{dr^2} + \frac{du}{dr} - \frac{u}{r} = \frac{3\varkappa}{3\varkappa + 4} \frac{\alpha \Delta T}{\ln \lambda} \tag{14}$$

and is given by

$$\frac{u}{r} = C_1 + C_2 \left(\frac{r_1}{r}\right)^2 + \frac{3\varkappa}{3\varkappa + 4} \frac{\alpha \Delta T \ln \frac{r}{r_1}}{2 \ln \lambda}. \tag{15}$$

Here $\varkappa$ stands for the ratio of the bulk modulus of elasticity C and the shear modulus G, $\varkappa = C/G$. The constants C_1 and C_2 in (15) and the axial strain ε_a follow from the boundary conditions (10) after they have been integrated with respect to time. For the constants C_1, C_2 and ε_a the following expressions are obtained

$$\begin{aligned} C_1 &= \frac{(3\varkappa + 4)(p_1 - \lambda^2 p_2) - (3\varkappa - 2) p_a}{18\, C(\lambda^2 - 1)} + \frac{\alpha}{3}\left[T_1 - T_0 + \frac{\Delta T}{2 \ln \lambda}\right] \\ &\quad + \frac{\alpha}{6} \frac{3\varkappa - 8}{3\varkappa + 4} \frac{\lambda^2}{\lambda^2 - 1} \Delta T, \\ C_2 &= \frac{\lambda^2}{\lambda^2 - 1}\left[\frac{p_1 - p_2}{2G} - \frac{\alpha}{2} \frac{3\varkappa}{3\varkappa + 4} \Delta T\right], \\ \varepsilon_a &= \frac{-(3\varkappa - 2)(p_1 - \lambda^2 p_2) + (3\varkappa + 1) p_a}{9\, C(\lambda^2 - 1)} + \frac{\alpha}{3}\left[T_1 - T_0 + \frac{\Delta T}{2 \ln \lambda}\right] \\ &\quad - \frac{\alpha}{3} \frac{\lambda^2}{\lambda^2 - 1} \Delta T. \end{aligned} \tag{16}$$

The strain tensor is completely defined by the components $3\varepsilon = \varepsilon_a + \varepsilon_t + \varepsilon_r$, e_r, and e_t. Hence according to (12) and (15):

$$\begin{aligned} 3\varepsilon &= 2C_1 + \varepsilon_a - \frac{3\varkappa}{3\varkappa + 4} \frac{\alpha \Delta T \left(2 \ln \frac{r}{r_1} + 1\right)}{2 \ln \lambda}, \\ e_r &= \frac{1}{3} C_1 - \frac{1}{3} \varepsilon_a - C_2 \left(\frac{r_1}{r}\right)^2 - \frac{3\varkappa}{3\varkappa + 4} \frac{\alpha \Delta T \left(\ln \frac{r}{r_1} + 2\right)}{6 \ln \lambda}, \\ e_t &= \frac{1}{3} C_1 - \frac{1}{3} \varepsilon_a + C_2 \left(\frac{r_1}{r}\right)^2 - \frac{3\varkappa}{3\varkappa + 4} \frac{\alpha \Delta T \left(\ln \frac{r}{r_1} - 1\right)}{6 \ln \lambda}. \end{aligned} \tag{17}$$

In many cases expressions (15), (16), and (17) can serve as initial conditions for the creep process defined by Eqs. (9) through (12). The expressions can also be used to determine the increments of the variables for abrupt changes in the loading condition for which the change of $\bar{e}_r^k$ and $\bar{e}_t^k$ may be neglected.

It is obvious that substitution of expressions (16) and (17) into the stress-strain relations leads directly to the wellknown Lamé equations for the stresses in a thick-walled tube.

(c) Steady State Creep Solution

When the loading condition and the temperature distribution are kept constant, the creep process in a thick-walled tube will tend asymptotically to a steady state characterized by

$$\dot{\bar{e}}_r^k = \dot{e}_r, \quad \dot{\bar{e}}_t^k = \dot{e}_t, \quad \dot{\varepsilon} = 0. \tag{18}$$

These are the necessary and sufficient conditions for the stresses to be constant in time. Since Eqs. (9) and (10) are in fact the equilibrium equations for the stress rates, these equations are automatically satisfied.

The condition $\dot{\varepsilon} = 0$ implies the differential equation for $\dot{u}$

$$\frac{d\dot{u}}{dr} + \frac{\dot{u}}{r} + \dot{\varepsilon}_a = 0 \tag{19}$$

of which the general solution is given by

$$\frac{\dot{u}}{r} = -\frac{1}{2}\dot{\varepsilon}_a + \dot{C}_2\left(\frac{r_1}{r}\right)^2. \tag{20}$$

The corresponding expressions for the strain rates are

$$\begin{aligned} \dot{e}_r &= -\frac{1}{2}\dot{\varepsilon}_a - \dot{C}_2\left(\frac{r_1}{r}\right)^2, \\ \dot{e}_t &= -\frac{1}{2}\dot{\varepsilon}_a + \dot{C}_2\left(\frac{r_1}{r}\right)^2. \end{aligned} \tag{21}$$

Because of (18) the invariants I_2^k (12) have values independent of time that can be expressed in terms of $\dot{e}_r$ and $\dot{e}_t$ by means of the quantity

$$J_2 = \frac{1}{2}\dot{e}_{\alpha\beta}^k\,\dot{e}_{\beta\alpha}^k = \dot{e}_r^2 + \dot{e}_r\,\dot{e}_t + \dot{e}_t^2 = \frac{3}{4}\dot{\varepsilon}_a^2 + \dot{C}_2^2\left(\frac{r_1}{r}\right)^4. \tag{22}$$

From (3) it follows that

$$J_2 = I_2^k\,\{g(T)\,f^k(I_2^k)\}^2. \tag{23}$$

Since $f^k(I_2^k)$ must be a monotonically increasing function of I_2^k, Eq. (23) defines implicitely I_2^k as a single-valued function of J_2

$$I_2^k = h^k(J_2, T). \tag{24}$$

Eqs. (2), (3), (21), and (24) then lead to the following expressions for s_r and s_t

$$\begin{aligned} s_r &= \frac{2G}{g(T)}\dot{e}_r\sum_1^N\frac{\psi^k}{f^k(I_2^k)} = -\frac{2G}{g(T)}\left\{\frac{1}{2}\dot{\varepsilon}_a + \dot{C}_2\left(\frac{r_1}{r}\right)^2\right\}\sum_1^N\frac{\psi^k}{f^k\{h^k(J_2, T)\}}, \\ s_t &= \frac{2G}{g(T)}\dot{e}_t\sum_1^N\frac{\psi^k}{f^k(I_2^k)} = -\frac{2G}{g(T)}\left\{\frac{1}{2}\dot{\varepsilon}_a - \dot{C}_2\left(\frac{r_1}{r}\right)^2\right\}\sum_1^N\frac{\psi^k}{f^k\{h^k(J_2, T)\}}. \end{aligned} \tag{25}$$

From (22) and (25) it is seen that s_r and s_t depend, for a given temperature distribution, only on the constants $\dot{\varepsilon}_a$ and $\dot{C}_2$. These constants are determined by the loading condition and can be expressed in terms of p_1, p_2, and p_a with the aid of the equilibrium equations

$$r\frac{d\sigma_r}{dr}+s_r-s_t=0 \quad \begin{cases} r=r_1:\sigma_r=-p_1, \\ r=r_2:\sigma_r=-p_2, \end{cases} \tag{26}$$

$$\int_{r_1}^{r_2}\sigma_a\, r\, dr=\int_{r_1}^{r_2}(\sigma_r-2\, s_r-s_t)\, r\, dr=\frac{p_a\, r_1^2}{2}. \tag{27}$$

Integration of Eq. (26) and partial integration of Eq. (26) followed by substitution of Eq. (26) lead to

$$\int_{r_1}^{r_2}\frac{s_t-s_r}{r}\, dr=p_1-p_2, \tag{28}$$

$$3\int_{r_1}^{r_2}(s_t+s_r)\, r\, dr=(p_1-p_a)\, r_1^2-p_2\, r_2^2. \tag{29}$$

Eqs. (28) and (29) with the expressions (25) for s_t and s_r furnish two equations for $\dot{\varepsilon}_a$ and $\dot{C}_2$ that determine these constants uniquely in terms of p_1, p_2, and p_a. However in general an explicit expression for $\dot{\varepsilon}_a$ and $\dot{C}_2$ cannot be given. Only in the case $\dot{\varepsilon}_a=0$ can Eqs. (28) and (29) be integrated readily if the creep law is a simple power law as defined by (6).

For $\dot{\varepsilon}_a=0$ it follows from (25) that $s_t=-s_r$, and thus Eq. (29) reduces to

$$(p_1-p_a)\, r_1^2-p_2\, r_2^2=0. \tag{30}$$

It can be observed that for $p_2=0$ one has $p_1=p_a$ according to (30). Hence for $p_2=0$ these case $\dot{\varepsilon}_a=0$ applies to a tube under internal pressure with free but closed ends.

Before Eq. (28) can be integrated first the function $g(T)$ and the temperature distribution have to be specified. If in expression (13) for the temperature distribution corresponding to steady heat flow the temperature difference ΔT between inner and outer wall of the tube is small as compared to $T_1 \ln\lambda$, as will be nearly always the case, then $\frac{1}{T}$ may be written as

$$\frac{1}{T}=\frac{1}{T_1}\left(1+\frac{\Delta T}{T_1}\frac{\ln\frac{r}{r_1}}{\ln\lambda}\right).$$

The function $g(T)$ given by (5) then reads

$$g(T)=g(T_1)\left(\frac{r_1}{r}\right)^{\frac{T_c\Delta T}{T_1^2\ln\lambda}}. \tag{31}$$

The creep law defined by (6) together with Eqs. (22), (25), (31), and (28) defines $\dot{C}_2$ in terms of p_1 and p_2 as follows ($\dot{\varepsilon}_a=0$):

$$\dot{C}_2=2^{-\frac{q+1}{2}}g(T_1)\left[\frac{q'}{2}\left(1-\lambda^{-\frac{2}{q'}}\right)\sum_1^N\psi^k\, p_k^{\frac{q+1}{q}}\right]^{-q}\left(\frac{p_1-p_2}{2G}\right)^q, \tag{32}$$

where

$$\frac{1}{q'} = \frac{1}{q}\left(1 - \frac{T_c\, \Delta T}{2\, T_1^2 \ln \lambda}\right). \tag{33}$$

The deviator components of the stress tensor now follow from (25) after substitution of (32) and $\dot{\varepsilon}_a = 0$:

$$s_t = -\, s_r = \left[q'\left(1 - \lambda^{-\frac{2}{q'}}\right)\right]^{-1} (p_1 - p_2) \left(\frac{r_1}{r}\right)^{\frac{2}{q'}}. \tag{34}$$

For the axial stress component

$$\begin{aligned} \sigma_a = \sigma = \sigma_r - s_r &= -\, s_r - p_1 + \int_{r_1}^{r} \frac{s_t - s_r}{r}\, dr \\ = -\left(1 - \lambda^{-\frac{2}{q'}}\right)^{-1} &\left[\frac{q' - 1}{q'} (p_1 - p_2) \left(\frac{r_1}{r}\right)^{\frac{2}{q'}} + \lambda^{-\frac{2}{q'}} p_1 - p_2\right]. \end{aligned} \tag{35}$$

Since

$$3\, \varepsilon = \frac{du}{dr} + \frac{u}{r} + \varepsilon = \frac{\sigma}{C} + \alpha (T - T_0)$$

the general expression for the displacement u under steady state creep conditions is given by

$$\begin{aligned} u = &\left[\left(1 - \lambda^{-\frac{2}{q'}}\right)^{-1} \lambda^{-\frac{2}{q'}} \frac{p_1 - p_2}{2\, C} - \frac{\varepsilon_a}{2} + \frac{\alpha}{2} (T_1 - T_0) + \frac{\alpha}{4} \frac{\Delta T}{\ln \lambda}\right] r \\ &+ C_2\, r_1 \left(\frac{r_1}{r}\right) - \frac{\alpha}{2} \Delta T\, \frac{r \ln \frac{r}{r_1}}{\ln \lambda} - \left(1 - \lambda^{-\frac{2}{q'}}\right)^{-1} \frac{p_1 - p_2}{2\, C}\, r_1 \left(\frac{r}{r_1}\right)^{\frac{q' - 2}{q'}}, \end{aligned} \tag{36}$$

where the values of ε_a and C_2 can only be determined by numerical integration over the creep process that has taken place before a steady state was reached.

The solution for a thick-walled tube under internal pressure with free but closed ends in the case of steady state creep governed by a power law in the stress was first given in Ref. [*1*] for $\Delta T = 0$, while the solution for $\Delta T \neq 0$ is found in Ref. [*12*].

(d) Approximate Solution of Transient Problem

In general solutions of Eqs. (25), (28), (29), and (35) for steady-state creep of a thick-walled tube under axially symmetric loading can only be determined numerically. The solutions for steady state creep do not furnish reliable information on the total deformation; moreover in many instances it will be desirable to know the state of stress and deformation as a function of time under varying loading and temperature conditions. A convenient numerical method for the solution of the problem of non-stationary creep is therefore of practical interest. In view of the complexity of the problem this method should be given in a form adapted to the use of automatic digital computers. The application of the extremum principle of Ref. [*10*] was found to be most suitable for this end.

According to Eqs. (9) through (12) the dependent variables in the problem are $u = u(r, t)$, $\varepsilon_a = \varepsilon_a(t)$, $\bar{e}_r^k = \bar{e}_r^k(r, t)$, and $\bar{e}_t^k = \bar{e}_t^k(r, t)$. The functions $p_1 = p_1(t)$, $p_2 = p_2(t)$, $p_a = p_a(t)$, and $T = T(r, t)$ are prescribed. First it can be observed that the dependent variables will be continuous differentiable functions in r in the interval $r_1 \leq r \leq r_2$. Hence the displacement u can be expanded in the following power series that converges for all r in this interval

$$u = r_1 \sum_0^\infty \mu_n(t) \left(1 - \frac{r}{r_2}\right)^n. \tag{37}$$

For steady-state creep the values of $\dot{\mu}_n$ are according to expression (20), apart from the linear term $-\frac{1}{2}\dot{\varepsilon}_a r$ and a constant factor, the coefficients of the expansion of $(r/r_1)^{-1}$. This expansion converges slowly. In two special cases the exact dependence of u on r is known. In the solution (15) for u in the case of elastic deformation as well as in the general expression (36) for u in the case of the steady-state creep of a tube under internal pressure a term in $(r/r_1)^{-1}$ appears. Expressions (15) and (36) have in common terms in r and $r \ln r$, but expression (36) shows in addition a term in $(r/r_1)^{\frac{q'-2}{q'}}$. However the power series expansion (37) for this term converges rapidly for small values of λ and large values of q'. If the expansion is truncated at the term that is of the third degree in r, the error is at most of the order of 1% for $\lambda = 2$ and $q' = 3$. The dependence on r of the additional term in the general expression for u in the case of steady state creep is determined by the expression for σ. Since the elastic and the steady-state creep solution represent two limiting cases for the stress distribution, a good approximation of the dependence of u on r is obtained if in the general expression for u the terms are retained that the elastic and the steady state creep solutions have in common, and if the additional dependence on r for stationary and non-stationary creep is represented by a polynomial of the third degree in r. This reduces the problem of the determination of the infinitely many functions $\mu_n = \mu_n(t)$ in (37) to a problem in which the deformation is defined by only seven unknown functions of time. These are the deformation parameters a_1 through a_6 defined as follows

$$\begin{aligned} \varepsilon_a &= a_1, \\ \frac{u}{r_1} &= a_2 \varrho^{-1} + a_3 + a_4 \varrho + a_5 \varrho^2 + a_6 \varrho^3 + a_7 \varrho \ln \varrho, \\ \varrho &= \frac{r}{r_1}. \end{aligned} \tag{38}$$

Table 1. *Elements of the Symmetric Matrix A*

$\frac{3\varkappa+4}{12}(\lambda^2-1)$	0	$\frac{3\varkappa-2}{6}(\lambda-1)$	$\frac{3\varkappa-2}{6}(\lambda^2-1)$	$\frac{3\varkappa-2}{6}(\lambda^3-1)$	$\frac{3\varkappa-2}{6}(\lambda^4-1)$	$\frac{3\varkappa-2}{6}\lambda^2\ln\lambda$
	$1-\lambda^{-2}$	$1-\lambda^{-1}$	0	$-(\lambda-1)$	$-(\lambda^2-1)$	$-\ln\lambda$
		$\frac{3\varkappa+4}{6}\ln\lambda$	$\frac{3\varkappa+1}{3}(\lambda-1)$	$\frac{3\varkappa}{4}(\lambda^2-1)$	$\frac{6\varkappa-1}{9}(\lambda^3-1)$	$\frac{3\varkappa+1}{3}\lambda\ln\lambda-\frac{3\varkappa+4}{6}(\lambda-1)$
			$\frac{3\varkappa+1}{3}(\lambda^2-1)$	$\frac{3\varkappa+1}{3}(\lambda^3-1)$	$\frac{3\varkappa+1}{3}(\lambda^4-1)$	$\frac{3\varkappa+1}{3}\lambda^2\ln\lambda$
				$\frac{9\varkappa+4}{8}(\lambda^4-1)$	$\frac{6\varkappa+3}{5}(\lambda^5-1)$	$\frac{3\varkappa+1}{3}\lambda^3\ln\lambda+\frac{3\varkappa+4}{18}(\lambda^3-1)$
					$\frac{12\varkappa+7}{9}(\lambda^6-1)$	$\frac{3\varkappa+1}{3}\lambda^4\ln\lambda+\frac{3\varkappa+4}{12}(\lambda^4-1)$
						$\frac{3\varkappa+1}{3}\lambda^2(\ln\lambda)^2+\frac{3\varkappa+4}{12}(\lambda^2-1)$

According to the extremum principle of Ref. [*10*] the approximate solution for the rate of deformation in the general creep problem is characterized by a maximum value of the integral expression (7) with respect to arbitrary variations of $\dot{a}_1$ through $\dot{a}_7$. The condition for a stationary value of expression (7) gives rise to a system of simultaneous integrodifferential equations of the type

$$\sum_{j=1}^{7} A_{ij}\,\dot{a}_j = A_{i0}^{(1)}\,\dot{p}_1 + A_{i0}^{(2)}\,\dot{p}_2 + A_{i0}^{(a)}\,\dot{p}_a + \varkappa\,\alpha\int_1^{\lambda} b_i(\varrho)\,\dot{T}\,d\varrho + \sum_{k=1}^{N}\psi^k\int_1^{\lambda}\{c_i(\varrho)\,\dot{e}_r^k + d_i(\varrho)\,\dot{e}_t^k\}\,d\varrho,\quad i=1,2,\ldots,7. \tag{39}$$

A_{ij} is an element of the matrix of constant coefficients given in Table 1. Table 2 contains the column vectors of the numerical factors A_{i0} as well as the column vectors b, c, and d of which the elements are functions of ϱ.

With the aid of the notation

$$x_k = e_r - \bar{e}_r^k,\qquad y_k = e_t - \bar{e}_t^k. \tag{40}$$

the expressions (11) for $\dot{e}_r^k$ and $\dot{e}_t^k$ are written

$$\dot{e}_r^k = g(T)\,x_k\,f^k(I_2^k),\qquad \dot{e}_t^k = g(T)\,y_k\,f^k(I_2^k),\qquad I_2^k = x_k^2 + x_k\,y_k + y_k^2. \tag{41}$$

Table 2. *Column vectors and row vectors occurring in integro-differential Eq.* (44)

$$A_0^{(1)} = \frac{1}{2G}\begin{vmatrix} 0 \\ 1 \\ 1 \\ 1 \\ 1 \\ 1 \\ 0 \end{vmatrix}, \qquad A_0^{(2)} = \frac{1}{2G}\begin{vmatrix} 0 \\ -1 \\ -\lambda \\ -\lambda^2 \\ -\lambda^3 \\ -\lambda^4 \\ -\lambda^2 \ln \lambda \end{vmatrix}, \qquad A_0^{(a)} = \frac{1}{2}\begin{vmatrix} \frac{1}{2} \\ 0 \\ 0 \\ 0 \\ 0 \\ 0 \\ 0 \end{vmatrix}$$

$$b = \begin{vmatrix} \frac{1}{2}\varrho \\ 0 \\ \frac{1}{2} \\ \varrho \\ \frac{3}{2}\varrho^2 \\ 2\varrho^3 \\ 2\varrho\ln\varrho + \varrho \end{vmatrix}, \qquad c = \begin{vmatrix} \varrho \\ -\varrho^{-1} \\ 0 \\ \varrho \\ 2\varrho^2 \\ 3\varrho^3 \\ \varrho\ln\varrho + \varrho \end{vmatrix}, \qquad d = \begin{vmatrix} \varrho \\ \varrho^{-1} \\ 1 \\ \varrho \\ \varrho^2 \\ \varrho^3 \\ \varrho\ln\varrho \end{vmatrix}$$

$$v' = \left| -\frac{1}{3}, -\varrho^{-2}, -\frac{1}{3}\varrho^{-1}, \frac{1}{3}, \varrho, \frac{5}{3}\varrho^2, \frac{1}{3}(\ln\varrho + 2) \right|$$

$$w' = \left| -\frac{1}{3}, \varrho^{-2}, \frac{2}{3}\varrho^{-1}, \frac{1}{3}, 0, -\frac{1}{3}\varrho^2, \frac{1}{3}(\ln\varrho - 1) \right|$$

Since

$$e_r = \sum_{i=1}^{7} v_i(\varrho)\, a_i, \quad e_t = \sum_{i=1}^{7} w_i(\varrho)\, a_i, \tag{42}$$

where $v_i(\varrho)$ and $w_i(\varrho)$ are known functions of ϱ determined by (12) and (38), it follows that

$$\begin{aligned} \dot{x}_k &= \sum_{i=1}^{7} v_i(\varrho)\,\dot{a}_i - g(T)\, x_k\, f^k(I_2^k), \\ \dot{y}_k &= \sum_{i=1}^{7} w_i(\varrho)\,\dot{a}_i - g(T)\, y_k\, f^k(I_2^k). \end{aligned} \tag{43}$$

The row vectors v' and w' are given in Table 2. If the matrix A is inverted and if the solution for the $\dot{a}_i$'s is substituted in (43), the problem is reduced to a problem of numerical integration of the following integro-differential equations for given initial values of x_k and y_k and for prescribed $p_1(t)$, $p_2(t)$, $p_a(t)$, and $T(r, t)$:

$$
\begin{aligned}
\dot{x}_k &= v' A^{-1}\Bigg[A_0^{(1)}\dot{p}_1 + A_0^{(2)}\dot{p}_2 + A_0^{(a)}\dot{p}_a + \varkappa\,\alpha \int_1^\lambda b\,\dot{T}\,d\varrho \\
&\quad + \sum_{l=1}^{N}\psi^k \int_1^\lambda \{c\,x_l + d y_l\}\,g(T)\,f^l(I_2^l)\,d\varrho\Bigg] - g(T)\,x_k\,f^k(I_2^k), \\
\dot{y}_k &= w' A^{-1}\Bigg[A_0^{(1)}\dot{p}_1 + A_0^{(2)}\dot{p}_2 + A_0^{(a)}\dot{p}_a + \varkappa\,\alpha \int_1^\lambda b\,\dot{T}\,d\varrho \\
&\quad + \sum_{l=1}^{N}\psi^l \int_1^\lambda \{c\,x_l + d y_l\}\,g(T)\,f^l(I_2^l)\,d\varrho\Bigg] - g(T)\,y_k\,f^k(I_2^k).
\end{aligned}
\tag{44}
$$

The components of the stress deviator at any instant are determined by

$$
\begin{aligned}
s_r &= 2\,G \sum_1^N x_k\,, \\
s_t &= 2\,G \sum_1^N y_k,
\end{aligned}
\tag{45}
$$

while the isotropic stress σ follows from

$$
\sigma = \sigma_r - s_r = -\,s_r - p_1 + \int_{r_1}^{r} \frac{s_t - s_r}{r}\,dr\,. \tag{46}
$$

The state of deformation as a function of time is obtained by numerical integration of the equations

$$
\begin{aligned}
\dot{a} &= A^{-1}\Bigg[A_0^{(1)}\dot{p}_1 + A_0^{(2)}\dot{p}_2 + A_0^{(a)}\dot{p}_a + \varkappa\,\alpha \int_1^\lambda b\,\dot{T}\,d\varrho \\
&\quad + \sum_{k=1}^{N}\psi^k \int_1^\lambda \{c\,x_k + d y_k\}\,g(T)\,f^k(I_2^k)\,d\varrho\Bigg].
\end{aligned}
\tag{47}
$$

Eqs. (44) and (47) can be conveniently handled by means of an automatic computer. The integrations over the interval $1 \leq \varrho \leq \lambda$ may then for instance be carried out by SIMPSON's rule which reduces the equations to a system of first order nonlinear ordinary differential equations. The computer program should be sufficiently flexible so as to allow variations of the parameters $\varkappa$, α, λ, ψ^k and of the functions $f^k(I_2^k)$ and $g(T)$ between different cases. The number of equations to be solved is reduced by a factor N if primary creep is neglected. If furthermore the temperature variation over the wall-thickness is negligible, then the integration can be carried out with respect to a temperature corrected time parameter like τ in Fig. 1.

It should be mentioned that especially under variable loading and temperature conditions it may prove necessary to correct at regular intervals the numerical integration process for accumulated errors with the aid of the equilibrium Eqs. (28) and (29). These determine at any instant the actual loading condition and small fictitious pressure increments may have to be applied to bring the actual loading condition in concordance with the prescribed loading condition.

If p_1, p_2, p_a, and T are held constant, the solution of Eqs. (44) and (47) will converge to the solution for steady state creep characterized by

$$\begin{aligned} \dot{e}_r^k &= -\frac{1}{2}\dot{a}_1 - \dot{a}_2\,\varrho^{-2}, \\ \dot{e}_t^k &= -\frac{1}{2}\dot{a}_1 + \dot{a}_2\,\varrho^{-2}, \\ \dot{a}_4 &= -\frac{1}{2}\dot{a}_1,\ \dot{a}_3 = \dot{a}_5 = \dot{a}_6 = \dot{a}_7 = 0. \end{aligned} \tag{48}$$

It is easily verified that substitution of (48) leads to $\dot{x}_k = \dot{y}_k = 0$.

In many cases the initial conditions will be given in the form of a state of elastic deformation as defined by expressions (15), (16), and (17)

$$\begin{aligned} &a_1 = \varepsilon_a,\ a_2 = C_2,\ a_4 = C_1,\ a_7 = \frac{3\,\varkappa}{3\,\varkappa + 4}\,\frac{\alpha\,\Delta T}{2\ln\lambda}, \\ &a_3 = a_5 = a_6 = 0, \\ &x_k = e_r,\ y_k = e_t. \end{aligned} \tag{49}$$

As a numerical illustration of the method here presented the transition from a state of elastic deformation to steady state creep has been analysed for a tube under internal pressure and constant temperature in the simple case of an incompressible material ($\varepsilon = 0$) with only secondary creep ($N = 1$) governed by a power law in the stress (6).

For $\varepsilon = 0$ the parameters a_3, a_5, a_6, and a_7 are identically zero; moreover $a_4 = -\frac{1}{2}a_1$ (see (20)). This reduces the Eqs. (44) and (47) to

$$\begin{aligned} &(\dot{p}_1 = \dot{p}_2 = \dot{p}_a = 0),\ \tau' = 2^{\frac{q-1}{2}}\, p^{-(q+1)} \int_0^t g(T)\,dt' \\ &\frac{da_1}{d\tau'} = +\frac{2}{\lambda^2 - 1}\int_1^\lambda (x+y)\,(x^2 + x\,y + y^2)^{\frac{q-1}{2}}\,\varrho\,d\varrho, \\ &\frac{da_2}{d\tau'} = -\frac{1}{1-\lambda^{-2}}\int_1^\lambda (x-y)\,(x^2 + x\,y + y^2)^{\frac{q-1}{2}}\,\varrho^{-1}\,d\varrho, \\ &\frac{dx}{d\tau'} = -\frac{1}{2}\frac{da_1}{d\tau'} - \varrho^{-2}\frac{da_2}{d\tau'} - x\,(x^2 + x\,y + y^2)^{\frac{q-1}{2}}, \\ &\frac{dy}{d\tau'} = -\frac{1}{2}\frac{da_1}{d\tau'} + \varrho^{-2}\frac{da_2}{d\tau'} - y\,(x^2 + x\,y + y^2)^{\frac{q-1}{2}}. \end{aligned} \tag{50}$$

The initial state of elastic deformation for the tube under internal pressure, $p_1 = p_a$, $p_2 = 0$, is according to (15), (16) and (17)

$$\begin{aligned} \tau' = 0: a_1 = 0,\ a_2 &= \frac{1}{1-\lambda^{-2}} \frac{p_1}{2G}, \\ x = -y &= \frac{-p_1}{2G} \frac{\varrho^{-2}}{1-\lambda^{-2}}. \end{aligned} \tag{51}$$

It follows from (50) that the transition from the elastic to the steady state creep solution is governed in this case by the equations

$$\begin{aligned} \frac{dy}{d\tau'} &= -\frac{dx}{d\tau'} = \frac{2\varrho^{-2}}{1-\lambda^{-2}} \int_1^\lambda y^q \varrho^{-1} d\varrho - y^q, \\ \frac{da_2}{d\tau'} &= \frac{2}{1-\lambda^{-2}} \int_1^\lambda y^q \varrho^{-1} d\varrho, \quad \frac{da_1}{d\tau} = 0. \end{aligned} \tag{52}$$

Eqs. (52) have been integrated by means of an automatic computer for two different values of λ and q each. The range of integration $1 \leq \varrho \leq \lambda$ was divided into ten intervals for the application of Simpson's rule. The integro-differential equation in (52) is then replaced by eleven first order non-linear ordinary differential equations. Figs. 2 through 5 give a graphic presentation of the results of the computations.

$$(\tau = (p_1/2G)^{q-1} \tau').$$

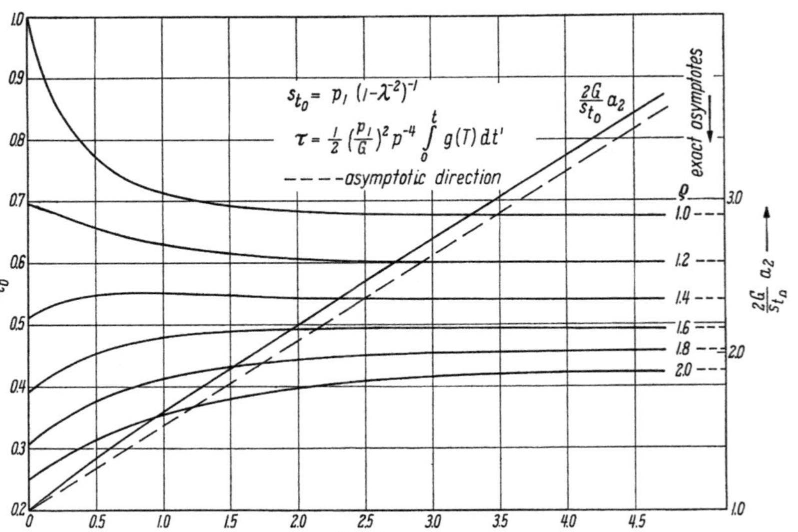

Fig. 2. Transient creep in tube under constant internal pressure ($\lambda = 2$, $q = 3$)

It was found that though Eqs. (44) and (47) will converge to a steady state creep solution ($\dot{x}_k = \dot{y}_k = 0$, $a =$ constant) independent of

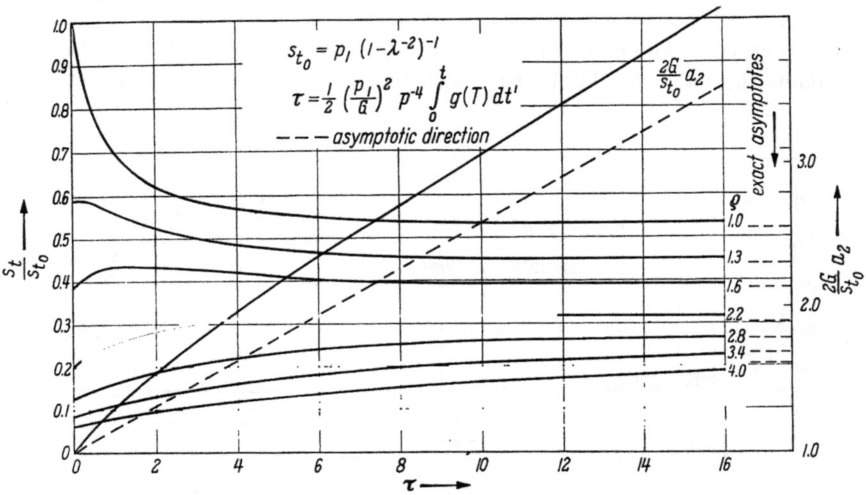

Fig. 3. Transient creep in tube under constant internal pressure ($\lambda = 4$, $q = 3$)

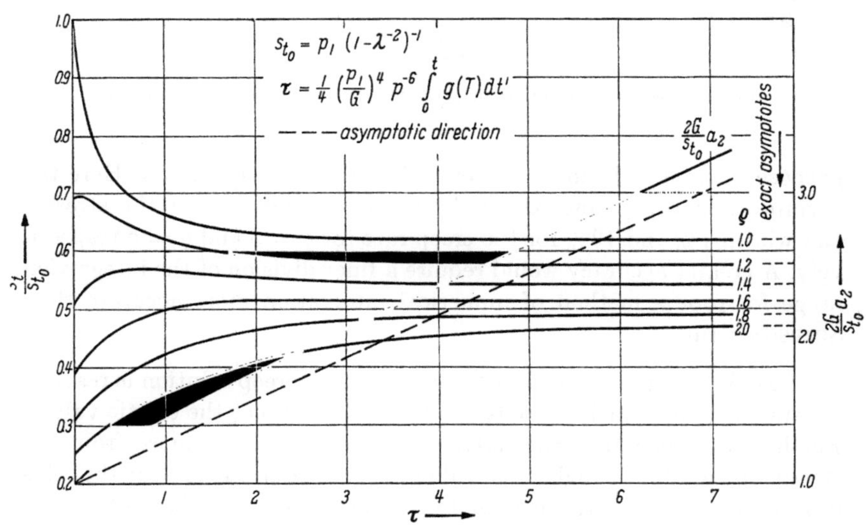

Fig. 4. Transient creep in tube under constant internal pressure ($\lambda = 2$, $q = 5$)

the initial condition and deformation history once the loading and temperature conditions are held constant, the numerical integration of Eqs. (52) did not lead exactly to the steady state creep solution defined by (32) and (34). This is mainly due to the accumulated error produced by the application of SIMPSON's rule which is much larger than the rounding off errors inherent in any numerical process. For $\lambda = 2$ the asymptotic state is reached within 0.04% for the stresses and within 0.2% for the creep rate. The error in the creep rate does not exceed 0.6% when τ reaches the value 50. While this is sufficient accuracy for any practical

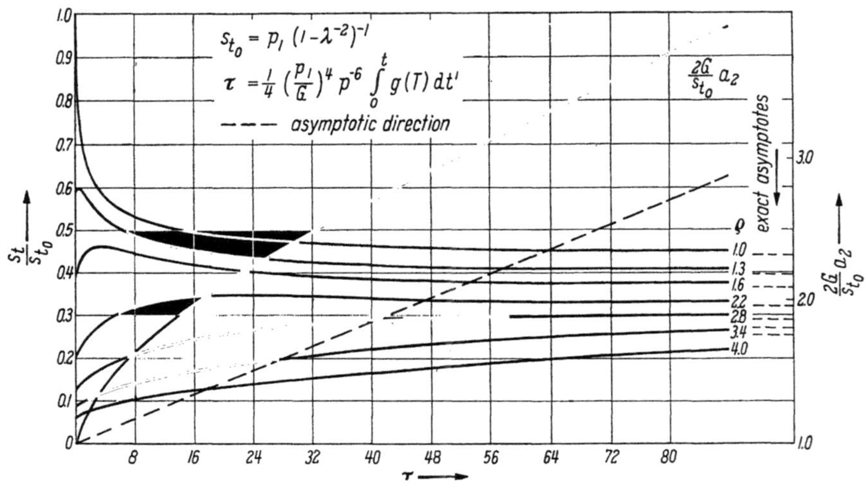

Fig. 5. Transient creep in tube under constant internal pressure ($\lambda = 4$, $q = 5$)

purpose, more significant is the error for the case with $\lambda = 4$. Here the asymptotic state is approximated within 4% and 6% for the stresses and within 7% and 10% for the creep rate for $q = 3$ and $q = 5$ respectively. A greater accuracy would require a finer division of the integration range for ϱ or a method of numerical integration more accurate than Simpson's rule.

Figs. 2 through 5 show that the steady-state creep solution is reached at strains values which are only a small number times the elastic values. Further it is seen that the initial strain rate deviates more from the steady creep rate the higher the values of q and λ are. It is clear that this deviation will be greatly increased if the primary creep phase of the material is also taken into account.

Finally it is of interest to remember that for a tube of uniform temperature under axially symmetric loading the elastic solution for the stresses does not depend on the elastic constants. The stress distribution for steady-state creep is always independent of these constants. Hence if the temperature difference across the wall may be neglected these two limiting states of stress are identical for the compressible and incompressible material.

4. Concluding Remarks

Since lengthy computations seem unavoidable in the analysis of non-stationary creep problems the purpose of this paper was to outline a numerical method for the solution of these problems that is particularly suited to the use of automatic digital computers.

It was shown that sufficient information on the elastic as well as on the steady-state creep solutions was of prime importance in the application of this method to thick-walled tubes under radially symmetric loading. Other problems where similar information is available may be open to numerical analysis by the method presented here.

Acknowledgement

The author gratefully acknowledges the co-operation of the staff of the Institute for Applied Mathematics of the Technical University at Delft. He is particularly indebted to Mr. D. H. Wolbers who supervised the computational phase of the investigation, and to Mr. F. P. Bremer who did the coding for the ZEBRA automatic computer.

References

[1] Bailey, R. W.: The Utilization of Creep Test Data in Engineering Design Proc. Inst. Mech. Engrs., **131**, 131—349 (1935).

[2] Soderberg, C. R.: Interpretation of Creep Tests for Machine Design, Trans. ASME, **58**, 733 (1936).

[3] Coffin, L. E., P. R. Shepler and G. S. Cherniak: Primary Creep in the Design of Internal Pressure Vessels, J. Appl. Mechanics, **16**, no. 3, 229—241 (1949).

[4] Voorhees, H. R., C. M. Sliepcevitch and J. W. Freeman: Thick-Walled Pressure Vessels, Ind. Eng. Chem. **48**, 872 (1956).

[5] Besseling, J. F.: A Theory of Elastic, Plastic, and Creep Deformations of an Initially Isotropic Material, J. Appl. Mechanics, **25**, no. 4, 529—536 (1958).

[6] Besseling, J. F.: Thermodynamic Foundations of the Theory of Deformation, Proc. Durand Centennial, Oxford: Pergamon Press Ltd., 1960.

[7] Odqvist, F. K. G.: Theory of Creep under the Action of Combined Stresses with Applications to High-Temperature Machinery, Proc. Roy. Swed. Inst. Eng. Res., no. 141, 1—31 (1936).

[8] Dorn, J. E.: Some Fundamental Experiments of High-Temperature Creep, Journal of Mechanics and Physics of Solids, **3**, no. 2, 85 (1954).

[9] Sherby, O. D., J. L. Lytton and J. E. Dorn: Activation Energies for Creep of High-Purity Aluminium, Acta Metallurgica, **5**, 219—227 (1957).

[10] Wang, A. J., and W. Prager: Thermal and Creep Effects in Work-Hardening Elastic-Plastic Solids, Journal of the Aeronautical Sciences, **21**, no. 5, 343—344, 360 (1954).

[11] Besseling, J. F.: A Theory of Small Deformations of Solid Bodies, SUDAER no. 84, AFOSR TN-59-605, ASTIA no. 217.172, Stanford University, February 1959.

[12] Finnie, I., and W. R. Heller: Creep of Engineering Materials, New York, Toronto, London: McGraw Hill Book Comp., Inc. 1959, p. 185.

Discussion

E. H. Lee, Brown University: I wish to comment on the question brought up by Professor Onat of the relative merits of solving the differential equations, representing the problem, directly by numerical methods, or making use of the variational principle. It seems to me that the choice must depend on the accuracy that can be achieved with the same expenditure of computing effort. In this connection, it is interesting to observe that, in the theory of work-hardening plasticity, variational principles also refer to the strain and stress rates at a given loading situation. Approximate application of these give functions which must be added incrementally, and to my knowledge this has not been attempted because of the loss of accuracy to be expected. Application of variational principles has been limited to such problems as ideal plastic flow, where the principle applies to the final solution, and not merely incrementally. I would appreciate comments from Professor Besseling on this aspect of his work.

F. K. G. Odqvist, Royal Institute of Technology, Stockholm: Without underrating the importance of work of the kind carried out by the author, I would like to point out the desirability of having only few free constants in the body relations of engineering theories of creep.

Author's concluding remarks: In answer to the question raised by Professor Lee I would like to mention that the application of the variational principle for the displacement rates to plasticity problems meets with two essential difficulties: (1) The domain of integration has to be divided into continuously changing elastic and elastic-plastic domains; (2) The coefficients in the resulting equations for the unknown rates of the deformation parameters depend on the stress distribution in the elastic-plastic domains and thus on the history of the deformation process. These difficulties are not present in creep problems where the variational principle leads to a system of integro-differential equations, valid throughout the whole deformation process, that can be solved by means of an automatic computer to any required degree of accuracy. The final numerical process is a step-by-step method of integrating a semilinear first order system of ordinary differential equations where the step length is determined by the computing machine from the required accuracy of the results. I do not know of any other numerical method for solving the differential equations of the problem considered in my paper that matches the variational approach in simplicity and flexibility.

Though I fully agree with the comment made by Professor Odqvist it seems to me that we can only hope that further experiments may lead to a theory that is both simple and adequate.

A Comparison of Flow Criteria Applied to Elevated Temperature Creep of Rotating Disks with Consideration of the Transient Condition

By

A. M. Wahl

Westinghouse Research Laboratories, Pittsburgh, Pennsylvania, USA

Abstract

A study is made of stress distributions in rotating disks with holes, under steady-state creep at elevated temperature, based on various flow criteria (Tresca, Tresca-Mises or Mises), and assuming a creep rate-stress-time function of the form $\dot{\varepsilon} = K\,\sigma^n f(t)$. On this basis, comparisons are given between maximum stresses and creep deformations obtained using the various flow rules for disks having different ratios of outside to inside diameters and composed of material with different n values. Results of these show generally lower peak stresses and creep strains if the Mises criterion is used. Spin tests on steel disks at 1000 F and on lead disks at 85 F also indicate that the commonly used Mises criterion gives low creep deformations compared to average test values; better results are usually obtained using the Tresca or Tresca-Mises criteria. Further experimental data, however, are required before definite conclusions can be drawn.

Using the various flow rules and both strain-hardening and time-hardening assumptions, methods are given for taking into account the additional creep resulting during the initial transient period after starting, during which the stress distributon in the disk changes from an initial elastic-plastic one to the steady state. These indicate that, for the spin tests reported, a negligible error is made by assuming the steady-state stress distribution during the transient period, first stage creep being taken into account by the time function in the expression for strain rate. In certain cases in practice it may, however, be necessary to consider this transient condition.

1. Introduction

In 1954 and 1956, the results of some long-time spin tests on chrome-steel disks at 1000° F were reported [*1*, *2*]. These tests indicated that the use of the MISES criterion, as commonly applied in creep problems, may give too low results compared to actual spin test values. Better agreement between test and calculated creep values was obtained if the maximum shear or TRESCA criterion were used, but it was concluded that further test data would be required before definite conclusions could be drawn. Since that time, additional spin tests [3] have been made on lead disks under creep conditions which give results very similar to those previously obtained on the steel disks, the use of the MISES criterion again giving low creep values compared to test results.

The analyses reported on previously [1, 2] were carried out assuming a *steady-state condition*, i. e., that stresses in the disk, although functions of the disk radius, did not change with time. It was also assumed that the creep rate $\dot{\varepsilon}$ for the material could be taken as a product of a function of stress times a function of time, t i. e., that $\dot{\varepsilon} = F(\sigma) \cdot f(t)$. Thus primary or first stage creep was taken into account by the time function $f(t)$ but the transient change in stress distribution from an initial value at zero time to the steady-state value, during a period of time defined here as the "transient period", was neglected. An approximate analysis [2] indicated that no appreciable error was involved by assuming the steady state stress distribution during this transient period.

Recently attempts [4] have been made to show that an appreciable error was made in [1] by neglecting this transient change in stress and that the discrepancy between test results and theoretical values based on the MISES criterion could be nearly eliminated by taking into account these transient effects. However, as shown in a recent discussion [5], the comparisons made in [4] between test and calculated results using the MISES criterion are not valid since tension creep data were used which did not apply for the material used. Also the comparisons made in the closure of [4] between the MISES "transient" and the MISES "steady state" have no significance since the "transient" curves were calculated using an empirical relation giving creep rates around 15 to 30 percent higher than those used for the "steady state". Thus the two curves are not directly comparable. Actually, as shown in the present paper, if the "transient" and "steady state" solutions are based on the same tensile creep data, no appreciable difference in creep deformations is found for strains of the order obtained in [1].

Since the question of the effect of the transient period of stress has been raised, some further studies of this have been carried out and the results are reported here. In addition, some comparisons are given between various flow criteria for disks having various ratios of hole diameter to outer diameter; also additional comparisons will be made with available spin test data. In the present paper, only disks with central holes are considered and for simplicity only power function stress-creep rate relations of the forms

$$\dot{\varepsilon} = K\, \sigma^n\, t^m, \tag{1}$$

$$\dot{\varepsilon} = K\, \sigma^n\, \varepsilon^m \tag{1a}$$

are considered, where ε = strain, $\dot{\varepsilon} = d\varepsilon/dt$ = creep rate, σ = stress, t = time and m and n are constants. Although these equations do not always apply to actual materials, in many cases they are sufficiently accurate for engineering use, particularly in view of the scatter commonly found in creep data.

2. Comparison of Steady-State Solutions using Various Flow Criteria

In [6] a comparison of steady-state stress distributions using various flow criteria was made for long rotating cylinders with holes having various diameter ratios and n values. In this section a similar comparison is made for disks with holes. For brevity as in [6] these criteria will be referred to in what follows as the "TRESCA", "TRESCA-MISES" and "MISES" criteria. The first of these is based on the TRESCA or maximum shear criterion and the associated flow rule as used in plasticity theory [7, 8], the second is based on the TRESCA criterion for stress intensity combined with the flow rule corresponding to the MISES criterion [2] and the third on the MISES criterion and the associated flow rule [9, 10, 11].

For a rotating disk of constant thickness (Fig. 1), the equilibrium condition to be satisfied in all cases is

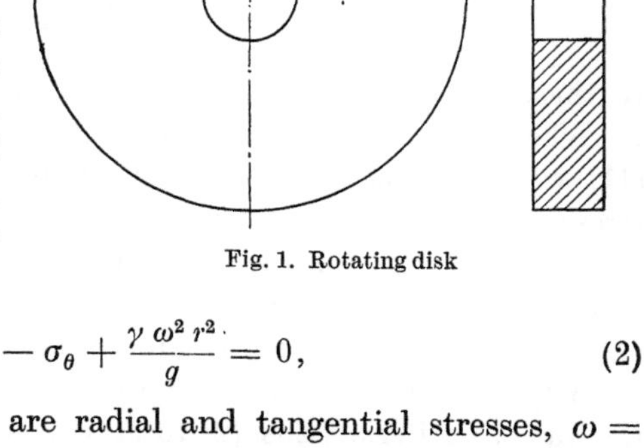

Fig. 1. Rotating disk

$$\frac{d}{dr}(r\,\sigma_r) - \sigma_\theta + \frac{\gamma\,\omega^2\, r^2}{g} = 0, \tag{2}$$

where r = radius, σ_r and σ_θ are radial and tangential stresses, ω = angular velocity, γ = weight per unit volume and g = acceleration of gravity.

For steady state conditions, the following additional equations must be satisfied for the various criteria [2]

Tresca Criterion:

$$\dot{\varepsilon}_\theta = \frac{\dot{w}}{r} = \frac{\dot{w}_i}{r} = K\,\sigma_\theta^n\, f(t), \tag{3}$$

$$\dot{\varepsilon}_r = \frac{d\dot{w}}{dr} = 0, \text{ for } \sigma_\theta > \sigma_r > 0, \tag{4}$$

where $\dot{w}$, $\dot{w}_i$ are radial deformation rates at r and r_i, respectively (Fig. 1), $\dot{\varepsilon}_\theta$ and $\dot{e}_r$ are tangential and radial strain rates respectively.

Tresca-Mises Criterion ($\sigma_\theta \gtrless \sigma_r > 0$)

$$\dot{\varepsilon}_\theta = \frac{\dot{w}}{r} = K\,\sigma_\theta^{n-1}\, f(t)\,(\sigma_\theta - \sigma_r/2), \tag{5}$$

$$\dot{\varepsilon}_r = \frac{d\dot{w}}{dr} = K\,\sigma_\theta^{n-1}\, f(t)\,(\sigma_r - \sigma_\theta/2). \tag{6}$$

Mises Criterion:

$$\dot{\varepsilon}_\theta = \frac{\dot{w}}{r} = K\,\sigma_e^{n-1}\, f(t)\,(\sigma_\theta - \sigma_r/2), \tag{7}$$

$$\dot{\varepsilon}_r = \frac{d\dot{w}}{dr} = K\,\sigma_e^{n-1}\, f(t)\,(\sigma_r - \sigma_\theta/2), \tag{8}$$

where the effective stress σ_e is:

$$\sigma_e = \sqrt{\sigma_\theta^2 - \sigma_\theta\,\sigma_r + \sigma_r^2}\,. \tag{9}$$

Using Eqs. (2), (3), (4) for the Tresca criterion [*2*] the expressions for tangential and radial stresses can be expressed in closed form as follows, for no loading at the periphery of the disk:

$$\sigma_\theta = \frac{\alpha\,(r_0 - r_i)\,\sigma_{av}}{\left(r_0^\alpha - r_i^\alpha\right) r^{1/n}}, \tag{10}$$

$$\sigma_r = \frac{(r_0 - r_i)\,(r^\alpha - r_i^\alpha)\,\sigma_{av}}{\left(r_0^\alpha - r_i^\alpha\right) r} - \frac{\gamma\,\omega^2\,(r^3 - r_i^3)}{3\,g\,r}, \tag{11}$$

where the average tangential stress over the cross section is

$$\sigma_{av} = \frac{\gamma\,\omega^2 (r_0^3 - r_i^3)}{3\,g\,(r_0 - r_i)}\,. \tag{12}$$

In these $\alpha = (n-1)/n$, r_i = inner radius, r_0 = outer radius.

Where an external peripheral tension stress is present Eqs. (10) and (11) will apply under certain restrictions, if σ_{av} is the average tangential stress [*12*].

The equations for the Tresca-Mises and Mises criteria may be solved for the steady-state condition using methods given in [*10*] and [*1*], which involve numerical integration of Eqs. (2), (5), (6), (7) and (8).

A typical plot of tangential and radial stress distributions based on the various criteria [Eqs. (2) to (11)] is shown on Fig. 2 for $r_i/r_0 = 0.1$ and $n = 6$.[1] Values here are plotted in terms of ratios $\sigma_\theta/\sigma_{av}$ and σ_r/σ_{av}.

[1] Distributions for several other n and r_i/r_0 values based on the Tresca criterion are given in [*12*].

(In all cases in Figs. 2—10 which follow, the TRESCA criterion is represented by the full line, the TRESCA-MISES by the short dashes and the MISES by the long dashes.) In Fig. 2, the dot-dash line represents the elastic tangential stress distribution for POISSON's ratio = 0.3. Also since the steady-state distributions of radial stress showed little variation for the various criteria only a mean curve of σ_r/σ_{av} is shown. It may be noted that lower stresses at the inside and outside diameters of the disk are obtained using the MISES criterion as compared with those found using the other two criteria.

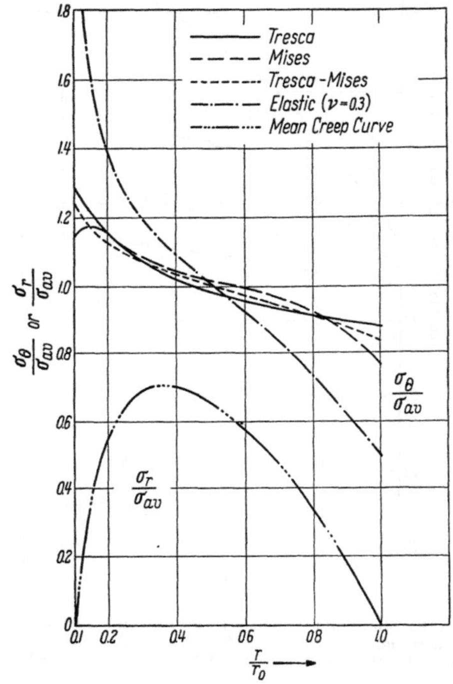

Fig. 2. Typical stress distributions in rotating disks under steady state creep for various how criteria ($r_i/r_0 = 0.1$, $n = 6$)

Curves showing values of maximum tangential stress in the disk, plotted against the ratio r_0/r_i for n values varying from 4 to 10, are shown on Fig. 3 based on the various flow rules. For the TRESCA and TRESCA-MISES criteria these maximum stresses occur at the bore while in most cases for the MISES criterion they occur at a radius slightly greater than the inside radius (see Fig. 2, for example). It may be noted that the ratio $\sigma_{\max}/\sigma_{av}$ is greater for the larger r_0/r_i values and for the smaller n values. Also, other things being equal, the peak stresses are generally largest for the TRESCA, and smallest for the MISES criterion. Similar curves showing values of the tangential stress σ_{r0} at the outside diameter as a function of r_0/r_i are shown on Fig. 4 for various n values. Here again, for the same n values, lower stresses are found for the MISES than for the other two criteria.

Since creep rates are usually proportional to some rather high power of the stress, much greater relative differences are found between the various criteria if creep rates rather than stresses are used as a basis for comparison. In Fig. 5, the ratio $\dot{\varepsilon}_{\theta i}/\dot{\varepsilon}_n$ between creep rates at the inside diameter and the nominal rate $\dot{\varepsilon}_n$ based on the average tangential stress σ_{av} is plotted against r_0/r_i for various n values based on the various criteria

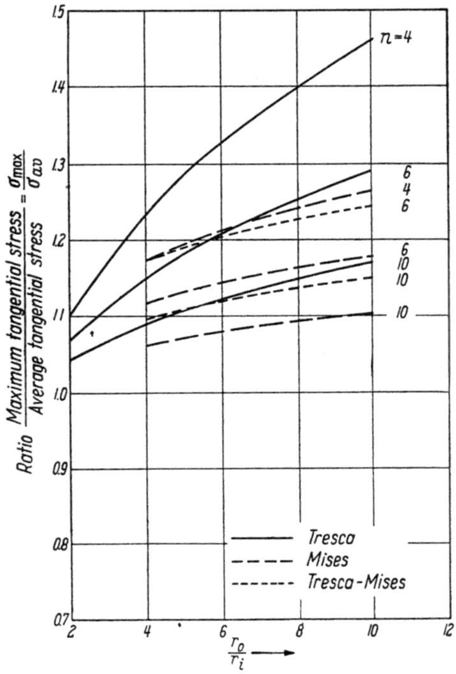

Fig. 3. Maximum tangential stresses for various n values and flow criteria. (Note: for mises criterion maximum stress may Occur at radius larger than r_i)

[Eqs. (2) to (10)]. In this figure,

$$\dot{\varepsilon}_n = K\,\sigma_{av}^n\,f(t). \qquad (13)$$

It may be seen from Fig. 5 that in general the calculated creep rates at the inside diameter are much lower for the Mises than for either the Tresca or Tresca-Mises criteria; this difference is also more pronounced for the higher values of n and of r_0/r_i. For example for $n = 10$ and $r_0/r_i = 10$ the creep rate based on the Mises criterion is only about onethird those found for the other two criteria.

Similar curves for the outside diameter of the disk are shown on Fig. 6 where the ratios $\dot{\varepsilon}_{\theta 0}/\dot{\varepsilon}_n$ between creep rate at the outside

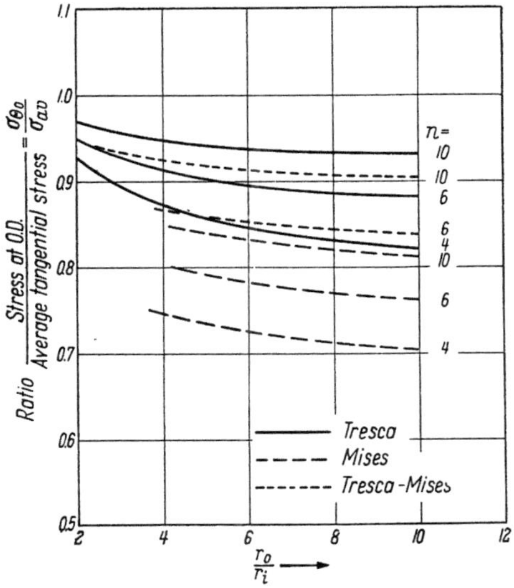

Fig. 4. Tangential stresses at outside diameter for various n values and flow criteria

diameter and nominal creep rate are plotted versus r_0/r_i for various n values using Eqs. (2) to (10). Again it may be seen that the creep rates using the MISES criterion are much lower than those found using the other two criteria.

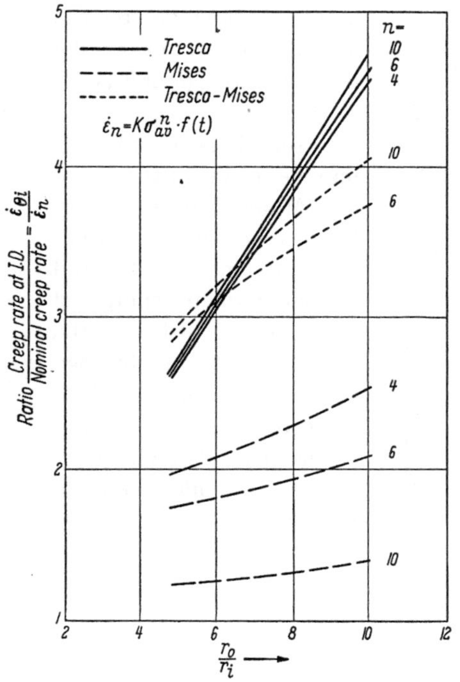

Fig. 5. Creep rate at inside diameter for various n values and flow criteria (steady-state condition)

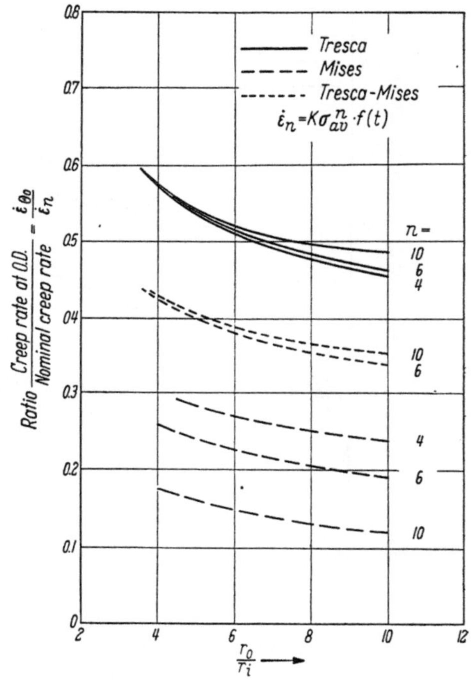

Fig. 6. Creep rate at outside diameter for various n values and flow criteria (steady-state condition)

3. Comparison of Theoretical and Test Results

(a) Tests on Chrome Steel Disks at 1000 F

In Figs. 7 and 8 are shown results of tests [*1*] on 12 inch diameter chrome steel disks with 2—1/2 inch diameter central holes, tested at 1000 F for periods up to about 1000 hours at 15000 rpm. (The scatter bands shown represent test results on 3 disks). For comparison the calculated curves using average tensile creep test data and Eqs. (2) to (11), based on steady-state conditions, are shown for the three criteria, the mean test curve being shown by the dot-dash line. In this case, as discussed in [*1*] the calculated curves were based on true stress, the effects of disk expansion being taken into account. Also the average creep rates used for calculating disk expansion were based on true stresses obtained from tensile creep data [1]. The fact that in these spin tests the disks were slowed down slightly to maintain approximately constant peripheral speed beyond about 180 hours was also considered. First stage creep is taken into account by the time function $f(t)$ of Eqs. (3) to (8).

From Figs. 7 and 8, it may be seen that the use of the Mises criterion gives low creep strain values both at the inside and outside diameter compared to average test results. Better results are obtained using the Tresca-Mises criterion while the Tresca criterion yields high values at the outside diameter but good results at the bore.

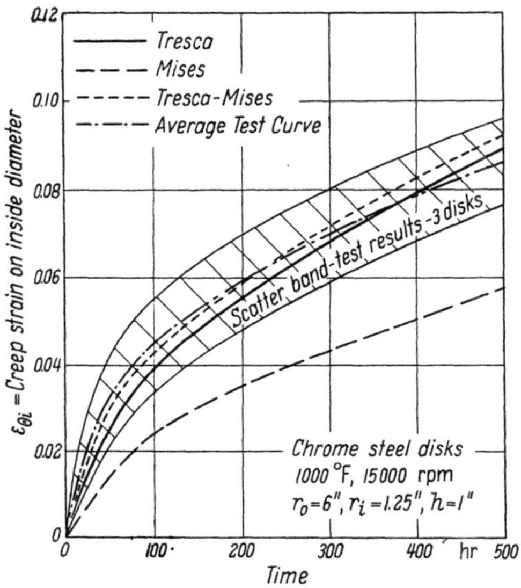

Fig. 7. Comparison of calculated and test values of creep strain at inside diameter of rotating steel disks

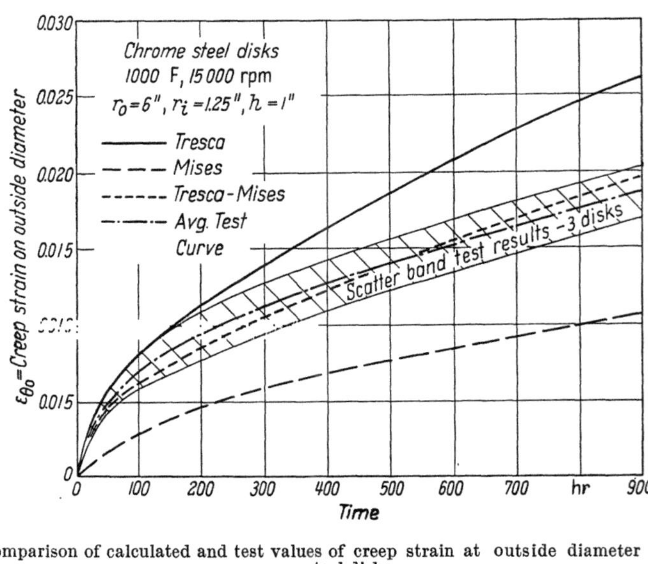

Fig. 8. Comparison of calculated and test values of creep strain at outside diameter of rotating steel disks

(b) Tests on Lead Disks at 85 F

Recently spin-creep tests [3] have been carried out at 85 F on lead disks having the same diameters as the steel disks tested in [1]. The results are shown in Figs. 9 and 10 for the inside and outside diameters,

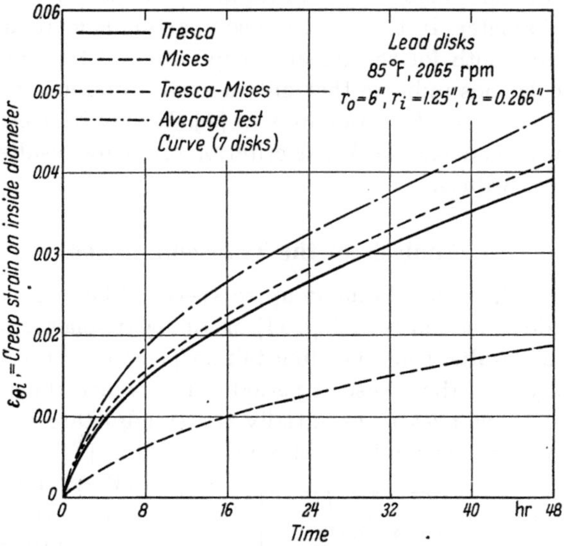

Fig. 9. Comparison of calculated and test values of creep strain at inside diameter lead disks (D'ISA, Ref. [3])

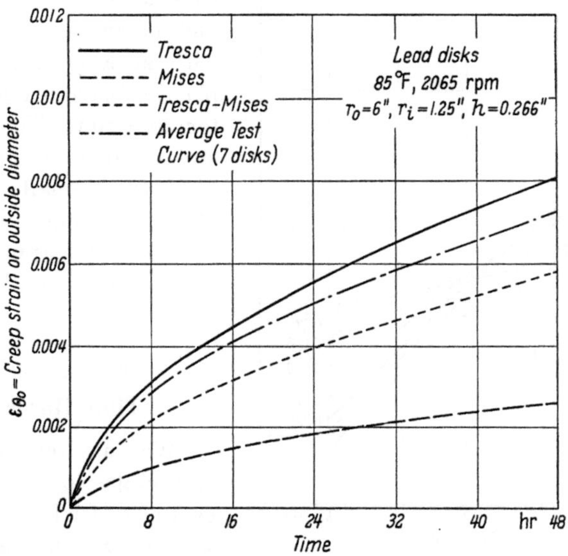

Fig. 10. Comparison of calculated and test values of creep strain at outside diameter of rotating lead disks (D'ISA, Ref. [3])

respectively, the theoretical results based on average tensile creep data being shown for the various criteria by either the full or dashed lines. The average test curves based on tests on 7 disks are shown by the dot-dash-lines. Considerable scatter was present in these test results, the standard deviation at 48 hours being .0023 strain at the outside and .0128 strain at the inside diameter. In this case, the comparison is made on the basis of nominal stress both for the tensile creep data and for the spin tests; however, calculations indicate that only a small difference would result if true stresses were used as was done in [1]. It may be seen that as in the case of the steel disks, the Mises criterion gives low results compared to average test values.

4. Analysis of the Transient Period

When the disk is first brought up to speed, either an elastic stress distribution (dot-dash curve of Fig. 11) or an elastic-plastic distribution (dashed curve) results, some yielding taking place near the bore of the disk. The analysis of the transient period during which the stress distribution gradually changes to essentially the steady-state value, is here carried out by methods similar to those described in [2] except that the Tresca-Mises and Mises criteria are also utilized. In addition, a strainhardening as well as a time-hardening rule is considered. Essentially, the method involves starting out with an approximate elastic-plastic distribution of stress for zero time. The change in stress distribution during successive small increments of time is found by a method of successive approximations as described in [2].

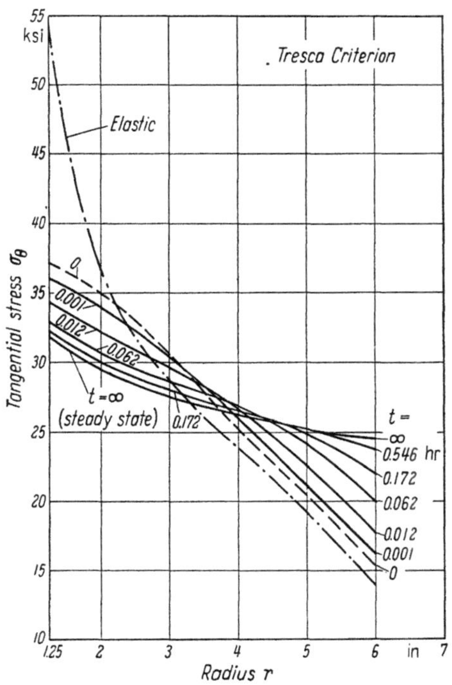

Fig. 11. Change in tangential stress distribution during transient period

(a) Tresca Criterion — Time-Hardening

We assume an initial elastic-plastic stress distribution of σ_θ such that represented by the dashed curve marked $t = 0$ in Fig. 11. Since for the disk $\sigma_\theta > \sigma_r > 0$, the change in stress $\Delta\sigma_\theta$ for successive time increments Δt is then calculated using the following fundamental equations based on the TRESCA criterion and given in [2].

$$\Delta\varepsilon_\theta = \frac{\Delta w}{r} = \frac{\Delta\sigma_\theta - \nu\,\Delta\sigma_r}{E} + K\,\sigma_{\theta a}^n \int_t^{t+\Delta t} t^m\,dt, \tag{14}$$

$$\Delta\varepsilon_r = \frac{d\Delta w}{dr} = \frac{\Delta\sigma_\theta - \nu\,\Delta\sigma_r}{E}. \tag{15}$$

In these, $\Delta\varepsilon_\theta$ and $\Delta\varepsilon_t$ are changes in tangential and radial strains during the time increment Δt; Δw is the increment of radial deflection at radius r in Δt; $\nu =$ POISSON's ratio, $E =$ modulus of elasticity and $\sigma_{\theta a} = \sigma_\theta + (1/2)\,\Delta\sigma_\theta$ is the average tangential stress during the increment Δt, σ_θ being the stress at the beginning of the increment. This amounts to assuming that the plastic creep rate $\dot\varepsilon_{\theta p}$ can be represented by using an average stress during the time increment.

$$\dot\varepsilon_{\theta p} = K\,\sigma_{\theta a}^n\,t^m. \tag{15a}$$

This equation involves the time hardening assumption and assumes that $\Delta\sigma_\theta$ is small compared to σ_θ. Also since no load acts at the inner or outer radius, from equilibrium using Eq. (2)

$$\Delta\sigma_r = \frac{1}{r}\int_{r_i}^{r} \Delta\sigma_\theta\,dr, \tag{16}$$

$$\int_{r_i}^{r_0} \Delta\sigma_\theta\,dr = 0. \tag{17}$$

It is assumed that Δt is taken small enough so that, approximately

$$\sigma_{\theta a}^n = \sigma_\theta^n \left(1 + \frac{n\,\Delta\sigma_\theta}{2\,\sigma_\theta}\right). \tag{17a}$$

On this basis, using Eqs. (14)—(17) the following equations given in [2] are obtained.

$$\Delta\sigma_\theta = \psi(r) + \psi_1(r), \tag{18}$$

where

$$\psi(r) = \frac{E\,\Delta\varepsilon_{\theta si}}{\Phi_1}\left[\frac{\Sigma_1}{r\,\Sigma_2} - \left(\frac{\sigma_\theta}{\sigma_{\theta si}}\right)^n\right], \tag{19}$$

$$\psi_1(r) = [B - \Sigma_3/\Sigma_2]/(r\,\Phi_1), \tag{20}$$

$$B = \int_{r_i}^{r} \frac{dr}{r} \int_{r_i}^{r} \Delta\sigma_\theta\,dr, \tag{21}$$

$$\Phi_1 = \frac{n\,E\,\Delta\varepsilon_{\theta si}}{2\,\sigma_{\theta si}}\left(\frac{\sigma_\theta}{\sigma_{\theta si}}\right)^{n-1} + 1, \tag{22}$$

$$\Sigma_1 = \int_{r_1}^{r_0} \frac{1}{\Phi_1}\left(\frac{\sigma_\theta}{\sigma_{\theta si}}\right)^n dr, \tag{23}$$

$$\Sigma_2 = \int_{r_i}^{r_0} \frac{dr}{r\,\Phi_1}, \tag{24}$$

$$\Sigma_3 = \int_{r_i}^{r_0} \frac{B\,dr}{r\,\Phi_1}, \tag{25}$$

$$\Delta\varepsilon_{\theta s i} = \frac{K\,\sigma_{\theta s i}^n}{m+1}\,[(t+\Delta t)^{m+1} - t^{m+1}]. \tag{26}$$

In these equations σ_θ is the tangential stress at the beginning of the time increment Δt while $\sigma_{\theta s i}$ is the steady-state stress at $r = r_i$ calculated from Eq. (10).

The stress increment $\Delta\sigma_\theta$ during Δt is determined from Eq. (18) as follows [2], starting from an initial elastic-plastic distribution of σ_0, estimated approximately from tensile stress-strain data. The function $\psi(r)$ is first calculated from Eq. (19) using Σ_1, Σ_2, and Φ_1 values from Eqs. (23), (24) and (22). Using this value of $\psi(r)$ as a first approximation for $\Delta\sigma_\theta$, values of $\psi_1(r)$ are calculated from Eq. (20) using values of B and Σ_3 found from Eqs. (21) and (25). Now using these values of $\psi_1(r)$ a new approximation for $\Delta\sigma_\theta$ is found from Eq. (18) and using this a new $\psi_1(r)$ is found. This process converges rapidly. Having found $\Delta\sigma_\theta$, $\Delta\sigma_r$ may be calculated from Eq. (16). The results of this method as applied to the disk tested in [1] are shown on Fig. 11 where σ_θ is plotted against disk radius r for various time increments based on the following constants based on tensile creep data: $K = 1.67 \times 10^{-30}$, $n = 6$ and $m = -0.5$.

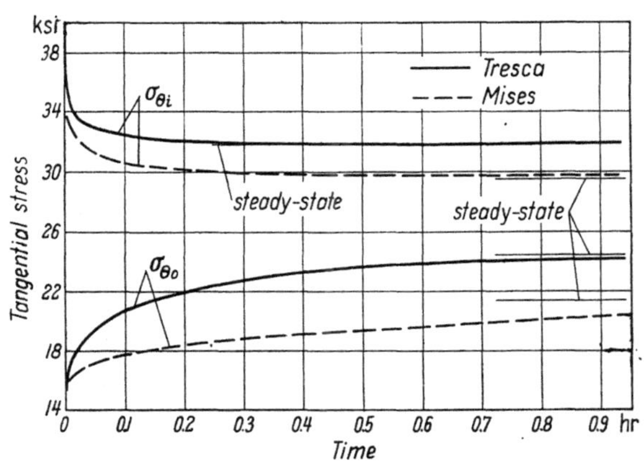

Fig. 12. Tangential stresses $\delta_{\theta i}$ and δ_{θ_0} at inside and outside of disks during transient period

(Applied to Eq. (15a) these give creep rate per hour if stress is in psi.) Stresses $\sigma_{\theta i}$ and $\sigma_{\theta 0}$ at the inside and outside of the disk are plotted versus time in Fig. 12. It may be seen that, in 0.3 hour, the tangential stress at the inside of the disk drops to within about 100 psi of the steady-state stress value of 31,850 psi; thus for all practical purposes the steady state value is reached at the inside diameter in this time.

In Fig. 13 the strain at the inside diameter of the disk obtained using Eq. (14) for the various time increments is shown on the upper full line curve, while the lower full line represents the values found using the steady-state solution and taking $\sigma_{\theta a} = \sigma_{\theta si} = 31{,}850$ psi and the same n, m and K values in Eq. (15a). After about 0.6 hour the difference between the transient and steady-state solutions remains constant at a strain of about 0.0004, a value which is negligible compared to creep strains reached in the test. Thus no appreciable error is here made using the steady state solution and neglecting the transient.

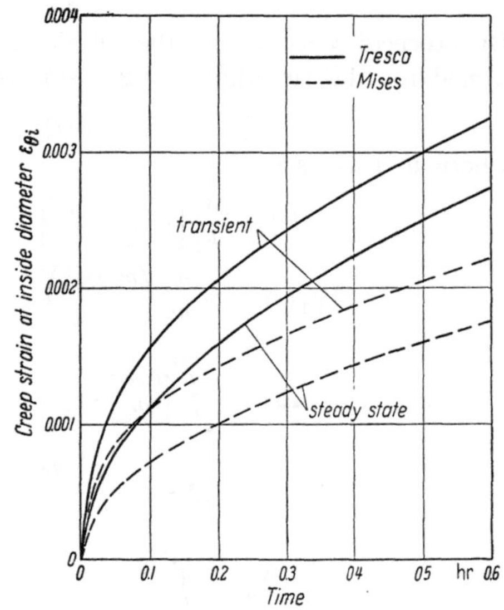

Fig. 13. Strain at inside diameter with and without consideration of transient condition

(b) Tresca Criterion — Strain Hardening

In this case it is assumed that the average plastic strain rate $\dot{\varepsilon}_{\theta pa}$ is the product of a power function of $\sigma_{\theta a}$ times a power function of the average plastic strain $\varepsilon_{\theta pa}$ during the time increment Δt. Thus

$$\dot{\varepsilon}_{\theta pa} = K\,\sigma_{\theta a}^{n}\,\varepsilon_{\theta pa}^{m}, \tag{26a}$$

where

$$\sigma_{\theta a} = \sigma_\theta + (\Delta\sigma_\theta/2) \text{ and } \varepsilon_{\theta pa} = \varepsilon_{\theta p} + (\Delta\varepsilon_{\theta p}/2).$$

It is also assumed that Δt is taken small enough so that Eq. (17a) holds and that $\Delta\varepsilon_{\theta p} = \dot{\varepsilon}_{\theta pa}\,\Delta t = K\,\sigma_{\theta a}^{n}\,\varepsilon_{\theta pa}^{m}\,\Delta t$. Proceeding as before, the total strain increments become (based on the Tresca criterion where $\sigma_\theta > \sigma_r > 0$):

$$\Delta\varepsilon_\theta = \frac{\Delta w}{r} = \frac{\Delta\sigma_\theta - \nu\,\Delta\sigma_r}{E} + K\,\sigma_{\theta a}^{n}\,\varepsilon_{\theta pa}^{m}\,\Delta t, \tag{27}$$

$$\Delta\varepsilon_r = \frac{d(\Delta w)}{dr} = \frac{\Delta\sigma_r - \nu\,\Delta\sigma_\theta}{E}. \tag{28}$$

We again use Eqs. (16) and (17) and assume further that $\Delta\varepsilon_{\theta p}/\varepsilon_{\theta p}$ is small so that, approximately

$$\varepsilon_{\theta p a}^{m} = \varepsilon_{\theta p}^{m}\left(1 + \frac{m\,\Delta\varepsilon_{\theta p}}{2\,\varepsilon_{\theta p}}\right). \tag{29}$$

By proceeding as before, the following equations (similar to those obtained for the time-hardening case) are found:

$$\Delta\sigma_\theta = \psi(r) + \psi_1(r), \tag{30}$$

where in this case,

$$\psi(r) = \frac{1}{\Phi_3}\left[\frac{\Sigma_1}{r\,\Sigma_2} - \Phi_2\right], \tag{31}$$

$$\psi_1(r) = -\frac{\Phi_2}{\Phi_3}\left(\frac{m\,\Delta\varepsilon_{\theta p}}{2\,\varepsilon_{\theta p}}\right) + \frac{1}{r\,\Phi_3}\left[B + \frac{\Sigma_4 - \Sigma_3}{\Sigma_2}\right], \tag{32}$$

$$\Sigma_1 = \int_{r_i}^{r_0} \frac{\Phi_2\,dr}{\Phi_3}, \tag{33}$$

$$\Sigma_2 = \int_{r_i}^{r_0} \frac{dr}{r\,\Phi_3}, \tag{34}$$

$$\Sigma_3 = \int_{r_i}^{r_0} \frac{B\,dr}{r\,\Phi_3}, \tag{35}$$

$$B = \int_{r_i}^{r} \frac{dr}{r} \int_{r_i}^{r} \Delta\sigma_\theta\,dr, \tag{36}$$

$$\Sigma_4 = \int_{r_i}^{r_0} \frac{m\,\Phi_2\,\Delta\varepsilon_{\theta p}\,dr}{2\,\Phi_3\,\varepsilon_{\theta p}}, \tag{37}$$

$$\Phi_2 = E\,K\,\sigma_\theta^n\,\varepsilon_{\theta p}^m\,\Delta t, \tag{38}$$

$$\Phi_3 = 1 + \frac{n\,\Phi_2}{2\,\sigma_\theta}, \tag{39}$$

$$\Delta\varepsilon_{\theta p} = \frac{\Phi_2}{E}\left(1 + \frac{n\,\Delta\sigma_\theta}{2\,\sigma_\theta}\right)\Bigg/\left(1 - \frac{m\,\Phi_2}{2\,E\,\varepsilon_{\theta p}}\right). \tag{40}$$

These equations may be solved by a procedure similar to that used for the time-hardening case. To start out, the time-hardening equations are used until the plastic strains are large enough so that the condition represented by Eq. (29) is fulfilled. As before $\psi(r)$ is first calculated using Eqs. (31), (33), (34), (38) and (39) and taking values of σ_θ and $\varepsilon_{\theta p}$ at the beginning of the time interval Δt. This value of $\psi(r)$ is then taken as a first approximation for $\Delta\sigma_\theta$. Now B and Σ_3 can be found from Eqs. (36) and (35), and $\Delta\varepsilon_{\theta p}$ from Eq. (40). Using the latter value, Σ_4 is found from Eq. (37) and from this a new value $\Delta\sigma_\theta$ using Eq. (30). Using this value, a second approximation for $\psi_1(r)$ is found by proceeding as before.

This procedure has been applied to the rotating disks tested in [1] and it was found that the stresses at the inside diameter do not drop quite as rapidly as was the case when the time-hardening rule was used. For example, at 0.052 hour, $\sigma_\theta = 33{,}600$ psi for the strain-hardening case compared to 33,000 psi for the time-hardening assumption[1]. Also the steady-state condition is still practically reached within about an hour. This indicates that there is little difference in this case between the two assumptions. However, in other cases where stresses are lower, this may not be true.

(c) Tresca-Mises Criterion (Transient Period)

If the TRESCA-MISES criterion is used, together with the time-hardening assumption, the following equations apply:

$$\Delta\varepsilon_\theta = \frac{\Delta w}{r} = \frac{\Delta\sigma_0 - \nu\,\Delta\sigma_r}{E} + K\,\sigma_{\theta a}^n\left(1-\frac{x}{2}\right)\int\limits_t^{t+\Delta t} t^m\,dt\,, \tag{41}$$

$$\Delta\varepsilon_t = \frac{d(\Delta w)}{dr} = \frac{\Delta\sigma_r - \nu\,\Delta\sigma_\theta}{E} + K\,\sigma_{\theta a}^n\left(x-\frac{1}{2}\right)\int\limits_t^{t+\Delta t} t^m\,dt\,, \tag{42}$$

where

$$x = \frac{\sigma_{ra}}{\sigma_{\theta a}} = \frac{\sigma_r + \Delta\sigma_r/2}{\sigma_\theta + \Delta\sigma_\theta/2}\,. \tag{43}$$

x = ratio of average radial to average tangential stress at radius r during the time increment Δt.

Using Eqs. (16) and (17), and assuming that as before that $\Delta\sigma_\theta$ is small compared to σ_θ, we again obtain the following expression:

$$\Delta\sigma_\theta = \psi(r) + \psi_1(r)\,, \tag{44}$$

where in this case $\psi(r)$ and $\psi_1(r)$ are given by:

$$\psi(r) = \frac{E\,\Delta\varepsilon_{\theta si}}{\Phi_1}\left[\frac{\Sigma_1 - \Sigma_4}{r\,\Sigma_2} - \left(\frac{\sigma_\theta}{\sigma_{\theta si}}\right)^n\left(1-\frac{x}{2}\right) + \frac{B_1}{r}\right], \tag{45}$$

$$\psi_1(r) = \frac{1}{r\,\Phi_1}\left[B - \frac{\Sigma_3 + \Sigma_5}{\Sigma_2} + A\right], \tag{46}$$

$$A = \frac{n\,E\,\Delta\varepsilon_{\theta si}}{2\,\sigma_{\theta si}}\int\limits_{r_i}^{r}\left(\frac{\sigma_\theta}{\sigma_{\theta si}}\right)^{n-1}\left(x-\frac{1}{2}\right)\Delta\sigma_\theta\,dr\,, \tag{47}$$

$$\Phi_1 = 1 + \frac{n\,E\,\Delta\varepsilon_{\theta si}}{2\,\sigma_{\theta si}}\left(\frac{\sigma_\theta}{\sigma_{\theta si}}\right)^{n-1}\left(1-\frac{x}{2}\right), \tag{48}$$

---

[1] The writer is indebted to Mr. S. GREEN for calculating these results.

$$\Sigma_1 = \int_{r_i}^{r_0} \frac{1}{\Phi_1} \left(\frac{\sigma_\theta}{\sigma_{\theta s i}}\right)^n \left(1 - \frac{x}{2}\right) dr, \tag{49}$$

$$B_1 = \int_{r_i}^{r} \left(\frac{\sigma_\theta}{\sigma_{\theta s i}}\right)^n \left(x - \frac{1}{2}\right) dr, \tag{50}$$

$$\Sigma_4 = \int_{r_i}^{r_0} \frac{B_1\, dr}{r\, \Phi_1}, \tag{51}$$

$$\Sigma_5 = \int_{r_i}^{r_0} \frac{A\, dr}{r\, \Phi_1}. \tag{52}$$

In these equations $\Delta \varepsilon_{\theta s i}$ is obtained from Eq. (26) using the value of $\sigma_{\theta s i}$ obtained for the steady-state condition based on the Tresca-Mises criterion by the method described in [1].

The quantities B, Σ_2 and Σ_3 appearing in Eqs. (45) and (46) are given in Eqs. (21), (24) and (25) respectively using the value of Φ_1 given by Eq. (48).

For this case, $\Delta\sigma_\theta$ may be found for each time increment Δt, by using a method of successive approximations similar to that discussed for the Tresca criterion. To calculate $\psi(r)$, values of x are first assumed equal to the σ_r/σ_θ ratios at the start of the time interval. Later when first approximations for $\Delta\sigma_\theta$ and $\Delta\sigma_r$ are found, new values of $x = \sigma_{ra}/\sigma_{\theta a}$ can be obtained. Values of $\Delta\sigma_r$ are again found from Eq. (16). Generally it will be found that the new values of x will not differ significantly from those first assumed. Results thus obtained for $\sigma_{\theta i}$ at the inside radius fall slightly above the full line of Fig. 12 for the Tresca case while the value of stress at the outside radius falls between the $\sigma_{\theta 0}$ lines for the Tresca and the Mises criteria. The time taken to practically reach the steady-state at the inside diameter is somewhat greater than that found for the Tresca criterion but still less than one hour.

(d) Mises Criterion (Transient Period)

In this case, to calculate the plastic strain increment during the time Δt, it is assumed that an average effective stress σ_{ea} during the increment may be used, equal to

$$\sigma_{ea} = \sqrt{\sigma_{\theta a}^2 - \sigma_{\theta a}\,\sigma_{ra} + \sigma_{ra}^2} = \sigma_{\theta a}\sqrt{1 - x + x^2}. \tag{53}$$

In this $x = \sigma_{ra}/\sigma_{\theta a}$ as before. Based on the Mises criterion and associated flow rule, and using the time-hardening relation, the fundamental equations become:

$$\Delta\varepsilon_\theta = \frac{\Delta w}{r} = \frac{\Delta\sigma_\theta - \nu\,\Delta\sigma_r}{E} + K\,\sigma_{ea}^{n-1}\left(\sigma_{\theta a} - \frac{\sigma_{ra}}{2}\right)\int\limits_t^{t+\Delta t} t^m\,dt, \tag{54}$$

$$\Delta\varepsilon_r = \frac{d(\Delta w)}{dr} = \frac{\Delta\sigma_r - \nu\,\Delta\sigma_\theta}{E} + K\,\sigma_{ea}^{n-1}\left(\sigma_{ra} - \frac{\sigma_{\theta a}}{2}\right)\int\limits_t^{t+\Delta t} t^m\,dt. \tag{55}$$

These assume a plastic creep rate-stress-time relation of the form $\dot\varepsilon_p = K\,\sigma_{ea}^n\,t^m$ during the time increment Δt.

Taking

$$\beta = (1 - x + x^2)^{(n-1)/2} \tag{56}$$

these equations can be written:

$$\Delta\varepsilon_\theta = \frac{\Delta w}{r} = \frac{\Delta\sigma_\theta - \nu\,\Delta\sigma_r}{E} + K\,\sigma_{\theta a}^n\,\beta\left(1 - \frac{x}{2}\right)\int\limits_t^{t+\Delta t} t^m\,dt, \tag{57}$$

$$\Delta\varepsilon_r = \frac{d(\Delta w)}{dr} = \frac{\Delta\sigma_r - \nu\,\Delta\sigma_\theta}{E} + K\,\sigma_{\theta a}^n\,\beta\left(x - \frac{1}{2}\right)\int\limits_t^{t+\Delta t} t^m\,dt. \tag{58}$$

Eqs. (57) and (58) are essentially the same as Eqs. (41) and (42) for the TRESCA-MISES criterion, except that $\sigma_{\theta a}^n\,\beta$ is now used instead of $\sigma_{\theta a}^n$. Proceeding in the same manner the final equations reduce the same form used for the previous cases, i. e.,

$$\Delta\sigma_\theta = \psi(r) + \psi_1(r), \tag{59}$$

where $\psi(r)$ and $\psi_1(r)$ are now given by

$$\psi(r) = \frac{E\,\Delta\varepsilon_{\theta si}}{\Phi_1}\left[\frac{\Sigma_1 - \Sigma_4}{r\,\Sigma_2} - \left(\frac{\sigma_\theta}{\sigma_{\theta si}}\right)^n \beta\left(1 - \frac{x}{2}\right) + \frac{B_1}{r}\right], \tag{60}$$

$$\psi_1(r) = \frac{1}{r\,\Phi_1}\left[B - \frac{\Sigma_3 + \Sigma_5}{\Sigma_2} + A\right], \tag{61}$$

$$A = \frac{n\,E\,\Delta\varepsilon_{\theta si}}{2\,\sigma_{\theta si}}\int\limits_{r_i}^{r}\left(\frac{\sigma_\theta}{\sigma_{\theta si}}\right)^{n-1}\beta\left(x - \frac{1}{2}\right)\Delta\sigma_\theta\,dr, \tag{62}$$

$$\Phi_1 = 1 + \frac{n\,E\,\Delta\varepsilon_{\theta si}}{2\,\sigma_{\theta si}}\left(\frac{\sigma_\theta}{\sigma_{\theta si}}\right)^{n-1}\beta\left(1 - \frac{x}{2}\right), \tag{63}$$

$$\Sigma_1 = \int\limits_{r_i}^{r_0}\frac{1}{\Phi_1}\left(\frac{\sigma_\theta}{\sigma_{\theta si}}\right)^n\beta\left(1 - \frac{x}{2}\right)dr, \tag{64}$$

$$B_1 = \int\limits_{r_i}^{r}\left(\frac{\sigma_\theta}{\sigma_{\theta si}}\right)^n\beta\left(x - \frac{1}{2}\right)dr, \tag{65}$$

$$\Sigma_4 = \int_{r_i}^{r_0} \frac{B_1\, dr}{r\,\Phi_1}, \tag{66}$$

$$\Sigma_5 = \int_{r_i}^{r_0} \frac{A\, dr}{r\,\Phi_1}. \tag{67}$$

The quantities B, Σ_1 and Σ_3 appearing in Eqs. (60) and (61) are again given by Eqs. (21), (23) and (25), respectively, provided the value of Φ_1 given by Eq. (63) is used. As will be noted, except for the β term, these equations are essentially the same as those used for the TRESCA-MISES case. The method of finding $\Delta\sigma_\theta$ and $\Delta\sigma_r$ is essentially the same as that described for the latter case. Tangential stress distributions for various times based on Eq. (59) for the MISES case are shown in Fig. 14 for the disks tested in [1] the same values of K, n and m being used as for the TRESCA case of Fig. 11. (In this case the same approximate curve representing the elastic-plastic stress distribution at zero time was assumed as for the TRESCA case.) Values of stresses $\sigma_{\theta i}$ and $\sigma_{\theta 0}$ at the inside and outside diameters are shown versus time by the dashed curves of Fig. 12, the steady state value at the inside being reached (within 100 psi) in about one hour. Although the stress relaxes somewhat more slowly than for the TRESCA case, the steady state value is still reached in a relatively short time.

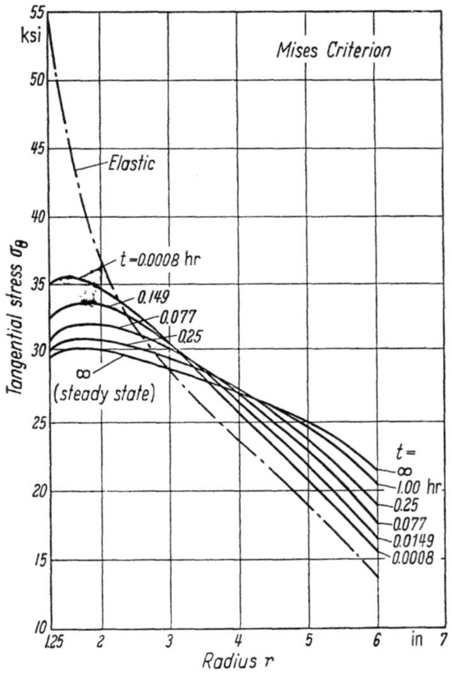

Fig. 14. Change in tangential stress distribution during transient period

Based on Eq. (57), values of creep strain at the inside diameter for the transient condition are represented by the upper dashed curve of Fig. 13. The lower dashed curve representing the steady state solution is based on Eq. (15a) taking $\sigma_{\theta a}$ equal to the steady state value of stress at the inside diameter, i. e., 29,600 psi. After about 0.5 hour the difference between the transient and steady state curves becomes constant at a strain of 0.00045, a value which is again negligible compared to the total strains reached during the test.

5. Summary and Conclusions

This study indicates that if creep rates can be expressed as $\dot{\varepsilon} = K\sigma^n f(t)$ the Mises criterion generally gives creep rates and deformations in rotating disks which are considerably lower than those obtained using the Tresca or Tresca-Mises criteria, steady state conditions being assumed. The difference is particularly pronounced for the higher values of n where creep rates using the Mises criterion may be only a fraction of those obtained using the Tresca criterion.

Spin tests on chrome steel disks at 1000 F and on lead disks at 85 F indicate that creep deformations as calculated by the commonly used Mises criterion give too low values compared to average test results. Better results are obtained in these cases using either the Tresca or Tresca-Mises criteria. However, further experimental data are required before definite conclusions can be drawn as to whether this is generally true.

Methods are described for taking into account the change in stress distribution and the additional strain occurring during the transient period as the stress relaxes from an initial elastic-plastic distribution to the steady state, based on the various flow criteria and using both time-hardening and strain-hardening rules. Comparisons of these various criteria indicates that a somewhat faster relaxation occurs if the Tresca criterion is used. Applied to the spin tests on chrome steel disks reported in [*1*], the results indicate that only a negligible error in calculating creep strain is made by using the steady-state stress distribution as a basis; first stage creep is, however, taken into account by the time function used. Thus the discrepancies shown by available test data between calculated and test results using the Mises criterion cannot be resolved by consideration of the transient period. It is realized, however, that there are cases in practice particularly where lower stresses are present where it may be necessary to consider the transient period to obtain satisfactory results.

Acknowledgement

The writer is indebted to Mr. R. E. Peterson for his support of this work and to Mr. W. Gabrys for assistance in carrying out the calculations.

References

[1] WAHL, A. M., G. O. SANKEY, M. J. MANJOINE and E. SHOEMAKER: Trans. ASME, **76**, (Jl. of Appl. Mech.) 225 (1954).
[2] WAHL, A. M.: Trans. ASME, **78**, (Jl. of Appl. Mech.) 231 (1956).
[3] D'ISA, FRANK A.: Dissertation, University of Pittsburgh, (1960).
[4] MENDELSON, A., M. H. HIRSCHBERG and S. S. MANSON: Trans. ASME, **81**, 585 (1959).
[5] WAHL, A. M., and M. J. MANJOINE: discussion, Trans. ASME, **81**, 595 (1959).
[6] WAHL, A. M.: Proc. Third US National Congress of Applied Mech., 685 (1958).
[7] KOITER, W.: Biezeno Anniversary Vol., Haarlem, Holland, 232 (1953).
[8] PRAGER, W.: Trans. ASME, **75**, (Jl. of Appl. Mech.) 317 (1953).
[9] SODERBERG, C. R.: Trans. ASME, **58**, 734 (1936).
[10] BAILEY, R. W.: Proc. Inst. of Mech. Eng., **131**, 260 (1935).
[11] MILLENSON, M. B., and S. S. MANSON: NACA Report No. 906, (1948).
[12] WAHL, A. M.: Trans. ASME, **80**, (Jl. of Appl. Mech.) 243 (1958); also **79**, 299 (1957).

Discussion

F. K. G. ODQVIST, Royal Institute of Technology, Stockholm: It is suggested that the author should use Eq. (1a) instead of (1) when generalizing from one- to three-dimensional cases in order to obtain a consistent theory for the creep of rotating disks. Such generalization may be made as proposed in a paper by the writer, see reference [6] of writer's paper at this symposium, Eq. (6).

Author's closing remarks: The author agrees with Professor ODQVIST's suggestion that it would be better to use the strain-hardening expression [Eq. (1a)] instead of the time-hardening expression [Eq. (1)]. However, if this is done, the resulting equations become much more complicated and so far no calculated results have been obtained. In this connection, it is of interest to note that, as mentioned in the paper, if the TRESCA criterion is applied, for the disks tested in reference [1] only a small difference is found between the results calculated by the strain-hardening and time-hardening relations.

Bending of a Plate with Nonlinear Strain Hardening Creep

By

T. H. Lin

University of California, Los Angeles, California, USA

Abstract

A method of analyzing plates subject to bending with arbitrary creep characteristics is derived. The creep strain is treated as equivalent loads and edge moments acting on the plate. This method is applicable to cases with varying temperature in the plane of the plate. The general equations are simplified for the cases of circular plates subject to axially symmetrical lateral loading. An illustrative example of a circular plate with nonlinear strain hardening creep is shown.

1. Introduction

The use of structures at elevated temperatures has recently much increased. The analyses of beams [*1*], columns [*2*] and beam columns [*3*] with arbitrary creep laws have been given by a number of investigators. Analyses of the bending of plates with arbitrary creep laws have not been published. Steady-state solutions of circular plates under lateral load with the creep rate assumed to be proportional to the nth power of the stress and independent of the existing amount of creep strain have been studied [*4, 5, 6*]. These do not include transient creep and are not applicable to materials with other creep characteristics. Extensive creep tests by Johnson [*7*] and others [*8, 9*] have shown that the creep rate depends on both the stress and the amount of creep strain present. The reduction of the creep rate with the amount of creep strain developed is called strain hardening creep. This strain hardening creep seems to be common to most engineering materials. The stress-strain-time relationship varies with temperature and material. This paper shows a method of analyzing plate bending with arbitrary creep characteristics.

Symbols used.

a radius of the circular plate

A a constant

B a constant

D flexural rigidity of the plate $= \dfrac{E\,h^3}{12(1-\mu^2)}$
e strain
E Young's modulus at temperature T
f_r radial stress
f_t tangential stress
F a function
h thickness of plate
i, j varies from 1 to 2
I_2 second creep strain invariant
$\dot{I}_2$ rate of increase of the second creep strain invariant
J_2 second stress invariant
$\bar{J}_2$ second deviatoric stress invariant
K a constant
M section moment per unit width
P vertical load
Q vertical shear per unit width
r radius
S_{ij} deviatoric stress component
S_r deviatoric radial stress
S_t deviatoric tangential stress
w vertical deflection
z vertical axis
μ Poisson's ratio
φ $= -\dfrac{dw}{dr}$
ζ $= \dfrac{z}{h/2}$

2. Equivalent Load and Moment Due to Creep

In the presence of creep, the total strain consists of elastic strain and creep strain which are denoted by single and double primes, respectively. Let x, y and z be a set of rectangular coordinates

$$f_x = \frac{E}{1-\mu^2}\,[e_x + \mu\,e_y - e_x'' - \mu\,e_y''],$$

$$f_y = \frac{E}{1-\mu^2}\,[e_y + \mu\,e_x - e_y'' - \mu\,e_x''],$$

$$f_{xy} = 2\,G(e_{xy} - e_{xy}'')$$

$$M_x = \int f_x\,z\,dz$$

$$= -D\left(\frac{\partial^2 w}{\partial x^2} + \mu\,\frac{\partial^2 w}{\partial y^2}\right) - \frac{E}{1-\mu^2}\int (e_x'' + \mu\,e_y'')\,z\,dz, \tag{1}$$

$$M_y = \int f_x\,z\,dz$$

$$= -D\left(\frac{\partial^2 w}{\partial y^2} + \mu\,\frac{\partial^2 w}{\partial x^2}\right) - \frac{E}{1-\mu^2}\int (e_y'' + \mu\,e_x'')\,z\,dz, \tag{2}$$

$$M_{xy} = \int f_{xy}\,z\,dz = D(1-\mu)\,\frac{\partial^2 w}{\partial x\,\partial y} + \frac{E}{1+\mu}\int e_{xy}''\,z\,dz, \tag{3}$$

$$\frac{\partial^2 M_x}{\partial x^2} - 2\,\frac{\partial^2 M_{xy}}{\partial x\,\partial y} + \frac{\partial^2 M_y}{\partial y^2} + q = 0,$$

$$- D\,\Delta^4\,w = q - \frac{E}{1-\mu^2}\,\frac{\partial^2}{\partial x^2}\int (e_x'' + \mu\,e_y'')\,z\,dz,$$

$$- \frac{E}{1-\mu^2}\,\frac{\partial^2}{\partial y^2}\int (e_y'' + \mu\,e_x'')\,z\,dz - \frac{2\,E}{1+\mu}\,\frac{\partial^2}{\partial x\,\partial y}\int e_{xy}''\,z\,dz. \tag{4}$$

It is seen from Eqs. (2) and (3) that the second terms on the right hand give the equivalent sectional bending moments M_x', M_y' and M_{xy}'. In Eq. (4) the 2nd, 3rd and 4th terms give the equivalent lateral loads q'. The deflection due to creep strains e_x'', e_y'' and e_{xy}'' can be calculated by applying the equivalent load q' on the plate and the equivalent moments M_x', M_y' and M_{xy}' on the edge.

The amounts of creep strain e_x'', e_y'' and e_{xy}'' depend on the history of stresses and temperature and the stresses depend on the current amounts of strain. The problem is simplified by considering incremental creep strains in a small time interval. Consider a particular point in the plate and let the stresses at this point be f_x, f_y, and f_{xy}. The stresses vary with time and are generally represented by smooth curves. Next the smooth curves are replaced by many incremental steps. Each step consists of a constant stress period Δt followed by an instantaneous increment of stress Δf as shown in Fig. 1. The incremental creep strains $\Delta e_x''$, $\Delta e_y''$ and $\Delta e_{xy}''$ in the period Δt are obtained from the polyaxial creep characteristics of the material at the particular temperature with the particular history of loading. These incremental strains can be obtained from tests following the same history of loading. The error introduced through keeping the stresses constant in Δt decreases to zero as the time interval Δt approaches zero.

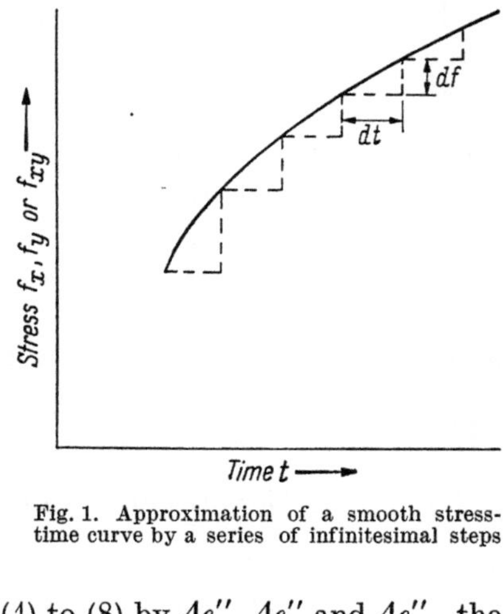

Fig. 1. Approximation of a smooth stress-time curve by a series of infinitesimal steps

Replacing e_x'', e_y'' and e_{xy}'' in Eqs. (4) to (8) by $\Delta e_x''$, $\Delta e_y''$ and $\Delta e_{xy}''$, the equivalent incremental sectional edge moments $\Delta M'$ and the equivalent incremental lateral load Δq for the time interval Δt are obtained. Then the creep deflection and stresses in Δt can be calculated accordingly.

This gives the equivalent loading and the equivalent boundary moments for plates of arbitrary shape. The method is readily applicable to the study of rectangular plates.

3. Circular Plate under Symmetrical Loading

For the cases of circular plates under symmetrical loading, the radial and tangential directions are those of the principal stresses and strains. The two principal curvatures [10] are $-\frac{d^2w}{dr^2}$ and $-\frac{1}{r}\frac{dw}{dr}$. The sectional moments M_r and M_t following the same derivation as given in Eqs. (3) to (4) are

$$M_r = -D\left(\frac{d^2w}{dr^2} + \frac{\mu}{r}\frac{dw}{dr}\right) - \frac{E}{1-\mu^2}\int (e_r'' + \mu\, e_t'')\, z\, dz, \tag{5}$$

$$M_t = -D\left(\frac{1}{r}\frac{dw}{dr} + \mu\frac{d^2w}{dr^2}\right) - \frac{E}{1-\mu^2}\int (e_t'' + \mu\, e_r'')\, z\, dz, \tag{6}$$

$$M_{rt} = 0.$$

Let

$$\varphi = -\frac{dw}{dr},$$

$$M_r = D\left(\frac{d\varphi}{dr} + \mu\frac{\varphi}{r}\right) - \frac{E}{1-\mu^2}\int (e_r'' + \mu\, e_t'')\, z\, dz, \tag{7}$$

$$M_t = D\left(\frac{\varphi}{r} + \mu\frac{d\varphi}{dr}\right) - \frac{E}{1-\mu^2}\int (e_t'' + \mu\, e_r'')\, z\, dz. \tag{8}$$

The condition of equilibrium of an element $dr\, d\theta$ of the circular plate is

$$\left(M_r + \frac{dM_r}{dr}dr\right)(r + dr)\, d\theta - M_r\, r\, d\theta - M_t\, dr\, d\theta + Q\, r\, d\theta\, dr = 0,$$

$$M_r + \frac{dM_r}{dr} r - M_t + Q\, r = 0. \tag{9}$$

Substituting Eqs. (7) and (8) into Eq. (9) we obtain

$$D\left[\frac{d^2\varphi}{dr^2} + \frac{1}{r}\frac{d\varphi}{dr} - \frac{\varphi}{r^2}\right] - \frac{E}{1-\mu^2}\left[\frac{d}{dr}\int (e_r'' + \mu\, e_t'')\, z\, dz + \frac{1}{r}(1-\mu)\int (e_r'' - e_t'')\, z\, dz\right] + Q = 0. \tag{10}$$

Multiplying Eq. (10) by $2\pi r$ we get

$$2\pi r D\left[\frac{d^2\varphi}{dr^2} + \frac{1}{r}\frac{d\varphi}{dr} - \frac{\varphi}{r^2}\right] - \frac{2\pi E}{1-\mu^2}\left[r\frac{d}{dr}\int (e_r'' + \mu\, e_t'')\, z\, dz + (1-\mu)\int (e_r'' - e_t'')\, z\, dz\right] + 2\pi Q\, r = 0. \tag{11}$$

Differentiation with respect to r yields

$$2\pi D\frac{d}{dr}\left\{r\left[\frac{d^2\varphi}{dr^2} + \frac{1}{r}\frac{d\varphi}{dr} - \frac{\varphi}{r^2}\right]\right\} dr$$
$$-\frac{2\pi E}{1-\mu^2}\frac{d}{dr}\left[r\frac{d}{dr}\int (e_r'' + \mu\, e_t'')\, z\, dz + (1-\mu)\int (e_r'' - e_t'')\, z\, dz\right] dr$$
$$+\frac{d}{dr}(2\pi Q\, r)\, dr = 0, \tag{12}$$

$$\frac{d}{dr}(2\pi Q\, r)\, dr = 2\pi q\, r\, dr. \tag{13}$$

It is seen from Eqs. (12) and (13) that

$$-\frac{2\pi E}{1-\mu^2}\frac{d}{dr}\left[r\frac{d}{dr}\int(e_r''+\mu e_t'')\,z\,dz+(1-\mu)\int(e_r''-e_t'')\,z\,dz\right]dr$$

is equivalent to $2\pi q r\,dr$ and is denoted by $2\pi q' r\,dr$ or P' in dr. Hence the increment

$$-\frac{2\pi E}{1-\mu^2}\Delta\left[r\frac{d}{dr}\int(e_r''-\mu e_t'')\,z\,dz+(1-\mu)\int(e_r''-e_t'')\,z\,dz\right] \tag{14}$$

gives the equivalent lateral loading P' in the interval Δr.

At the boundary of the plate, the equivalent moment $\Delta M_r'$ of $-\frac{E}{1-\mu^2}\int(\Delta e_r''+\mu\,\Delta e_t'')\,z\,dz$ is added. With a built-in edge, this additional creep equivalent moment is taken by the edge. With a simply supported edge M_r must be zero all the time. The edge moment $\Delta M_r'$ lost due to creep must be added to keep the edge free of radial moments. The deflections due to the creep strains $\Delta e_r''$ and $\Delta e_t''$ can be calculated by computing the deflections caused by concentric loads P' acting on various concentric circular arcs and those due to the equivalent moment M_r' at the edge.

4. Polyaxial Stress-Strain-Time Relationship

Tests have shown that there is no appreciable change in volume in creep deformation; hence

$$e_{11}''+e_{22}''+e_{33}''=0, \tag{15}$$

where the subscripts 1, 2, and 3 refer to a set of rectangular axes x_1, x_2 and x_3. The creep strain components are the same as the deviatoric strain components. The creep strain rates are assumed to be proportional to the corresponding deviatoric stress components:

$$\frac{\dot e_{11}''}{s_{11}}=\frac{\dot e_{22}''}{s_{22}}=\frac{\dot e_{33}''}{s_{33}}=\frac{\dot e_{12}''}{s_{12}}=\frac{\dot e_{23}''}{s_{23}}=\frac{\dot e_{31}''}{s_{31}}=K, \tag{16}$$

where s_{ij} are the deviatoric stress components. This is commonly assumed in incremental theories of plasticity [*11*]. JOHNSON's tests [*7*] have shown that for a number of engineering materials, the rate of increase of the second creep strain invariant I_2'' can be expressed with good accuracy as a function of the deviatoric second stress invariant $\bar J_2$ and of the strain invariant I_2'' itself. From uniaxial tests along the x_1-axis, the data are generally expressed as $\dot e_{11}''=F(\sigma_{11}, e_{11}'')$. In uniaxial tests $\bar J_2=\frac{2\sigma_{11}^2}{3}$ and $I_2''=\frac{3}{2}e_{11}''^2$. One may set

$$\sqrt{\frac{2}{3}}\frac{d\sqrt{I_2''}}{dt}=F\left(\sqrt{\frac{3}{2}J_2},\ \sqrt{\frac{2}{3}}\sqrt{I_2''}\right). \tag{17}$$

Under polyaxial loading

$$\bar{J}_2 = \frac{1}{3}\,[(\tau_{11}-\tau_{22})^2 + (\tau_{22}-\tau_{33})^2 + (\tau_{33}-\tau_{11})^2] \\ + 2[\tau_{12}^2 + \tau_{23}^2 + \tau_{31}^2], \tag{18}$$

$$I_2'' = -2[e_{11}'' e_{22}'' + e_{22}'' e_{33}'' + e_{33}'' e_{11}'' - e_{12}''^2 - e_{23}''^2 - e_{31}''^2]. \tag{19}$$

Under plane stress

$$\bar{J}_2 = \frac{1}{3}\,[(\tau_{11}-\tau_{22})^2 + \tau_{22}^2 + \tau_{11}^2] + 2\tau_{12}^2, \tag{20}$$

$$I_2'' = 2(e_{11}''^2 + e_{22}''^2 + e_{11}'' e_{22}'' + e_{12}''^2). \tag{21}$$

From Eq. (17) one obtains, for Δt

$$\Delta \sqrt{I_2''} = \sqrt{\frac{3}{2}}\, F\left(\sqrt{\frac{3}{2}}\,\bar{J}_2, \sqrt{\frac{2}{3}}\sqrt{I_2''}\right) \Delta t. \tag{22}$$

For the case of plane stress

$$\Delta\sqrt{\frac{I_2''}{2}} = [(e_{11}''+\Delta e_{11}'')^2 + (e_{22}''+\Delta e_{22}'')^2 + (e_{11}''+\Delta e_{11}'')\,(e_{22}''+\Delta e_{22}'') \\ + (e_{12}''+e_{12}'')^2]^{1/2} - [e_{11}''^2 + e_{22}''^2 + e_{11}'' e_{22}'' + e_{12}'']^{1/2}. \tag{23}$$

Since $\Delta e_{ij}'' = K\, s_{ij}$ where K is a constant

$$\Delta\sqrt{\frac{I_2''}{2}} = [(e_{11}''+K\,s_{11})^2 + (e_{22}''+K\,s_{22})^2 + (e_{11}''+K\,s_{11})\,(e_{22}''+K\,s_{22}) \\ + (e_{12}''+K\,s_{12})^2]^{1/2} - \sqrt{\frac{I_2''}{2}}\,. \tag{24}$$

Hence I_2'', s_{ij} and e_{ij}'' are known, the constant K is determined from Eq. (24) and $\Delta e_{ij}''$ can be computed. This gives $\Delta e_r''$ and $\Delta e_t''$. Inserting these in place of e_r'' and e_t'' in Eq. (14), the equivalent loads P' are obtained.

For a concentric loading *[10]* P' at $r = b$, for the portion of the plate characterized by $r > b$ we have

$$w = \frac{P}{8\pi D}\left[(a^2 - r^2)\left(1 + \frac{1-\mu}{1+\mu}\cdot\frac{a^2-b^2}{2a^2}\right) + (b^2+r^2)\log\frac{r}{a}\right]. \tag{25}$$

When $r < b$, the equation becomes

$$w = \frac{Pa^2}{8\pi D}\left[\left(\frac{b^2}{a^2}+\frac{r^2}{a^2}\right)\log\frac{b}{a} + \left(1-\frac{b^2}{a^2}\right)\left(\frac{3+\mu-(1-\mu)\frac{r^2}{a^2}}{2(1+\mu)}\right)\right]. \tag{26}$$

The moments M_r and M_t due to P are obtained by differentiating w:

$$M_r = -D\left(\frac{\partial^2 w}{\partial r^2} + \frac{\mu}{r}\frac{dw}{dr}\right), \tag{27}$$

$$M_t = -D\left(\frac{1}{r}\frac{dw}{dr} + \mu\frac{d^2w}{dr^2}\right). \tag{28}$$

The stresses due to P are obtained from these moments as

$$f_r = \frac{6\,M_r}{h^2}\,\zeta, \tag{29}$$

$$f_t = \frac{6\,M_t}{h^2}\,\zeta, \tag{30}$$

where $\zeta = \frac{z}{h/2}$.

The equivalent M_r' applied at the edge gives

$$w = -\frac{M_r'(a^2 - r^2)}{2\,D(1+\mu)} \tag{31}$$

and a constant f_r and f_t at different radii of the plate:

$$f_r = f_t = \frac{6\,M_r'}{h^2}\,\zeta. \tag{32}$$

From $\Delta e_r''$ and $\Delta e_t''$, the loss of stresses $\Delta f_r''$ and $\Delta f''_t$ is computed as

$$\Delta f_r'' = \frac{E}{1-\mu^2}\,(\Delta e_r'' + \mu\,\Delta e_t''), \tag{33}$$

$$\Delta f_t'' = \frac{E}{1-\mu^2}\,(\Delta e_t'' + \mu\,\Delta e_r''). \tag{34}$$

The stresses $\Delta f_r'$ and $\Delta f_t'$ due to the equivalent loads P' and the equivalent edge moment M_r' are obtained from Eqs. (25) to (32). At the end of the time interval, the radial stress will be $f_r - \Delta f_r'' + \Delta f_r'$ and the tangential stress, $f_t - \Delta f_t'' + \Delta f_t'$. These values will be the initial values of the next time interval and this process is repeated to cover the period desired. This method is illustrated in an example given in the appendix.

References

[1] Popov, E. P.: Bending of Beams With Creep, Journal of Applied Physics, **20**, 251—256 (1949).
[2] Lin, T. H.: Creep Stresses and Deflections of Columns, Journal of Applied Mechanics, **23**, 214 (1956).
[3] Lin, T. H.: Creep Deflections and Stresses of Beam-Columns, Journal of Applied Mechanics, **25**, 75 (1958).
[4] Finnie, I., and W. R. Heller: Creep of Engineering Materials, McGraw-Hill Book Co., Inc. (1959) pp. 190.
[5] Malinin, N. N.: Continuous Creep of Round Symmetrically Load Plates (in Russian), Moskov, Vysshe Technicheskoe Uchilische Truda 26: 221, 1953.
[6] Kachanov, L. M.: Certain Problems in the Theory of Creep (in Russian), Gos. Izdat. Tekh-Teor. Lit. Leningrad, 1949.
[7] Johnson, A. E., J. Henderson and V. Mather: Creep under changing complex stress systems, The Engineer, **206**, 209 (1958).
[8] Shanley, F. R.: Weight Strength Analysis of Aircraft Structures, McGraw-Hill Book Co., Inc. 1952, pp. 271, 275—282, 317.

[9] LIBOVE, C.: Creep Buckling of Columns, Journal of Aeronautical Sciences, **19**, 459 (1952).
[10] TIMOSHENKO, S., and S. WOINOWSKY-KRIEGER: Theory of Plates and Shells, McGraw-Hill Book Co., Inc. (1959) pp. 51—78.
[11] HILL, R.: Mathematical Theory of Plasticity, Oxford, England: The Clarendon Press (1950) pp. 33.

Appendix

Illustrative Example. A 75 ST aluminium alloy circular plate with simply supported edge at 600°F is subjected to a uniformly distributed load of 36 psi. The diameter of the plate is 10″ and the thickness is 0.5″. YOUNG's modulus of elasticity E at 600°F is 5.2×10^6 psi. POISSON's ratio is 0.3 for elastic strains and 0.5 for creep strains. The stress-strain-time relationship for constant stress is represented by [8]

$$e = \frac{\sigma}{E} + A\, e^{B\sigma}\, t^K,$$

$$e'' = A\, e^{B\sigma}\, t^K, \tag{A-1}$$

where e is the uniaxial strain and e'' the uniaxial creep strain, σ is the uniaxial stress and the values of the constants are $A = 2.64\times 10^{-7}$, $B = 1.92\times 10^{-3}$, $t =$ in hr., $K = 0.66$.

Expressing Eq. (A-1) in terms of stress and strain invariants

$$\sqrt{\frac{2}{3}}\sqrt{I_2''} = A\, e^{B\sqrt{\frac{3 J_2}{2}}}\, t^K. \tag{A-2}$$

The elastic lateral deflection under uniform load is given [10] by

$$w = \frac{q(a^2 - r^2)}{64\,D}\left(\frac{5 + \mu}{1+\mu}\, a^2 - r^2\right),$$

where

$$D = \frac{E\,h^3}{12(1-\mu^2)}, \quad a = 5'', \quad \mu = 0.3, \quad q = 36 \text{ psi},$$

$$\frac{w}{h} = 0{,}0118\left[1 - \left(\frac{r}{a}\right)^2\right]\left[4.07 - \left(\frac{r}{a}\right)^2\right]. \tag{A-3}$$

This is plotted in Fig. 2 as the zero time deflection curve. The moments [10] are

$$M_r = \frac{q}{16}(3+\mu)(a^2 - r^2),$$

$$M_t = \frac{q}{16}\left[a^2(3+\mu) - r^2(1+3\mu)\right],$$

$$f_r = \frac{6\,M_r}{h^2}\,\frac{z}{h/2} = \frac{3\,q}{8\,h^2}\,3.3(a^2 - r^2)\,\frac{z}{h/2}. \tag{A-4}$$

Similarly

$$f_t = \frac{3\,q}{8\,h^2}\,[3.3\,a^2 - 1.9\,r^2]\,\frac{z}{h/2}\,. \tag{A-5}$$

The first incremental time interval Δt is taken as 1/20 hr. For symmetrical loading

$$3\,\bar{J}_2 = [f_r^2 + f_t^2 + (f_r - f_t)^2]\,. \tag{A-6}$$

From $\bar{J}_2$ and t, $\sqrt{I_2''}$ at the end of 1/20 hr. is obtained from Eq. (A-2). The deviatoric stress components S_r and S_t are calculated from f_r and f_t. From Eq. (24), the constant K is determined and then e_r'' and e_t'' are calculated from Eq. (16). The relief of stresses due to creep $\Delta f_r''$ and $\Delta f_t''$ in this time interval is obtained from Eqs. (33) and (34). In the first time interval one has

$$\sqrt{\frac{3}{2}\,\bar{J}_2} \propto z\,,$$

$$\frac{(e_r'')_z}{(e_r'')_{\frac{h}{2}}} = \frac{\left(e^{B\sqrt{\frac{3\bar{J}_2}{2}}}\right)_z}{\left(e^{B\sqrt{\frac{3\bar{J}_2}{2}}}\right)_{\frac{h}{2}}}\,,$$

$$\int e_r''\,z\,dz = 2\int_0^{\frac{h}{2}} e_r''\,z\,dz$$

$$= 2\,(e_r'')_{\frac{h}{2}}\left(\frac{h}{2}\right)^2\left[\frac{1}{\left(B\sqrt{\frac{3\,\bar{J}_2}{2}}\right)_{\frac{h}{2}}} - \frac{1}{\left(B\sqrt{\frac{3\,\bar{J}_2}{2}}\right)^2_{\frac{h}{2}}}\left(1 - \frac{1}{e^{B\sqrt{\frac{3\bar{J}_2}{2}}}}\right)_{\frac{h}{2}}\right]. \tag{A-7}$$

For subsequent time intervals, the integrals are evaluated by assuming that the incremental creep strain varies as ζ^n. The value "n" is determined by the incremental creep strains at $\zeta = \frac{1}{2}$ and 1. The equivalent concentric loads due to creep are obtained from Eq. (14). With an equivalent load P' applied at $r = b$ [10], one gets for the region $r > b$

$$w = \frac{P'}{8\,\pi\,D}\left[(a^2 - r^2)\left(1 + \frac{.7}{2.6}\,\frac{a^2 - b^2}{a^2}\right) + (b^2 + r^2)\log\frac{r}{a}\right]$$

$$M_r = -D\left(\frac{d^2w}{dr^2} + \frac{1}{r}\,\frac{dw}{dr}\right)$$

$$= -\frac{P'}{8\,\pi}\left[.7 - \frac{.91}{1.3}\left(1 - \frac{b^2}{a^2}\right) + 2.6\log\frac{r}{a} - .7\,\frac{b^2}{r^2}\right], \tag{A-8}$$

$$M_t = -D\left(\frac{1}{r}\,\frac{dw}{dr} + \mu\,\frac{d^2w}{dr^2}\right)$$

$$= -\frac{P'}{8\pi}\left[-.7 - \frac{.91}{1.3}\left(1 - \frac{b^2}{a^2}\right) + .7\frac{b^2}{r^2} + 2.6 \log\frac{r}{a}\right], \qquad \text{(A-9)}$$

$$f_r = -\frac{3P'}{4\pi h^2}\left[-.7\frac{b^2}{a^2} + 2.6\log\frac{r}{a} - .7\frac{b^2}{r^2}\right], \qquad \text{(A-10)}$$

$$f_t = -\frac{3P'}{4\pi h^2}\left[-1.4 + .7\left(\frac{b^2}{a^2} + \frac{b^2}{r^2}\right) + 2.6\log\frac{r}{a}\right]. \qquad \text{(A-11)}$$

For $r < b$

$$w = \frac{P'}{8\pi D}\left[(b^2 + r^2)\log\frac{b}{a} + \frac{\left(3.3 - 0.7\frac{r^2}{a^2}\right)(a^2 - b^2)}{2.6}\right], \qquad \text{(A-12)}$$

$$M_r = M_t = \frac{P'}{8\pi}\left[2.6\log\frac{b}{a} - 0{,}7\left(\frac{a^2 - b^2}{a^2}\right)\right], \qquad \text{(A-13)}$$

$$f_r = f_t = \frac{-3P'}{4\pi h^2}\left[2.6\log\frac{b}{a} - 0{,}7\left(\frac{a^2 - b^2}{a^2}\right)\right]. \qquad \text{(A-14)}$$

The stresses due to these equivalent loads acting at different radii are calculated.

At the edge, the relief in bending moment $\Delta M_r'$ at $r = a$ is equal to $\frac{E}{1-\mu^2}\int (e_r'' + \mu e_t'')\, z\, dz$. In order to keep the simply supported edge free of M_r, this equivalent $\Delta M_r'$ is applied. This gives $\Delta M_t' = \Delta M_r'$ throughout, so

$$f_r = f_t = \frac{6\,\Delta M_r'}{h^2}\,\zeta. \qquad \text{(A-15)}$$

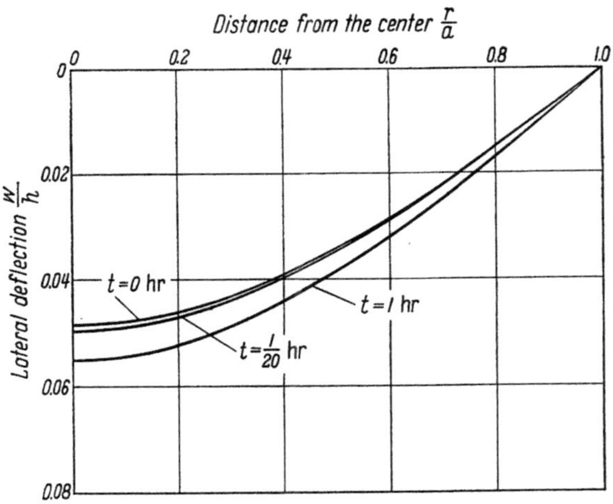

Fig. 2. Deflected surfaces at different time instants

The stresses at the end of the first time interval are equal to the initial stresses f_r and f_t minus the stresses relieved because of creep, $\Delta f_r''$ and

$\Delta f_t''$, plus the stresses caused by the equivalent concentric loads and edge moment $\Delta f_r'$ and $\Delta f_t'$.

Two subsequent time intervals 1/20 to 1 hr are considered. The stresses and deflections at the ends of these intervals and details of the computations are presented in Table 1. The results are plotted in Fig. 2.

It is seen that the rate of deflection decreases with time. This is partly due to the fact that the secondary creep of the material is slower than the primary creep and partly that the highly stressed regions have their stresses reduced because of creep.

Discussion

E. H. Lee, Brown University: The consideration of creep strain components to represent equivalent body forces in an elastic problem appears to provide an ingenious method of solution. It seems to me that this can only provide a satisfactory method of treatment for a limited time, since as the creep strains increase, the two parts on the right-hand sides of Eqs. (1), (2), (3) and (4) will be nearly equal in magnitude and opposite in sign, so that the computed quantities will be large, and the desired solution will be obtained from the difference between almost equal quantities. Did this difficulty arise in the problem considered?

Author's closing words: The author wishes to thank Professor Lee for his kind comment on this paper.

The equivalence of creep strain gradient to body force follows the general treatment of plastic strain gradient as an equivalent body force for a three dimensional medium derived by the author in his calculation of stress field caused by a uniform slip in an embedded crystal in an aggregate[1]. Considering the total strain as composed of the elastic strain and the creep strain, the stress components τ_{ij} are related to the elastic strain components by Hooke's law as

$$\tau_{ij} = \lambda\, \delta_{ij}\, \theta' + 2\, \mu\, e'_{ij}$$

where the subscripts $i\,j$ refer to a set of rectangular axes x_1, x_2 and x_3, θ' is the elastic dilatation, δ_{ij} the Kronecker delta, e'_{ij} the elastic strain component, and λ and μ are Lame's constants. The elastic strain is the difference between the total strain and the creep strain $\theta' = \theta - \theta''$, $e'_{ij} = e_{ij} - e''_{ij}$ where the double prime denotes creep strain. The equilibrium conditions are

$$\tau_{ij,j} + F_i = 0$$

where the repetition of the subscript denotes summation from one to three and the subscript after comma denotes differentiation. From these equations

$$\lambda\, \delta_{ij}(\theta_{,j} - \theta_{,j}'') + 2\, \mu (e_{ij,j} - e''_{ij,j}) + F_i = 0.$$

[1] Lin, T. H., and S. Uchiyama, Stress Field of Uniformly Plastically Deformed Crystal in an Aggregate, US Air Force Office of Scientific Research Rech. Note 59-1048, Oct. 1959, and Lin T. H., S. Uchiyama and D. Martin Stress Field in Metab at the Initial stage of Plastic Deformation, Journal of the Mechanics and Physics of Solids 1961 Vol 9 pp 200—209.

Table 1. *Creep Stresses and Deflections of a Circular Plate under Uniform Loading*

No.		(1)		(2)		(3)		(4)		(5)		(6)	
1	$\frac{r}{a}$	0		.2		.4		.6		.8		1.0	
2	$\zeta = \frac{z}{h/2}$	1/2	1	1/2	1	1/2	1	1/2	1	1/2	1	1/2	1
				$t = 0$ to 1/20 hour									
3	w/h	.0481		.0457		.0388		.0280		0.146		0	
4	f_r	2230	4460	2140	4280	1915	3830	1425	2850	800	1600	0	0
5	f_t	2230	4460	2180	4360	2020	4040	1765	3530	1410	2820	945	1890
6	S_r			700	1400	603	1207	362	724	64	128	−315	−630
7	S_t			740	1480	708	1417	702	1404	674	1328	630	1260
8	$\sqrt{3 J_2/2}$	2230	4460	2160	4320	1970	3940	1620	3240	1225	2450	945	1860
9	$e^{B\sqrt{3 J_2/2}}$	72	5200	63	3980	44	1920	31	500	105	109	6.1	37.6
10	$\sqrt{I_2'/2}$			19.6	1260	14	607	9.8	158	3.4	34.5	1.9	12
11	$[S_r^2 + S_t^2 + S_r S_t]^{1/2}$				2495		2276		1874		1452		1092
12	$\Delta e_r'' \times 10^{-7}$	13.2	950	11.0	710	7.4	323	3.8	61	.3	3.0	−1.1	−6.8
13	$\Delta e_t'' \times 10^{-7}$	13.2	950	11.7	747	8.7	379	7.4	118	3.2	32	2.2	13.7
14	$\Delta f_r''$	9.8	706	8.3	533	5.7	191	3.4	55	.7	7.3	−2.5	−1.6
15	$\Delta f_t''$	9.8	706	8.5	548	6.2	223	4.9	78	1.8	19	1.1	6.7
16	$\int \Delta e_r'' \, z \, dz / \left(\frac{h}{2}\right)^2$	196×10^{-7}		151×10^{-7}		74.1×10^{-7}		16.5×10^{-7}		$.50 \times 10^{-7}$		-2.78×10^{-7}	
17	$\int \Delta e_t'' \, z \, dz / \left(\frac{h}{2}\right)^2$	196×10^{-7}		158×10^{-7}		87.4×10^{-7}		32×10^{-7}		107×10^{-7}		5.59×10^{-7}	
18	$\int (\Delta e_r'' + \mu \Delta e_t'') \, z \, dz / \left(\frac{h}{2}\right)^2$	256×10^{-7}		200×10^{-7}		100×10^{-7}		26.7×10^{-7}		3.7×10^{-7}		$-.28 \times 10^{-7}$	
19	$\frac{\partial}{\partial r}$ (18)		-280×10^{-7}		-500×10^{-7}		-367×10^{-7}		-115×10^{-7}		-24×10^{-7}		
		0		-390×10^{-7}		-434×10^{-7}		-241×10^{-7}		-69.5×10^{-7}		-24×10^{-7}	
20	r (19)	0		-78×10^{-7}		-173×10^{-7}		-145×10^{-7}		-55.6×10^{-7}		-24×10^{-7}	
21	$(1-\mu) \int (\Delta e_r'' - \Delta e_t'') \, z \, dz / \left(\frac{h}{2}\right)^2$	0		-5.3×10^{-7}		-8.8×10^{-7}		-10.8×10^{-7}		-7.1×10^{-7}		-2.0×10^{-7}	
22	(20) + (21)	0		-83×10^{-7}		-182×10^{-7}		-156×10^{-7}		-62×10^{-7}		-26.0×10^{-7}	
23	$\frac{2 \pi r Q'}{(h/2)^2} \times 10^7$	0		300		652		558		216		78	
24	P'		18.9		22.1		−5.8		−21.4		−8.6		

26	$\Delta f'_t$	78	157	67	134	48	95	30	60	19	39	16	0
27	$f_r + \Delta f'_r - \Delta f''_r$	2298	3911	2190	3892	1938	3697	1431	2811	800	1594	0	0
28	$f_t + \Delta f'_t - \Delta f''_t$	2298	3911	2239	3946	2062	3912	1790	3512	1427	2840	960	1912
29	$\Delta w/h$	.00102		.00094		.00073		.00049		.00023		0	
30	w/h	.04912		.04664		.03953		.02849		.01483		0	
31	$\sqrt{3\bar{J}_2}$	2298	3911	2214	3910	2002	3810	1640	3220	1240	2470	960	1914
32	t_0 in hrs.	.041	.245	.0416	.164	.035	.074	.077	.0525	.0500	.0704	.0465	.0465
33	$t = t_0 + \Delta t$	.991	1.195	.992	1.114	.985	1.024	1.027	1.002	1.00	1.030	.9965	.9965
34	$t^{0.66}$	.994	1.125	.995	1.074	.990	1.016	1.018	1.002	1.00	1.013	.9976	.9976
35	$\sqrt{I''_2 + \Delta I''_2} \times 10^7$			183	5350	121	4000	63.1	1276	28.5	305	16.6	104
36	S_r			714	1258	604	1161	359	704	58	116		
37	S_t			763	1343	728	1376	720	1404	685	1362		
38	$\Delta e''_r \times 10^7$	111	2180	91	2440	57	1790	229	605	2.0	22	−8.5	−52
39	$\Delta e''_t \times 10^7$	111	2180	98	2606	66	2120	445	1208	23	259	17.0	104
40	$\Delta f''_r$	82	1620	69	1842	45	1390	207	553	5.1	56.6	−1.9	−12
41	$\Delta f''_t$	82	1620	72	1910	47	1616	294	794	13.5	152	8.3	51
42	$\int \Delta e''_r\, z\, dz / \left(\frac{h}{2}\right)^2$	692×10^{-7}		722×10^{-7}		514×10^{-7}		180×10^{-7}		8.0×10^{-7}		-23×10^{-7}	
43	$\int \Delta e''_t\, z\, dz / \left(\frac{h}{2}\right)^2$	692×10^{-7}		771×10^{-7}		605×10^{-7}		357×10^{-7}		95×10^{-7}		45×10^{-7}	
44	$[(42) + \mu\,(43)] \times 10^6$	900		953		696		287		36.5		−10	
45	$\frac{\partial}{\partial r}\,(44) \times 10^7$		265		−1285		−2045		−1250		−233		
				−510		−1665		−1648		−742		−233	
46	$r\,(45) \times 10^7$			−102		−666		−990		−593		−233	
47	$(1-\mu)\,[(42) - (43)] \times 10^7$			−34.3		−63.7		−124		−61		−47.6	
48	$[(46) + (47)] \times 10^7$			−136		−729.7		−1114		−654		−281	
49	$\frac{Q'\,r\,2\pi}{(h/2)^2} \times 10^7$			489		2620		4000		2340		1010	
50	$\Delta\,\frac{Q'\,r\,2\pi}{(h/2)} \times 10^7$		489		2131		1380		−1660		−1330		
51	P'		30.6		133		86.4		−104		−83.2		
52	$\Delta f'_r$	348	696	315	629	220	440	88	175	11.0	22.0	−1.9	−12
53	$\Delta f'_t$	348	696	330	660	277	554	199	397	133	265	94	187
54	$f_r = (27) - (40) + (50)$	2564	2987	2436	2656	2113	2747	1311	2434	806	1559	0	0
55	$f_t = (28) - (41) + (53)$	2564	2987	2497	7696	2292	2850	1695	3115	1546	2953	1046	2048
56	$\Delta w/h$	.006		.0056		.0046		.0031		.0015		0	
57	w/h	.0551		.0522		.0441		.0316		.0163		0	

It is seen that $-\lambda\, \delta_{ij}\, \theta''_{,i j} - 2\, \mu\, e''_{ij,\,j}$ is equivalent to the body force F_i in producing the stress field e_{ij}. The creep strain is assumed to cause no changes in volume, $\theta'' = 0$ throughout. Hence the equivalent body force equals $-2\, \mu\, e''_{ij,\,j}$. This derivation is quite general.

In applying this general equation to obtain the moment and curvature relationship given in Eq. (1) of this paper

$$M_x = -D\left(\frac{\partial^2 w}{\partial x^2} + \nu\,\frac{\partial^2 w}{\partial y^2}\right) - \frac{E}{1-\mu^2}\int (e''_x + \nu\, e''_y)\, z\, dz$$

the usual assumption that the normals of the middle plane before bending are deformed into the normals of the middle plane after bending is implied. Small deflection theory is used and the plane stresses caused by stretching of the plates due to lateral deflection are neglected. Within these limitations, the Eqs. (1) to (4) are valid.

Considering an element of plate under an applied moment M_x, the value of the integral $\frac{E}{1-\mu^2}\int (e''_x + \nu\, e''_y)\, zdz$ for an interval of time generally is of the same sign as M_x. So the change of curvature of the plate $-D\left(\frac{\partial^2 w}{\partial x^2} + \nu\,\frac{\partial^2 w}{\partial y^2}\right)$ corresponds to an applied moment of $M_x + \frac{E}{1-\mu^2}\int (e''_x + \nu\, e''_y)\, z\, dz$. These two terms are of the same sign, this holds also for Eqs. (2) and (3). For Eq. (4), the equivalent lateral load due to creep strain generally has the same sign as "q", so the deflection due to lateral load and that due to creep strain caused by the lateral load are of the same sign. This is shown by the increase in the lateral deflection as the creep strain increases. In the calculation of the numerical example, the deflections due to creep in successive intervals were calculated. No computational difficulty was encountered.

The stresses change with time. This change of stresses gives different creep strain rates. In order to calculate creep deflections and stresses at different instants more closely, shorter time increments are required.

Effect of Creep on Stresses in Cylindrical Shells

By

H. Poritsky

General Electric Company, Schenectady, New York, USA

1. Calculation of Creep

The ordinary elastic theory of shell deflections is based on considering stretching and bending of the shell. It assumes that only the stresses σ_x, σ_θ, $\tau_{x\theta}$ where x is the axial distance, θ the azimuthal angle along the cylindrical shell, are significant, while the remaining stress components σ_z, τ_{xz}, $\tau_{\theta z}$, where z is the distance along the normal to the middle surface of the shell, are negligible. This assumption will be adhered to in the following, both in the initial, elastic state, and as creep progresses.

In the original elastic stress distribution σ_x, σ_θ, $\tau_{x\theta}$ vary linearly across each line normal to the middle surface. As creep progresses these change from the linear distribution into a non-linear distribution as shown schematically in Fig. 1. This statement applies to all three stress components σ_x, σ_θ, $\tau_{x\theta}$. We proceed to formulate equations for the calculation of these stresses, and the elastic and plastic strains.

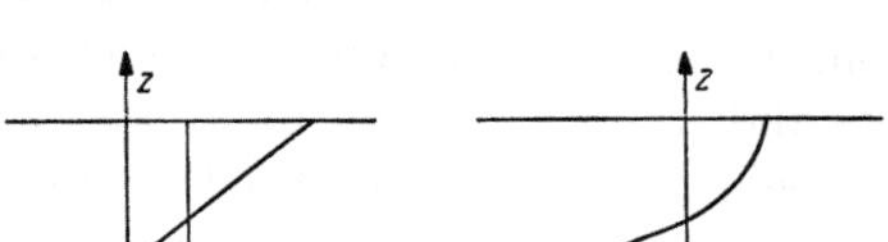

Fig. 1. Linear and non-linear stress distributions

Suppose that σ_x, σ_θ, $\tau_{x\theta}$ are known at a time t_0, both at each point (x, θ), along the middle surface, and across it, as a function of z, the distance from the middle surface along the normal, at each point. Then for each (x, θ, z) the rates of plastic creep $\dot{e}_x''$, $\dot{e}_\theta''$, $\dot{e}_{x\theta}''$ can be computed from the available creep data curves for the material. For definiteness we assume these are of the form

$$\dot{e}'' = f(e'', \sigma, T), \tag{1.1}$$

where e'' is the equivalent total plastic strain, σ the instantaneous equivalent stress, T the temperature, and dots denote time derivatives.

Eq. (1.1) is sometimes described as conforming to "strain hardening". In terms of the strain-elongation curves (for a fixed temperature), a change in σ is represented by a *horizontal* line on Fig. 2 from one (constant) σ-curve to another one, corresponding to the same plastic strain.

Another hypothesis regarding behavior of creep under changing stress, known as "time hardening", assumes that the variable rate of creep depends only upon time elapsed and upon the stress. A change of stress will now show up as indicated in Fig. 3, by a *vertical* transfer from one constant σ-curve to another one. Further hypotheses, combining both

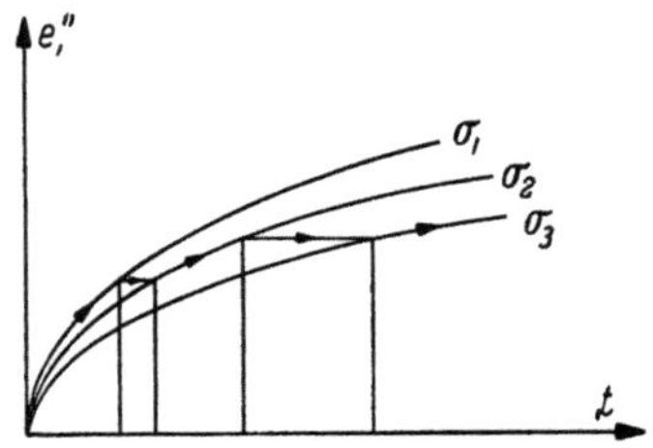

Fig. 2. Strain-hardening stress-elongation curves

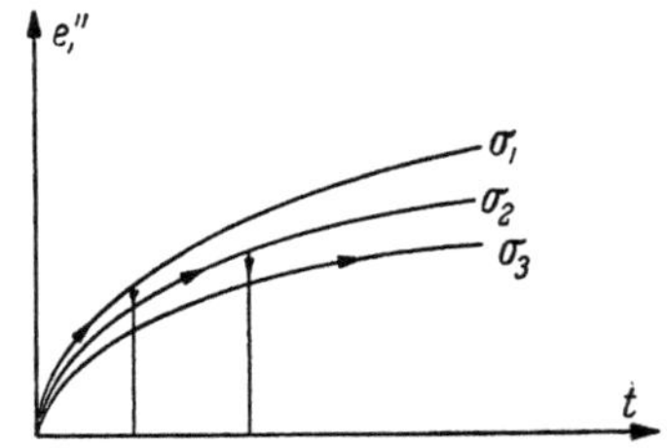

Fig. 3. Time-hardening stress-elongation curves

the strain and time hardening, are also sometimes used, as outlined in Ref. [*2*]. Dorn et al, Ref. [*3, 4*], have modified Eq. (1.1) in a manner to conform to the thermal and strain history of the material, as a result of tests carried out on specimen stressed partly at liquid air temperature, partly at room and elevated temperatures.

In the following we shall use the strain hardening hypothesis as explained in Ref. [*1*].

The curves of Fig. 2, for a virgin material (or freed from stress by a "heat stress relieving process") start with infinite slope at $t = 0$, $e'' = 0$. As the strain proceeds, the slope decreases and approaches a constant value over a fairly long period of time. Later the slope begins to increase again, and the material creeps to rupture. Over the nearly constant slope interval, the rate of creep is approximately given by

$$\dot{e}'' = K\,\sigma^n, \tag{1.2}$$

where $n \sim 5$ to 9 and K depends on the temperature.

As pointed out in [*1*], the equivalent stress and equivalent plastic strain are expressed in terms of the principal stresses and principal strains as follows:

$$\begin{aligned} \sigma &= \left[\frac{(\sigma_1-\sigma_2)^2+(\sigma_2-\sigma_3)^2+(\sigma_1-\sigma_3)^2}{2}\right]^{1/2}, \\ e'' &= \sqrt{\frac{2}{3}}\,[(e_1''-e_2'')^2+(e_2''-e_3'')^2+(e_1''-e_3'')^2]^{1/2}. \end{aligned} \tag{1.3}$$

The rates of plastic creep have principal (creep) directions coincident with the direction of principal stresses and these rates are given by

$$\begin{aligned}\dot{e}_1'' &= \frac{de_1''}{dr} = \frac{f}{\sigma}\left(\sigma_1 - \frac{\sigma_2}{2} - \frac{\sigma_3}{2}\right),\\ \dot{e}_2'' &= \frac{de_2''}{dt} = \frac{f}{\sigma}\left(\sigma_2 - \frac{\sigma_1}{2} - \frac{\sigma_3}{2}\right),\\ \dot{e}_3'' &= \frac{de_3''}{dt} = \frac{f}{\sigma}\left(\sigma_3 - \frac{\sigma_1}{2} - \frac{\sigma_2}{2}\right),\end{aligned} \tag{1.4}$$

where f is the function occurring in (1.1).

In the present case, the z-direction is one of the directions of principal stress, which stress, indeed, vanishes. As regards the principal stress directions normal to z, they correspond, in general, neither to the direction of increasing x nor θ. In the Mohr circle construction in a (σ, τ)-plane, indicated in Fig. 4, one plots the point A on Fig. 4 whose coordinates are (σ_x, τ_{xy}) and the point B with coordinates (σ_y, τ_{xy}). A circle of radius τ through these points with center at $(\sigma_x + \sigma_y)/2{,}0$, is now constructed: the end points of its diameter correspond to the two principal stresses, σ_1, σ_2. Moreover, the angle formed by the radius to the point (σ_x, τ_{xy}) with the σ-axis yields twice the angle between the direction of the principal stress σ_1 in the (x, θ)-plane and the x-axis. In equation form

Fig. 4. Mohr circle for stresses

$$\begin{aligned}\sigma_1 &= \frac{\sigma_x + \sigma_y}{2} + \sqrt{\frac{(\sigma_x - \sigma_y)^2}{4} + \tau_{xy}^2} = \sigma_m + \tau,\\ \sigma_2 &= \frac{\sigma_x + \sigma_y}{2} - \sqrt{\frac{(\sigma_x - \sigma_y)^2}{4} + \tau_{xy}^2} = \sigma_m - \tau.\end{aligned} \tag{1.5}$$

Application of Eqs. (1.3), (1.4) with vanishing σ_3 yields

$$\sigma = \sqrt{\sigma_1^2 + \sigma_2^2 - \sigma_1\sigma_2} = \sqrt{\sigma_m^2 + 3\tau^2} = \sqrt{(\sigma_x^2 + \sigma_y^2 - \sigma_x\sigma_y + 3\tau_{xy}^2)}, \tag{1.6}$$

$$\begin{aligned}\dot{e}_1'' &= \frac{f}{\sigma}\left[\sigma_1 - \frac{\sigma_2}{2}\right],\\ \dot{e}_2'' &= \frac{f}{\sigma}\left[\sigma_2 - \frac{\sigma_1}{2}\right].\end{aligned} \tag{1.7}$$

Similarly to σ, the strain rates $\dot{e}_1''$, $\dot{e}_2''$ may be expressed in terms of σ_x, σ_y, τ_{xy} by means of (1.5), (1.6). To the two Eqs. (1.7) may be added the equation for creep in the normal direction: $\dot{e}_3'' = -(\dot{e}_1'' + \dot{e}_2'')$.

Since $\sigma_z = 0$, one is led to a stress Mohr circle diagram as indicated in Fig. 5.

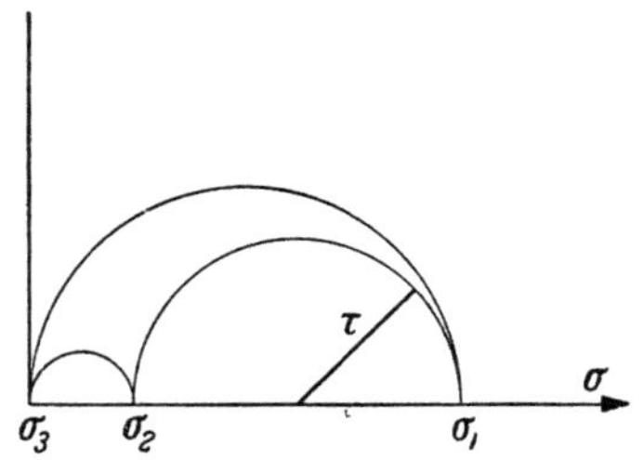

Fig. 5. Stress Mohr circle with $\sigma_3 = 0$

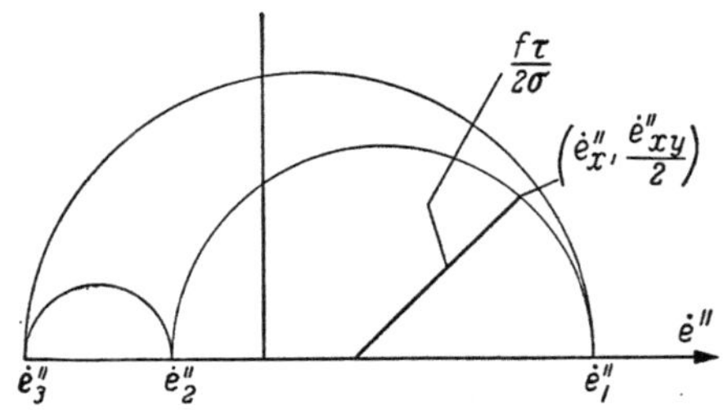

Fig. 6. Strain-rate Mohr circle

From (1.4) or (1.7) follows

$$\dot{e}_1'' - \dot{e}_2'' = \frac{3f}{2\sigma}(\sigma_1 - \sigma_2). \tag{1.8}$$

Hence the Mohr circles for the stresses and plastic strain rates have a ratio $3f/2\sigma$: there is also a shift in the horizontal direction. (See Fig. 5, and 6.) Half the sum of (1.7) yields for the center of the $(\dot{e}_1'', \dot{e}_2'')$-circle

$$\dot{e}_m'' = \frac{\dot{e}_1'' + \dot{e}_2''}{2} = \frac{f}{\sigma}\,\frac{\sigma_1 + \sigma_2}{4} = \frac{f\,\sigma_m}{2\sigma}. \tag{1.9}$$

From the similarity of the triangles in Figs. 5, 6, there results

$$\begin{aligned} \dot{e}_x'' &= \dot{e}_m'' + \frac{3f}{2\sigma}(\sigma_x - \sigma_m), \\ \dot{e}_y'' &= \dot{e}_m'' + \frac{3f}{2\sigma}(\sigma_\theta - \sigma_m), \\ \dot{e}_{xy}'' &= \frac{3}{2}\,\frac{f}{\sigma}\,\tau_{x\theta}, \end{aligned} \tag{1.10}$$

and substituting for $\sigma_m = (\sigma_x + \sigma_y)/2$,

$$\begin{aligned} \dot{e}_x'' &= \frac{f}{\sigma}\left(\sigma_x - \frac{\sigma_y}{2}\right), \\ \dot{e}_y'' &= \frac{f}{\sigma}\left(\sigma_y - \frac{\sigma_x}{2}\right), \\ \dot{e}_{xy}'' &= \frac{3f}{2\sigma}\,\tau_{xy}. \end{aligned} \tag{1.11}$$

Eqs. (1.11) reduce to (1.7) for $\tau_{xy} = 0$, while for $\tau_{xy} \neq 0$ the first two equations have the same form as (1.7) with σ_1, σ_2 replaced by σ_x, σ_y (though they differ in the meaning of σ).

This expresses the rates of plastic strain components in the terms of the stresses.

We write Eqs. (1.11) symbolically in the form

$$(\dot{e}_x'', \dot{e}_\theta'', \dot{e}_{x\theta}'') = f_x, f_\theta, f_{x\theta}, \tag{1.12}$$

where f_x, f_θ, $f_{x\theta}$ are given by the right-hand members of (1.11), with Eq. (1.1)—(1.8) supply the arguments of f.

Alternatively, the calculation may be carried out from (1.7) and converted to $\dot{e}_x''$, e_θ'', $e_{x\theta}''$.

2. Derivation of Elastic and Creep Equations

From the rates of change of the plastic strain components, their values may be found after a time interval Δt, if their values are known at the beginning of this time interval.

If each element of volume were separate, then the computed rates of change of the plastic strain (creep) components (1.8) at each point (x, θ, z) could be realized. However, due to contiguity, if creep took place in accordance with these equations, there would be either interference between adjacent elements, or gaps would open up between them: moreover, a line perpendicular to the middle plane will get distorted into a curve. Actually these tendencies are prevented from occurring (so long as no failure occurs) by *changes* in the *elastic strains* with corresponding changes in the stresses, so that the shell stays whole.

Unless conditions change rapidly with x, θ, it is proper to assume that lines originally normal to the middle surface remain normal to it, even as creep proceeds. This leads to the conclusion that the net strains e_x, e_θ, $e_{x\theta}$ (elastic *and* plastic), as functions of z, remain linear even as time progresses, as creep takes place, and the stresses σ_x, σ_θ, $\tau_{x\theta}$ change to non-linear functions of z as shown schematically in Fig. 1.

We assume further that the total or net strains can be expressed in terms of the displacements u, v, w in the axial, tangential, and inner normal directions of the middle surface, in a familiar manner as follows:

$$\begin{gathered}\bar{e}_x = \frac{\partial u}{\partial x}, \ \bar{e}_\theta = \frac{\partial v}{a\,\partial\theta} - \frac{w}{a}, \ \bar{e}_{x\theta} = \frac{\partial v}{\partial x} + \frac{\partial u}{a\,\partial\theta}, \\ \varkappa_x = \frac{\partial^2 w}{\partial x^2}, \ \varkappa_\theta = \frac{\partial^2 w}{a^2\,\partial\theta^2} + \frac{1}{a^2}\frac{\partial v}{\partial\theta}, \ \varkappa_{x\theta} = \frac{\partial^2 w}{a\,\partial x\,\partial\theta} + \frac{1}{a}\frac{\partial v}{\partial x}, \\ e_x = \bar{e}_x - \varkappa_x z, \ e_\theta = \bar{e}_\theta - \varkappa_\theta z, \ e_{x\theta} = \bar{e}_{x\theta} - 2\varkappa_{x\theta} z,\end{gathered} \tag{2.1}$$

where a is the radius of the (middle surface of the) shell: $\bar{e}_x$, $\bar{e}_\theta$, $\bar{e}_{x\theta}$ are the *mean* strains; $\varkappa_x$, $\varkappa_\theta$, $\varkappa_{x\theta}$ the changes in components of curvature; while e_x, e_y, $e_{x\theta}$ are the total or net strains.

The total strains e_x, e_y, $e_{x\theta}$ are the sum of the elastic components e_x', e_θ', $e_{x\theta}'$ and the plastic components e_x'', e_θ'', $e_{x\theta}''$:

$$e_x = e_x' + e_x'', \ e_\theta = e_\theta' + e_\theta'', \ e_{x\theta} = e_{x\theta}' + e_{x\theta}''. \tag{2.2}$$

The elastic components e_x', e_θ', $e_{x\theta}'$ are related to the stresses by means of HOOKE's law:

$$e_x' = \frac{\sigma_x - \nu\,\sigma_\theta}{E}, \quad e_\theta' = \frac{\sigma_\theta - \nu\,\sigma_x}{E}, \quad e_{x\theta}' = \frac{\tau_{x\theta}\,2(1+\nu)}{E} \tag{2.3}$$

whose solution yields

$$\sigma_x = \frac{E}{1-\nu^2}(e'_x + \nu\, e'_\theta), \quad \sigma_\theta = \frac{E}{1-\nu^2}(e'_\theta + \nu\, e'_x), \; \tau_{x\theta} = \frac{E}{2(1+\nu)} e'_{x\theta}. \quad (2.4)$$

By integrating σ_x, σ_θ, $\tau_{x\theta}$, with respect to z, one obtains the stress resultants N_x, N_θ, $M_{x\theta}$:

$$N_x = \int_{-h/2}^{h/2} \sigma_x\, dz, \; N_\theta = \int_{-h/2}^{h/2} \sigma_\theta\, dz, \; N_{x\theta} = \int_{-h/2}^{h/2} \tau_{x\theta}\, dz; \qquad (2.5)$$

by multiplying by z and integrating, the moments:

$$M_x = \int_{-h/2}^{h/2} z\,\sigma_x\, dz, \; M_\theta = \int_{-h/2}^{h/2} z\,\sigma_\theta\, dz, \; M_{\theta x} = \int_{-h/2}^{h/2} z\,\tau_{x\theta}\, dz = -M_{x\theta}. \; (2.6)$$

We next recall the equations of equilibrium for an element of area of the shell. These consist of three force and two moment equations with normal shears. Elimination of the latter leads to (TIMOSHENKO, Ref. [5], p. 440)

$$\begin{aligned}
&\frac{\partial N_x}{\partial x} + \frac{\partial N_{x\theta}}{a\,\partial\theta} + X = 0,\\
&\frac{\partial N_\theta}{a\,\partial\theta} + \frac{\partial N_{x\theta}}{\partial x} + \frac{\partial M_{x\theta}}{a\,\partial x} - \frac{1}{a^2}\frac{\partial M_\theta}{\partial\theta} + Y = 0,\\
&\frac{N_\theta}{a} + \frac{\partial^2 M_{\theta x}}{a\,\partial x\,\partial\theta} + \frac{\partial^2 M_x}{\partial x^2} - \frac{\partial^2 M_{x\theta}}{a\,\partial x\,\partial\theta} + \frac{1}{a^2}\frac{\partial^2 M_\theta}{\partial\theta^2} + Z = 0,
\end{aligned} \qquad (2.7)$$

where X, Y, Z are the external forces per unit area of the shell in the axial, tangential and inner normal directions.

Suppose now that at some time $t = t_0$ we know the stress distribution, all the extensions, both elastic and plastic, and the displacements u, v, w. Then their values are found at a later time, $t = t_1 = t_0 + (\Delta t)_1$, as follows.

Differentiating Eqs. (2.2) with respect to time and using Eqs. (1.12) (i. e., Eqs. (1.11)), we obtain

$$\begin{aligned}
\frac{de_x}{dt} &= \frac{de'_x}{dt} + f_x(e'', \sigma, T),\\
\frac{de_\theta}{dt} &= \frac{de'_\theta}{dt} + f_\theta(e'', \sigma, T),\\
\frac{de_{x\theta}}{dt} &= \frac{de'_{x\theta}}{dt} + f_{x\theta}(e'', \sigma, T).
\end{aligned} \qquad (2.8)$$

We now first take means of both sides of each Eq. (2.8) over the shell thickness, by integrating with respect to z and dividing by the total thickness $2h$. Upon recalling Eqs. (2.3), (2.5), there results

$$\begin{aligned}
\dot{e}_x &= \frac{1}{E\,h}(\dot{N}_x - \nu\,\dot{N}_\theta) + \frac{1}{h}\int_{-h/2}^{h/2} f_x(e'', \sigma, T)\,dz, \\
\dot{e}_\theta &= \frac{1}{E\,h}(-\nu\,\dot{N}_x + \dot{N}_\theta) + \frac{1}{h}\int_{-h/2}^{h/2} f_\theta(e'', \sigma, T)\,dz, \qquad (2.9)\\
\dot{e}_{x\theta} &= \frac{(1+\nu)}{2\,E\,h}\dot{N}_{x\theta} + \frac{1}{h}\int_{-h/2}^{h/2} f_{x\theta}(e'', \sigma, T)\,dz.
\end{aligned}$$

Similarly, we multiply both sides of each Eq. (2.8) by z and integrate with respect to z, at the same time utilizing Eqs. (2.3), (2.6). There results

$$\begin{aligned}
-I\,\dot{\varkappa}_x &= \frac{1}{E}(\dot{M}_x - \nu\,\dot{M}_\theta) + \int_{-h/2}^{h/2} z\,f_x(e'', \sigma, T)\,dz, \\
-I\,\dot{\varkappa}_\theta &= \frac{1}{E}(\dot{M}_\theta - \nu\,\dot{M}_x) + \int_{-h/2}^{h/2} z\,f_\theta(e'', \sigma, T)\,dz, \qquad (2.10)\\
-I\,\dot{\varkappa}_{x\theta} &= \frac{1+\nu}{E}\dot{M}_{\theta x} + \frac{1}{2}\int_{-h/2}^{h/2} z\,f_{x\theta}\,dz,
\end{aligned}$$

where

$$I = h^2/12 \qquad (2.11)$$

is the moment of inertia of the shell section, per unit length, about its neutral axis.

It is now possible to proceed in the same manner as in the derivation of the elastic shell equations. First, one solves Eqs. (2.9), (2.10) for $\dot{N}_x, \ldots, \dot{M}_x, \ldots$ obtaining

$$\begin{aligned}
\dot{N}_x &= \frac{E}{1-\nu^2}\left[(\dot{e}_x + \nu\,\dot{e}_\theta)\,h - \int f_x\,dz - \nu\int f_\theta\,dz\right], \\
\dot{N}_\theta &= \frac{E}{1-\nu^2}\left[(\dot{e}_\theta + \nu\,\dot{e}_x)\,h - \int f_\theta\,dz - \nu\int f_x\,dz\right], \qquad (2.12)\\
\dot{N}_{x\theta} &= \frac{E}{2(1+\nu)}\left[\dot{e}_{x\theta}\,h - \int f_{x\theta}\,dz\right],
\end{aligned}$$

and

$$\begin{aligned}
-\dot{M}_x &= \frac{E}{1-\nu^2}\left[I(\dot{\varkappa}_x + \nu\,\dot{\varkappa}_\theta) + \int z\,f_x\,dz + \nu\int z\,f_\theta\,dz\right], \\
-\dot{M}_\theta &= \frac{E}{1-\nu^2}\left[I(\dot{\varkappa}_\theta + \nu\,\dot{\varkappa}_x) + \int z\,f_\theta\,dz + \nu\int z\,f_x\,dz\right], \qquad (2.13)\\
-\dot{M}_{\theta x} &= \frac{E}{(1+\nu)}\left[I\,\dot{\varkappa}_{x\theta} + \frac{1}{2}\int z\,f_{x\theta}\,dz\right].
\end{aligned}$$

By applying Eqs. (2.1), differentiated with respect to time, to $e_x, \ldots, \varkappa_x, \ldots$, the left-hand members of (2.12), (2.13) can be expressed in terms of $\dot{u}$, $\dot{v}$, $\dot{w}$ and their x- and θ-derivatives. One obtains

$$\dot{N}_x = \frac{E}{1-\nu^2}\left[h\left\{\dot{u}_x + \nu\left(\frac{\dot{v}_\theta}{a} - \frac{\dot{w}}{a}\right)\right\} - \int f_x\, dz - \nu \int f_\theta\, dz\right],$$

$$\cdots\cdots\cdots\cdots\cdots\cdots \qquad (2.14)$$

$$-\dot{M}_{\theta x} = \frac{E}{1+\nu}\left[I\left(\frac{\dot{w}_{x\theta}}{a} + \frac{\dot{v}_x}{a}\right) + \frac{1}{2}\int z\, f_{x\theta}\, dz\right].$$

The resulting expressions are then substituted into the equilibrium Eqs. (1.9) after they have been differentiated with respect to time. In this way one obtains the final differential equations in terms of $\dot{u}$, $\dot{v}$, $\dot{w}$ and their x- and θ-derivatives:

$$\left(\frac{E\,h}{1-\nu^2}\right)\left[\dot{u}_{xx} + \frac{1+\nu}{2\,a}\dot{v}_{x\theta} - \frac{\nu}{a}\dot{w}_x + \frac{1-\nu}{2\,a^2}\dot{u}_{\theta^2}\right] + \dot{X}$$
$$= \frac{E}{1-\nu^2}\left\{\frac{\partial}{\partial x}\left[\int f_x\, dz + \nu\int f_\theta\, dz\right] + \frac{1}{a}\frac{\partial}{\partial\theta}\left[2(1-\nu)\int f_{x\theta}\, dz\right]\right\},$$

$$\left(\frac{E\,h}{1-\nu^2}\right)\left\{\frac{\dot{v}_{\theta\theta}}{a^2} + \frac{1+\nu}{2}\frac{\dot{u}_{x\theta}}{a} + \frac{(1-\nu)}{2}\dot{v}_{x^2} - \frac{\dot{w}_\theta}{a^2}\right.$$
$$\left. + \frac{h^2}{12\,a^2}\left[\dot{w}_{x^2\theta} + \dot{w}_{\theta^3}/a + \frac{(1-\nu)}{2}\dot{v}_{x^2} + \dot{v}_{\theta\theta}/a^2\right]\right\} + \dot{Y}$$
$$= \frac{E}{1-\nu^2}\left[\frac{1}{a}\frac{\partial}{\partial\theta}\left(f_\theta\, dz + \nu\int f_x\, dz\right) - \frac{\partial}{\partial\theta}\left(\int z\, f_\theta\, dz/a - \nu\int z\, f_x\, dz/a\right)\right. \qquad (2.15)$$
$$\left. + \frac{1-\nu}{2}\left(\frac{\partial}{\partial x}\int f_{x\theta}\, dz - \int z\, f_{x\theta}\, dz/a\right)\right],$$

$$\left(\frac{E\,h}{1-\nu^2}\right)\left\{\frac{\nu}{a}\frac{\partial\dot{u}}{\partial x} + \frac{\partial\dot{v}}{a^2\,\partial\theta} - \frac{\dot{w}}{a^2} - \frac{h^2}{12}(\nabla^4 \dot{w}) - \frac{h^2}{12}\left[\left(\frac{2-\nu}{a^2}\right)\frac{\partial^3\dot{v}}{\partial x^2\,\partial\theta} + \frac{\partial^3\dot{v}}{a^4\,\partial\theta^3}\right]\right\}$$
$$= \dot{Z} + \frac{E}{1-\nu^2}\left[\frac{\int f_\theta\, dz}{a} + \frac{\nu\int f_x\, dz}{a} + \frac{(1-\nu)}{a}\frac{\partial^2}{\partial x\,\partial\theta}\int z\, f_{x\theta}\, dz\right.$$
$$\left. + \frac{\partial^2}{\partial x^2}\left(\int z\, f_x\, dz + \nu\int z\, f_\theta\, dz\right) + \frac{1}{a^2}\frac{\partial^2}{\partial\theta^2}\left(\int z\, f_\theta\, dz + \nu\int z\, f_x\, dz\right)\right].$$

3. Discussion of Solutions of Eqs. (2.15)

Eqs. (2.15) differ from the elastic equations for cylindrical shells in two respects. In the first place, instead of the displacements u, v, w and their spacial (i. e. x- and θ-) derivatives, we have throughout $\dot{u}$, $\dot{v}$, $\dot{w}$ and their spacial derivatives. Furthermore, on the right-hand side there are contributions resulting from the integrated or resultant of the creep rates across the thickness, and the monents of the creep rates. The creep rates must be calculated from (1.1)—(1.7) at each point (x, θ, z), integrated over z from $-h$ to h, evaluated, the proper x- and θ-differentiations carried out; similarly, for the moments. These added terms may be regarded as added fictitious distributed loads (or rather their time rates) derived from resultants and moments of the creep rates.

Presumably, Eqs. (2.15) can be solved in a similar way to the solution of the elastic shell equation, except for the fact that one will be led only to the time rates of change u, v, w. By multiplying these by Δt over a small time interval, the net changes Δu, Δv, Δw in u, v, w will be obtained, these are added.

The process is now continued in the following fashion. From $\dot{u}$, $\dot{v}$, $\dot{w}$, the rates of net extensions $\dot{e}_x$ and the curvature rates $\dot{\varkappa}_x, \ldots$ are found by means of Eqs. (2.1). The left-hand members of Eqs. (2.8) now become known, and these equations can be used to solve for the rates of change of the elastic components of strain $\dot{e}'_x$, $\dot{e}'_\theta$, $\dot{e}'_{x\theta}$. Eqs. (2.4) can now be used to obtain the stresses at the time $t = t_1 = t + \Delta t$. The calculation can now be resumed for the next time interval.

Since f_x, f_θ, $f_{x\theta}$ are highly non-linear functions of the stresses and the total equivalent plastic strain e'' and its components are in general available only from the curves of Figs. 2, 3, it is evident that the evaluation of $e''_x, \ldots$, of the integrals $\int f_x\, dz, \ldots$ occurring on the right-hand sides of Eqs. (2.15), and their derivatives is a very complex calculation which may form one of the major difficulties in applying the above equations.

Theoretically, the solutions of Eqs. (2.15) for $\dot{u}$, $\dot{v}$, $\dot{w}$ can be carried out by expanding their right-hand members, as well as u, v, w, in the Fourier series in θ, thus converting the partial differential Eqs. (2.15) to ordinary differential equations in x for each term of the Fourier series. However, since as indicated above, the calculation of the right-hand members of (2.15), that is of the fictitious distributed load derived from the creep rates, presumably can only de done numerically, the calculation of the Fourier expansions at each time instant will likewise have to be carried out numerically based on a finite number of mesh points of a grid spread over the cylindrical shell.

Summarizing, it appears that the problem of calculating the creep in the cylindrical shell leads to repeated solutions of the same equations as the static equations of an elastic shell, except that these equations apply to $\dot{u}$, $\dot{v}$, $\dot{w}$, the time rates of change of the displacement components, rather than to the components themselves. Furthermore, certain right-hand members appear, in these equations in the same manner as the applied external forces, these extra terms being derived from the instantaneous rates of creep and involve their integral over the thickness of the shell and the moments of these rates of creep.

4. Axially Symmetric Example

To illustrate the application of the above creep calculations on a relative simple example, we consider an axially symmetric problem consisting of a semi-infinite cylindrical shell $x \geqq 0$, at a uniform temperature, initially free from stress and subject for $t \geqq 0$ to axially symmetric terminal boundary conditions at $x = 0$, corresponding to the application of a moment and vanishing tensile traction and vanishing shear:

$$N_x = 0,\ M_x = M_0,\ s = 0 \text{ at } x = 0. \tag{4.1}$$

It is evident that for such axial symmetry the displacement v will vanish, while the displacements u and w will be independent of θ, depending only on x. The strains, curvature changes, and mean strains become

$$\begin{aligned} \bar{e}_x &= \frac{du}{dx},\ \bar{e}_\theta = -\frac{w}{a},\ \bar{e}_{x\theta} = 0, \\ \varkappa_x &= \frac{d^2w}{dx^2},\ \varkappa_\theta = 0,\ \varkappa_{x\theta} = 0, \\ e_x &= \frac{du}{dx} - \frac{d^2w}{dx^2} z,\ e_\theta = -\frac{w}{a},\ e_{x\theta} = 0. \end{aligned} \tag{4.2}$$

Moreover, $\sigma_{x\theta}$ vanishes and the principal stress (and strain) directions at each point correspond to the directions of increasing x and increasing θ. Eq. (1.6) now yields

$$\sigma = \sqrt{\sigma_x^2 + \sigma_\theta^2 - \sigma_x \sigma_\theta}, \tag{4.3}$$

while Eqs. (1.11) result in

$$\begin{aligned} \dot{e}_x'' &= \frac{f}{\sigma}(\sigma_x - \sigma_\theta/2), \\ \dot{e}_\theta'' &= \frac{f}{\sigma}(\sigma_\theta - \sigma_x/2). \end{aligned} \tag{4.4}$$

The first and third Eqs. (2.15), in absence of applied loads, reduce to

$$\dot{u}_{xx} - \frac{\nu}{a}\dot{w}_x = \frac{d}{dx}\left[\int f_x\, dz + \nu \int f_\theta\, dz\right]/h, \tag{4.5}$$

$$\begin{aligned} \frac{\nu}{a}\dot{u}_x - \frac{\dot{w}}{a^2} - \frac{h^2}{12}\dot{w}_{x^4} &= \left(\int f_\theta\, dz + \nu \int f_x\, dz\right)/ha \\ &+ \frac{d^2}{dx^2}\left(\int z f_x\, dz + \nu \int z f_\theta\, dz\right)/h, \end{aligned} \tag{4.6}$$

while the second Eq. (2.15) is satisfied identically.

It will be noted that Eq. (4.5) can be integrated with respect to x into

$$\left(\dot{u}_x - \frac{\nu}{a}\dot{w}\right) = \left(\int f_x\, dz + \nu \int f_\theta\, dz\right)/h \tag{4.7}$$

without any additive constant, since at $x = \infty$ both members vanish.

We note further that the first equilibrium Eq. (2.7) simplifies to

$$\frac{\partial N_x}{\partial x} = 0, \; N_x = \text{const.} \tag{4.8}$$

Since $N_x = 0$ at $x = 0$, then (4.8) yields, for all x,

$$N_x = 0. \tag{4.9}$$

Eq. (4.7) allows one to solve for $\dot{u}_x$ in terms of $\dot{w}$:

$$\dot{u}_x = \frac{\nu}{a}\,\dot{w} + \frac{1}{h}\left(\int f_x\,dz + \nu \int f_\theta\,dz\right). \tag{4.10}$$

Substitution into (4.6) yields

$$\dot{w}_{x^4} + 4\,\beta^4\,\dot{w} = R(x), \quad \beta^4 = \frac{3(1-\nu^2)}{h^2\,a^2}, \tag{4.11}$$

where

$$R(x) = -\frac{12(1-\nu)}{h^3\,a}\int f_\theta\,dz - \frac{12}{h^3}\frac{d^2}{dx^2}\left(\int z\,f_x\,dz + \nu\int z\,f_\theta\,dz\right). \tag{4.12}$$

To simplify the calculations and still retain some of the features of non-linearity, we shall assume that the rate of creep is given by Eq. (1.2), which, however, we put in the form

$$\dot{e}'' = f(\sigma) = K\,\sigma^{n+1}, \tag{4.13}$$

where K depends only on the temperature. With the latter assumed constant, K is a constant. Eqs. (4.4) take on the form

$$\begin{aligned} \dot{e}'' &= K\,\sigma^n\,(\sigma_x - \sigma_\theta/2) = f_x, \\ \dot{e}'' &= K\,\sigma^n\,(\sigma_\theta - \sigma_x/2) = f_\theta. \end{aligned} \tag{4.14}$$

Eqs. (4.8) and (4.10) automatically include the boundary condition $N_x = 0$. The application of $M_x = M_0$ is carried out by time differentiation and recalling the first Eq. (2.13). There results

$$\left.\frac{(1-\nu^2)\,\dot{M}_x}{E}\right|_{x=0} = 0 = \left[\frac{h^3}{12}\,\dot{w}_{xx} + \int z\,f_x\,dz + \nu\int z\,f_\theta\,dz\right]_{x=0}. \tag{4.15}$$

To apply $S = 0$, we recall the moment equilibrium equation of an element, which yields

$$S = \frac{dM_x}{dx}. \tag{4.16}$$

Differentiation of the first Eq. (2.13) with respect to x and t yields

$$\left[\frac{h^3}{12}\,\dot{w}_{xxx} + \frac{d}{dx}\left(\int z\,f_x\,dz + \nu\int z\,f_\theta\,dz\right)\right]_{x=0} = 0. \tag{4.17}$$

Thus, the application of (4.1) leads to prescribed values of $\dot{w}_{xx}$ and $\dot{w}_{xxx}$ at $x = 0$.

If u, w, the stresses σ_x, σ_θ, and plastic strains e_x'', e_θ'' are known at a time t_n, they can be found at the time $t_{n+1} = t_n + (\Delta t)_n$ as follows.

A "grid" of points in the (x, z)-plane, corresponding to equidistant values of z, say

$$\frac{z}{h} = 0, \pm .1, \pm 0{,}2, \ldots, \pm 0{,}5 \tag{4.18}$$

and to equidistant values of x is considered. Assume that the values of $u, \ldots, e''_\theta$ are known at these "points" at the time t_n. Eqs. (4.3), (4.14) are now applied to calculate $\dot{e}''_x$, $\dot{e}''_\theta$ each each point. The integrals

$$I_1 = \int f_x\, dz,\ I_2 = \int f_\theta\, dz,\ I_3 = \int z\, f_x\, dz,\ I_4 = \int z\, f_\theta\, dz \tag{4.19}$$

are next evaluated, say by means of SIMPSON's rule, at each location x, and the function $R(x)$ in [(3.12) computed:

$$R(x) = -\frac{12(1-\nu)}{h^3\, a} I_2 - \frac{12}{h^3}\frac{d^2}{dx^2}(I_3 + \nu\, I_4). \tag{4.20}$$

The differential Eq. (4.11) is then solved for $\dot{w}$ as a function of x, subject to the boundary conditions (4.15), (4.17). Next $\dot{u}_x$ is evaluated from (4.10).

From (4.14) and the known values of e''_x, e''_θ at the time t_n, one may determine their values at the time t_{n+1} as follows

$$\begin{aligned} e''_x\big|_{t_{n+1}} &= e''_x\big|_{t_n} + \dot{e}''_x (\Delta t)_n, \\ e''_\theta\big|_{t_{n+1}} &= e''_\theta\big|_{t_n} + \dot{e}''_\theta (\Delta t)_n. \end{aligned} \tag{4.21}$$

Similarly, u_x, w may be obtained at the time t_{n+1} from their values at t_n and $\dot{w}$, $\dot{u}_x$ at t_n. From u_x, w the total strains e_x, e_θ and curvatures $\varkappa_x$, $\varkappa_\theta$ can be calculated by means of (4.1). Eqs. (2.2) can now be used to calculate the elastic strains e'_x, e'_θ at the time t_{n+1}, and Eqs. (2.4) the new stresses σ_x, σ_θ. As a check upon these calculations, the relation

$$\int_{-h/2}^{h/2} \sigma_x\, dz = N_x = 0 \tag{4.22}$$

may be noted.

At the beginning, at the time $t_0 = 0$, the values of u, w correspond to the elastic solution for the applied moment in question, namely

$$w = \frac{M_0\, e^{-\beta x}}{2\,\beta^2\, D}(\sin\beta\, x - \cos\beta\, x),\ D = \frac{E\, h^3}{12(1-\nu^2)},\ u_x = \nu\, w/a, \tag{4.23}$$

corresponding to the moments

$$M_x = D\, w_{xx} = M_0\, e^{-\beta x}[\sin\beta\, x + \cos\beta\, x],\quad M_\theta = \nu\, M_x. \tag{4.24}$$

The initial plastic strains vanish, and the total strains which are elastic reduce to

$$\begin{aligned} e_x &= e'_x = u_x - z\, w_{xx} = \frac{\nu\, w}{a} - z\, w_{xx}, \\ e_\theta &= e'_\theta = -\frac{w}{a}. \end{aligned} \tag{4.25}$$

The initial stresses are

$$\begin{aligned} \sigma_x &= -\frac{E}{1-\nu^2}\, z\, w_{xx}\,, \\ \sigma_\theta &= -E\left(\frac{w}{a} + \frac{\nu}{1-\nu^2}\, z\, w_{xx}\right), \end{aligned} \tag{4.26}$$

where w, w_{xx} are given by (4.23), (4.24).

With the value of $u, w, \ldots, \sigma_\theta$ available at the time t_{n+1}, the process is now repeated for the next time interval $(\Delta t)_{n+1}, \ldots$

The integrations of the differential Eq. (4.11) can be carried out either by means of numerical finite difference approximations or by means of GREEN's function which could be obtained even to satisfy proper boundary conditions at $x = 0$. It turns out simpler, however, to use a GREEN's function G which vanishes at infinity, satisfies the differential Eq. (4.11) in x except at $x = s$, at which point it is continuous with its first and second derivatives, but has a discontinuity in its third derivative of amount 1. It can be shown that G is a function of $x - s$, $G = G(x - s)$. For positive values of its argument it is given by

$$G(x) = \frac{1}{8\beta^3}(\cos\beta x + \sin\beta x)\, e^{-\beta x} \quad \text{for } x > 0\,, \tag{4.27}$$

for negative values it may be obtained from symmetry in x:

$$G(-x) = G(x) \quad \text{for } x = 0\,. \tag{4.28}$$

With G defined as above, a solution of (4.11) is given by

$$\dot{w}(x) = \int_0^\infty R(s)\, G(x-s)\, ds\,. \tag{4.29}$$

The integration here is presumably to be carried out numerically, breaking up the integral into two parts, from 0 to x and from x to infinity, using (4.27) for the former and (4.28) for the latter. This solution, however, does not satisfy the desired boundary conditions at $x = 0$. To this end the values of w and its derivatives are calculated at $x = 0$ by means of the following equations derived from (4.29) and its differentiations with respect to x, then equating x to 0:

$$\begin{aligned} \dot{w}(0) &= \int_0^\infty R(s)\, G(-s)\, ds = \int_0^\infty R(s)\, G(s)\, ds\,, \\ \dot{w}_x(0) &= \int_0^\infty R(s)\, G'(-s)\, ds = -\int_0^\infty R(s)\, G'(s)\, ds\,, \\ \dot{w}_{xx}(0) &= \int_0^\infty R(s)\, G''(-s)\, ds = \int_0^\infty R(s)\, G''(s)\, ds\,, \\ \dot{w}_{xxx}(0) &= \int_0^\infty R(s)\, G'''(-s)\, ds = -\int_0^\infty R(s)\, G'''(s)\, ds\,. \end{aligned} \tag{4.30}$$

Differentiation of (4.27) yields for positive x

$$G'(x) = -\frac{1}{4\beta^2} e^{-\beta x} \sin \beta x,$$
$$G''(x) = \frac{e^{-\beta x}}{4\beta} (\sin \beta x - \cos \beta x), \qquad (4.31)$$
$$G'''(x) = \frac{e^{-\beta x} \cos \beta x}{2},$$

while for negative x these functions are determined by noting that G', G'' are odd functions of their argument, while G'' is even.

Thus Eqs. (4.30) yield

$$\begin{aligned} \dot{w}(0) &= (J_1 + J_2)/8\beta^2, \\ \dot{w}_x(0) &= J_2/4\beta^2, \\ \dot{w}_{xx}(0) &= (J_2 - J_1)/4\beta, \\ \dot{w}_{xxx}(0) &= J_1/2, \end{aligned} \qquad (4.32)$$

where

$$\begin{aligned} J_1 &= \int_0^\infty R(s)\, e^{-\beta s} \cos \beta s \, ds, \\ J_2 &= \int_0^\infty R(s)\, e^{-\beta s} \sin \beta s \, ds. \end{aligned} \qquad (4.33)$$

To satisfy the boundary conditions (4.15), (4.17) the proper values of $\dot{w}_{xx}$, $\dot{w}_{xxx}$ at $x = 0$ are obtained from these equations, and the needed increments $\Delta\dot{w}_{xx}$, $\Delta\dot{w}_{xxx}$ to the values of these dervatives to be added to the values (4.32) of the special solution (4.29) computed. The following solution of the homogenized Eq. (4.11) is then added to the special solution (4.29):

$$(\Delta\dot{w}_{xx}|_0)\, e^{-\beta x} (\cos \beta x - \sin \beta x)/2\beta^2 + (\Delta\dot{w}_{xxx}|_0)\, e^{-\beta x} \cos \beta x/2\beta^3. \qquad (4.34)$$

The calculation of the plastic strains by means of Eqs. (4.21), as well as the corresponding calculations of u_x, w assume that the time rates of change of these quantities at the beginning of the interval (Δt) remain constant throughout that interval. Evidently a gross inaccuracy may be due to this assumption. This inaccuracy can be diminished by means of various methods familiar from the theory of numerical integration of ordinary differential equations, such as polynomial extrapolation from preceding rates of change to obtain a tentative value at the time t_{n+1}; repeating the calculation after it has been completed, using mean values of the time derivatives at the beginning and the end of the time interval, and so forth. The inaccuracy can likewise be diminished by taking smaller time intervals.

References

[1] PORITSKY, H., and F. A. FEND: Relief of Thermal Stresses Trough Creep, Trans. of ASME, **80**, 589—597 (1958).

[2] MENDELSON, A., M. H. HIRSCHBERG and S. S. MANSON: A General Approach to the Practical Solution of Creep Problem, Trans. of ASME, Journal of Basic Engineering, **81**, 585—598 (1959).

[3] SHERBY, O. D., R. L. ORR and J. E. DORN: Creep Correlation of Metals at Elevated Temperatures, Transactions of American Institute of Mining and Metallurgical Engineers, **200**, 71—80 (1954).

[4] DORN, J. E.: Fundamental Experiments on High Temperature Creep, Journal of Mechanics and Physics of Solids, **3**, 85—116 (1954).

[5] TIMOSHENKO, S.: Theory of Plates and Shells, 1st Edition, McGraw-Hill, 1940.

Discussion

C. R. CALLADINE, The English Electric Company: In his paper the author has set up equations, the solution of which would enable the behaviour of a cylindrical shell to be followed as the stress distribution changed from an elastic to a steady creep state.

As the author points out, solution of the equations would probably involve some difficulty.

An alternative approach to the problem, which would almost certainly require a shorter and a considerably less involved calculation, would be to divide the work into two parts, as follows.

1. Find a relationship between the creep deformation rates of a plate element and arbitrary applied stress resultants for a wide variety of loading conditions. ONAT and YÜKSEL (Ref. [1]) have shown how this may be done for a sandwich-plate made of a material obeying an n-power "Tresca" type of creep law. For "solid" plates and more general types of creep law, numerical work is probably necessary to establish the required relationship, but it seems plausible that a simple algebraic expression may be found to describe the cycle of numerical results with sufficient accuracy.

2. Use this load-deformation rate relationship combined with the elastic load-deformation relationship, in conjunction with the shell equilibrium equations (in terms of stress resultants) and compatibility equations, to obtain a numerical solution to the problem.

In this two-stage approach (which is, of course, directly analogous to the approach used in elastic shell theory) the shell element load-deformation relations are worked out once for all, thus eliminating the necessity for continual re-calculation, as implied in the method proposed in the paper.

E. H. LEE, Brown University: In connection with the question of the mathematical form of the equations presented in the paper, discussed by Professors BESSELING and ONAT, it may be worthwhile to point out that there are no real characteristics in the space directions, but that lines of increasing time for fixed spacial position are characteristics. Since three or more dimensions are involved, at least two space dimensions and time, characteristic lines do not divide the region as in two dimensions, and characteristics are curves along which purely "inner" differential

[1] ONAT, E. T., and H. YÜKSEL: "On the steady creep of shells". Proc. 3rd. US Nat. Congr. Appl. Mech., pp. 625—630, 1958.

relations exist, rather than curves across which discontinuities occur. "Inner" means that the partial differential equations reduce to an expression with derivatives only along the curve, which in effect becomes an ordinary differential relation. In the present case, this is the creep relation for a fixed material element. The method of integration suggested by Dr. PORITSKY is in accordance with this structure. Essentially, elliptic type boundary value problems are solved in the space dimensions, and characteristic integration along time increments in the nature of an initial value problem.

Author's closing words: Mr. CALLADINE proposes that the calculation of the behavior of a cylindrical shell should be carried out by a simplified calculation of creep in a shell, for instance, as described by ONAT and YÜKSEL. The latter propose a law of creep based on rate of work $\sigma_1 \dot{\varepsilon}_1 + \sigma_2 \dot{\varepsilon}_2 = F(\sigma)$ where σ is determined by the Tresca conditions for a shell in bending and tension. This is applied at the outer and inner shell boundary.

An alternative possible simplification is furnished by CALLADINE's own paper in this Colloquium, "On Creep of a Wrinkle". He presents in his Fig. 6 a set of relations for $\dot{\varkappa}$ as a function of the applied moment and traction for a beam of rectangular cross section based on various power laws of creep. This figure is reproduced from the work of A. G. YOUNG. A similar set of relations can, no doubt, be calculated for a plate under the action of tractions and moments.

The use of such simplified "universal" relations for $\dot{\varkappa}_x$ will, no doubt, reduce the amount of computational labor involved. Yet, the ONAT and YÜKSEL procedure is probably insufficient to treat general, three-dimensional creep. Nor is it clear whether the assumption of a universal relation between $\dot{\varkappa}_x$ and the moments and tractions will apply to the case where the load is first applied to a stress-free shell. Both the treatment by ONAT and YÜKSEL, and the further alternative proposed above are based on the assumption that the elastic component of strain is either negligible or does not change with time. With a constant $\dot{\varkappa}_x$, the plastic strain rate varies linearly with z, and the stress distribution with z adjusts itself so as to conform with the linear variation of the creep rate. It is not always true that in engineering applications the plastic creep becomes so large that the elastic component can be completely neglected.

It is to be expected that when the load is first applied to a stress-free shell, the initial stress distribution will correspond to the *elastic solution*, and the *stress* components at each x, θ will vary linearly with z. As time goes on, however, as a result of the creep, the stress distribution will change. It is not clear whether the stress distribution ever approaches the particular stress distribution which corresponds to constant $\dot{\varkappa}_x$. The changes in stress will cause a redistribution of the elastic strain components and thus vitiate some of the recommended simple procedures for calculating creep.

It may be added that the employment of high speed computing machines enables one nowadays to carry out calculations which formerly might have been too onerous to carry out with the aid of desk computers.

On the Creep of a Wrinkle

By

C. R. Calladine

English Electric Company Atomic Power Division, Whetstone, Leicester, England

Synopsis

The problem of the creep pressurisation of a thin circular cylindrical shell with closed ends onto a slightly smaller rod is introduced, and a qualitative description is given of the behaviour of the wrinkle which forms after the tube touches down onto the rod.

The equations governing the behaviour of the shell are set up in linear form (as a result of approximations which restrict the range of some of the variables) and are solved by a finite-difference technique, thus enabling the progress of the wrinkle to be followed.

A "strain intensification factor" I is defined as

$$I = \frac{\text{Maximum strain due to bending of shell}}{\text{Strain resulting from uniform collapse of shell}}$$

and the results of the calculations are given in a graph showing I as a function of various parameters defining the shell-rod system.

List of Symbols

When a lower case letter denotes a non-dimensional variable, the corresponding upper case letter denotes the original, dimensional variable.

A	Radius of rod + half thickness of shell
C, c	Curvature relative to circle radius A
H	Thickness of shell
I	Strain intensification factor
K_n	Parameter in creep law
M, m	Bending moment per unit axial length of shell
n	Index in creep law
P	External excess pressure on shell
Q, q	Reaction between rod and shell per unit axial length of shell
R, r	Direct stress resultant per unit length of shell
S	Perimeter of centre-line of shell
T, t	Time
U, u	Radial displacement of shell
V, v	Radial velocity of shell

W	Direct stress resultant per unit length of shell
x	Radial displacement of centre of wrinkle ($= U_0/U_0^*$)
Δ	Initial radial clearance between shell and rod
α	Angle subtended at centre of rod by part of shell under compression
β	Parameter describing pre-touchdown behaviour of shell
γ	Shell geometry-material parameter
δ	Finite-difference angle
$\dot{\varepsilon}$	Strain-rate
ν	Shell-rod system geometry-material parameter
θ	Circumferential angle co-ordinate
Φ	Value of θ at which reaction occurs

Subscripts

1, 2, 3	Refer to axes shown in Fig. 5
$0, 1, \ldots, 9$	Refer to finite-difference nodal points shown in Fig. 11
0	Refer to $\theta = 0$
Φ	Refer to $\theta = \Phi$

Superscripts

′	Denotes differentiation with respect to time
*	Denotes values of variables at first touch-down of shell

1. Introduction

(a) Background

In some advanced types of nuclear power reactor burning uranium dioxide fuel, it appears to be economically attractive to use fuel elements consisting of circular cylindrical fuel pellets enclosed in plain thin tubes with welded ends. For such systems operating at moderate heat ratings there appear to be no strong heat-transfer arguments for maintaining a tight fit between the fuel and the can: in most cases the fuel would not be operating near its maximum allowable temperature and, further, close contact between the fuel and the can is not particularly beneficial from the point of view of reducing maximum fuel temperature (as it is with uranium (metal) fuel) provided that the space between the fuel and the can is filled with a gas of high thermal conductivity (which should not be difficult to achieve). There is an obvious economic advantage in relaxing manufacturing tolerances on fuel pellet and can dimensions; however, some of this potential advantage may not be available if, on account of a substantial gap between the fuel and its can, the can under the action of the coolant fluid pressure creeps down onto the fuel irregularly, and forms a "wrinkle", as shown in Fig. 1.

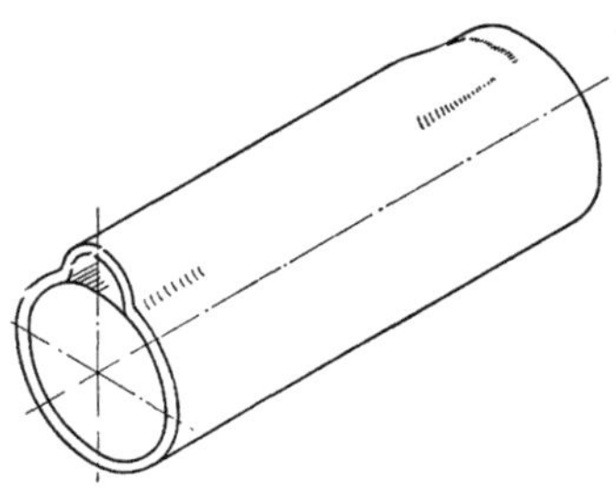

Fig. 1. Wrinkle

There are several reasons for wanting to minimise wrinkling. A pronounced wrinkle such as that shown in Fig. 1 would cause local overheating of the can due to increased interface heat resistance between the fuel and can at the wrinkle. Overheating in a region of high local strain could be a cause of premature can failure, especially if there were a significant amount of differential expansion between the can and the fuel caused by temperature cycling. Even for a much less pronounced wrinkle than that shown in Fig. 1, (which might possibly pass undetected by the naked eye) there would be a possibility of local strain levels in the can becoming sufficiently high to cause embarrasment if the can were made of a brittle material.

(b) Purpose of the Paper

The purpose of this paper is to investigate, by means of approximate mathematical techniques, the development of wrinkles in an idealised representation of a fuel/can system capable of describing a variety of fuel radius/can thickness/radial gap configurations.

(c) System Considered

The system to be studied is shown in Fig. 2. The shell of thickness H is originally circular, with a radial gap Δ between it and the rod to which the chain of pellets is considered to be equivalent. The mean radius of a tube which would touch the rod at all points is A. The rod is considered to be rigid: this assumption is valid if deflections of the rod are much smaller than those of the shell when both are subjected to similar forces — as would appear to be the case. The material of the tube is such that at stresses equal to those caused by the external pressure P, significant creep occurs at the working temperature. The pressure P is, however, sufficiently small for there to be no appreciable elastic buckling effects. Both of these conditions are common in a wide range of reactors.

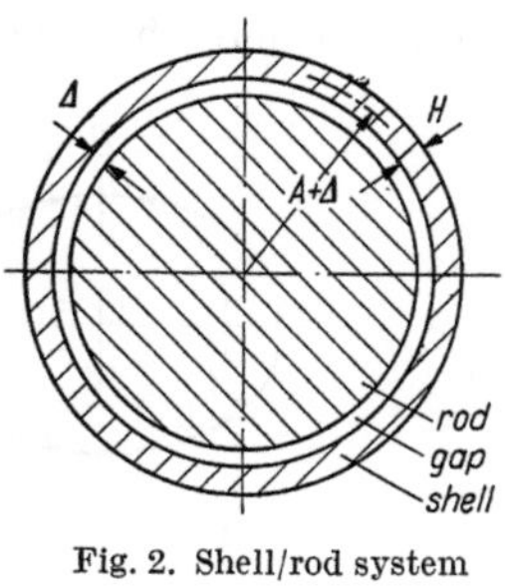

Fig. 2. Shell/rod system

Changes of temperature with time, causing differential expansion between the fuel and can, are ignored, as they would lead to complications unwarranted at this early stage of the analysis.

The section of tube considered is, for simplicity, taken as being free from effects due to the proximity of (virtually rigid) end caps. It is likely that the end caps prevent to some extent the development of wrinkles at the ends of the tube, and hence that the case considered is the most pessimistic.

(d) Qualitative Description of Creep Wrinkling

Consider the action of the differential coolant pressure P upon the shell.

As has been shown by HOFF et al [1], any initial small deviations of the tube from true circularity will be magnified roughly exponentially with time. This is because the more "oval" the tube becomes, the poorer is its disposition of material to resist external pressure. At the same time, a general shortening of the perimeter is taking place under the compressive circumferential stresses in the shell. If the primary creep stage of the material is past (as is generally assumed, to avoid complications) this shortening, and the associated change of mean radius of the shell, is linear with time. Thus the tube is subject to an exponentially-increasing ovality with time superimposed upon a linearly decreasing mean radius with time. Therefore, (sooner or later, depending upon the initial out-of-circularity) the shell will come into contact with the rod along two generators. If the rod was not originally disposed centrally with respect to the tube, these lines of contact will not, in general, be diametrically opposite. Whether or not subsequent deformation of the shell squeezes the rod into a more central position depends largely upon the coefficent of friction between the rod and the shell.

Assuming for the moment that by virtue of a small coefficient of friction between the rod and the tube, or by central location of the rod in the tube, the first reaction between the rod and the shell occurs at diametrically opposite generators, we arrive at the condition shown in Fig. 3a, where there is two-point contact at points B between the shell

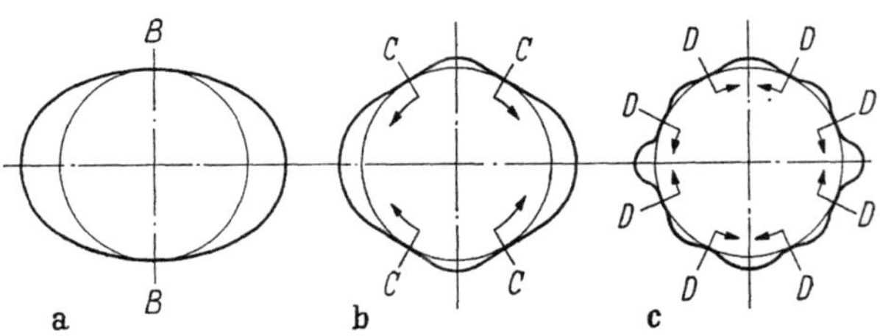

Fig. 3. Schematic development of wrinkle

and the rod. Further inward movement of the shell at points B is prevented by the presence of the rod, and reactions develop at B which, by causing "peaks" in the bending-moment distribution in the shell, make the shape of the shell at BB conform more closely to the shape of the rod. After a time, four-point contact is developed at BB, and the original points of contact divide into pairs of points of contact (as shown in Fig. 3b at CC) which proceed to move around the perimeter of the rod, lifting up the short length of shell between them. The points B on the shell thus begin to move outwards from the rod until (as we shall see later) points C have moved so far that the material between them begins to move inward again. This seems to take place by means of another division of the four points of contact C into eight points of contact D which move as shown in Fig. 3c, and raise between them four more shallow subsidiary wrinkles. Thus, as time passes, the angle subtended by the main wrinkle decreases while in the remaining space a set of more or less insignificant wrinkles plays.

Several points arising from this qualitative description merit further discussion. Firstly, so long as there is shell material not in contact with the rod the total perimeter of the shell becomes progressively smaller; for, by virtue of the shape of the shell, there is always circumferential compression in such material. Therefore the set of wrinkles eventually flattens itself out, complete collapse occurring when the perimeter of the shell becomes equal to that required for contact all round the rod. This marks a point of contrast to wrinkling produced when an oversize tube of rigid-plastic material is pressurised onto a rod. In this case, when a pronounced wrinkle occurs, deformation on subsequent small increase of pressure takes place by means or rotation of plastic "hinges" in the shell. The material away from the hinges is not in a state of plastic collapse, and so the wrinkle simply becomes more and more pronounced as the pressure is increased. Further, it seems that with rigid-plastic wrinkling of a given shell, there may exist a "critical" initial radial gap, below which the shell settles down uniformly on application of increasing pressure, and above which a wrinkle forms. Such a criterion is not to be expected in creep wrinkling; rather, if (say) the maximum strain developed is taken as a measure of the severity of the wrinkle, a continuous relationship between maximum strain and initial radial clearance should be expected.

Secondly, in the description of the behaviour in creep of the wrinkle, it was assumed that equal wrinkles occurred on each side of an axis of symmetry BB in the rod. (See Fig. 3). It may be that experimental evidence will show that this is possible, but it also seems plausible at this stage that a kind of instability may occur in which, in a pair of slightly unequal wrinkles, the flatter of the two could flatten preferentially and, if the rod were sufficiently smooth, push the shell around the rod so as to enlarge the larger wrinkle.

In the analysis of the following sections it is assumed that wrinkles form symmetrically on each side of an axis which passes through the centre of the rod; it may well be possible to apply the results to non-symmetrical wrinkles by assuming that the wrinkle on one side is unaware, so to speak, of the happenings on the other side of the rod.

Thirdly, the effect of the coefficient of friction between the rod and shell has not been discussed. It is clear that, in a case where two adjacent unequal lobes are in a state of creep (as, for example in Fig. 3b and c), there is a possibility of the rolling point of contact actually sliding a little: thus the effect of friction between the rod and the shell will be to separate to some extent the behaviour of adjacent lobes. Also, when a state is reached in the collapse process when a portion of the shell is lying against the rod, the circumferential stress in the shell depends largely upon the coefficient of friction between rod and shell. In general, the larger the coefficient of friction, the smaller will be the angle over which shortening takes place, and the greater will be the severity of the wrinkle. An attempt is made in Section 2c to establish upper and lower bounds on the effect of friction, although no provision is made for a non-zero coefficient of friction between the shell and the rod in setting up the basic equations.

2. Analysis

(a) Equations of Equilibrium

Consider unit axial length of the shell and rod as shown in Fig. 4. As we are particularly interested in the bending of the shell (we neglect, as is customary in shell theory, deformation due to radial shear), we need to express the circumferential bending moment in the shell in terms of the external pressure, the deflections of the system, the reactions between the rod and the shell and two further redundancies. The radius $\theta = 0$ is a line of symmetry through the wrinkle so that there is no radial shear force on the shell at $\theta = 0$.

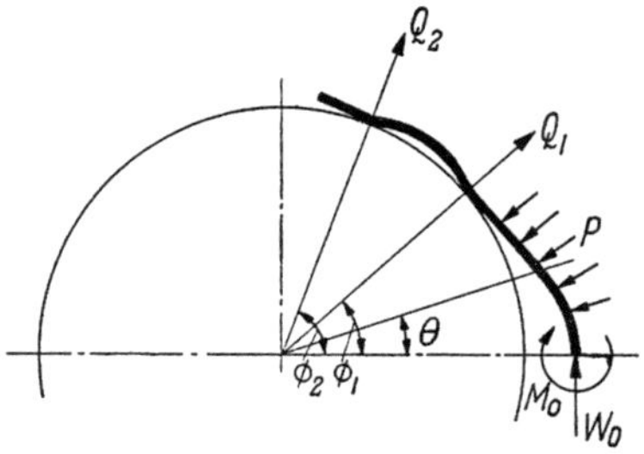

Fig. 4. Forces acting on shell

Frictional forces between the rod and the shell are omitted in order to simplify the analysis. It may be worth while at a later date to introduce a non-zero coefficient of friction, but it appears at this stage that such a complication is unnecessary from the point of view of gaining an understanding of the significant features of wrinkling.

Let us define the following symbols:

M: The circumferential bending moment per unit axial length of shell at angle θ; positive when tending to increase the original curvature of the shell.

W: The tangential force per unit axial length of shell; positive when compressive.

Q_i: The radial reaction per unit axial length of shell at point i. ($i = 1, 2, \ldots$ in order as θ increases from 0.) Q_i is always positive.

U: The radial displacement of the central surface of the shell, measured outwards from the position in which the shell touches the rod.

The subscript 0 denotes the value of the above variables at $\theta = 0$.

Taking moments about the central surface of the shell at θ, we find, if U/A is so small that $(U/A)^2$ is negligible compared to unity:

$$M = M_0 + P\,A\,[(A + U_0 + U)\,(1 - \cos\theta)] - W_0\,[A\,(1 - \cos\theta) + U_0 - U\cos\theta] - \Sigma\,A\,Q_i \sin(\theta - \Phi_i). \tag{1}$$

The summation is over all Φ_i less than θ. It seems likely that the force W_0 is nearly equal to PA[1], the value of W_0 when $U = 0$ and there is no contact with the rod: thus it is reasonable to replace W_0 by a variable representing the difference between W_0 and PA.

Thus, let

$$W_0 = P\,A + R_0.$$

Substituting for W_0 in Eq. (1) we obtain, if U/A is so small that $(1 + U/A)$ may be considered to be equal to unity for all values of θ;

$$M = M_0 + P\,A\,[U - U_0\cos\theta - (R_0/P)\,(1 - \cos\theta)] - \Sigma\,A\,Q_i \sin(\theta - \Phi_i). \tag{2}$$

It is here convenient to introduce non-dimensional variables, defined as follows:

$$\left.\begin{aligned} u &= \frac{U}{A}, \\ m &= \frac{M}{P\,A^2}, \\ r &= \frac{R}{P\,A}, \\ q &= \frac{Q}{P\,A}. \end{aligned}\right\} \tag{3}$$

The non-dimensional form of (2) is thus:

$$m = m_0 + u - u_0\cos\theta - r_0(1 - \cos\theta) - \Sigma\,q_i \sin(\theta - \Phi_i). \tag{4}$$

The size of u which can be tolerated depends largely upon the amount of inaccuracy which is acceptable. For the purposes of this paper values of u as large as 1/10 may be permissible.

---

[1] Assuming that the shell is thin.

The forces Q_i are linked by two equilibrium equations. For the case under consideration, $\theta = \pi/2$ is taken as an axis of symmetry, and the only relevant equilibrium equation is (in non-dimensional terms):

$$\Sigma\, q_i \sin \Phi_i = u_0 - r_0, \tag{5}$$

where the summation is over the whole range $0 \leq \Phi_i \leq \pi/2$.

(b) The Behaviour of a Shell-Element Subject to Circumferential Direct Stress and Bending Moment

The work of this section is based upon a VON MISES type of n-power tri-axial creep law and the assumption that, in all parts of the shell, the stresses associated with bending are small compared to those associated with direct compression. The n-power type of creep law (that is, a creep law of the type (strain-rate) = (material property) (stress)^n) is used primarily because of its simplicity. It presupposes that primary creep is unimportant (which may not, in fact, be justifiable) and that a value of n may be found to fit the law to some appropriate experimental results. Insofar as the order of magnitude of the stresses at all points in the major wrinkle does not change with time, this is justifiable. The assumption that bending stresses are small relative to direct stresses is introduced so that the stress/strain-rate law may be linearised. This will always be so if the value of m is sufficiently small compared to unity, but it is shown later that the linear moment/curvature rate relationship derived is valid for a larger range of bending moment. Let $\dot{\varepsilon}_1$, $\dot{\varepsilon}_2$ and $\dot{\varepsilon}_3$ represent creep strain rates on the principal axes 1, 2 and 3 in the material, and σ_1, σ_2 and σ_3 the direct stresses on the same axes. Let compressive stress and "contracting" strain be regarded as positive. The n-power VON MISES tri-axial creep law may be expressed as: (See Ref. [2], p. 536).

$$\left.\begin{aligned}
\dot{\varepsilon}_1 &= K_n \left(\frac{(\sigma_1-\sigma_2)^2+(\sigma_2-\sigma_3)^2+(\sigma_3-\sigma_1)^2}{2}\right)^{\frac{n-1}{2}} \left(\sigma_1 - \frac{1}{2}\sigma_2 - \frac{1}{2}\sigma_3\right),\\
\dot{\varepsilon}_2 &= K_n \left(\frac{(\sigma_1-\sigma_2)^2+(\sigma_2-\sigma_3)^2+(\sigma_3-\sigma_1)^2}{2}\right)^{\frac{n-1}{2}} \left(-\frac{1}{2}\sigma_1 + \sigma_2 - \frac{1}{2}\sigma_3\right),\\
\dot{\varepsilon}_3 &= K_n \left(\frac{(\sigma_1-\sigma_2)^2+(\sigma_2-\sigma_3)^2+(\sigma_3-\sigma_1)^2}{2}\right)^{\frac{n-1}{2}} \left(-\frac{1}{2}\sigma_1 - \frac{1}{2}\sigma_2 + \sigma_3\right),
\end{aligned}\right\} \tag{6}$$

where K_n and n typify the material at the working temperature.

Let $\dot{\bar{\varepsilon}}_1$ be the circumferential strain rate of the central surface of the shell, and $\dot{C}_1$ be the rate of change of curvature in the (1) direction (see Fig. 5), positive changes in curvature being caused by positive values of M_1. If we assume, as is reasonable, that "plane sections remain plane", that $\dot{\varepsilon}_2 = 0$ (which is so in the symmetrical case: see Eqs. (6)) and that generators of the shell remain straight during deformation, we may show, if the shell is so thin that σ_3 may be neglected and if "bending" strain rates are small compared to direct strain rates, that

$$\bar{\dot{\varepsilon}}_1 = K_n \left(\frac{3}{4}\right)^{\frac{n+1}{2}} \left(\frac{P A}{H}\right)^n \tag{7}$$

and

$$\dot{C}_1 = \bar{\dot{\varepsilon}}_1 \, n \, \frac{2}{H} \left(\frac{6 M_1}{H^2}\right) \left(\frac{H}{P A}\right). \tag{8}$$

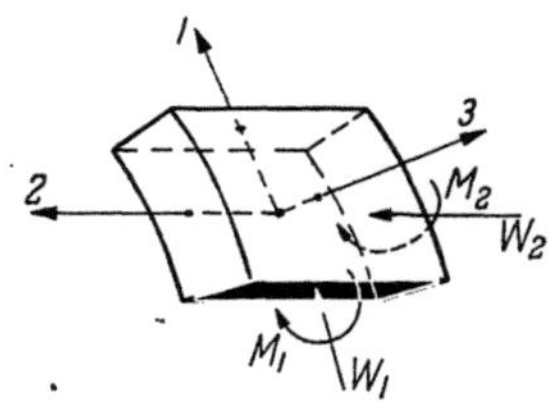

Fig. 5. Shell element

Introducing a shell parameter γ as follows:

$$\gamma = \frac{H^2}{12 \, n \, A^2}. \tag{9}$$

Eq. (8) becomes

$$\frac{A \, \dot{C}}{\bar{\dot{\varepsilon}}} = \frac{m}{\gamma}. \tag{10}$$

Now, although Eq. (10) is based upon the hypothesis that bending stresses are small relative to direct stresses, it is, in fact legitimate to use Eq. (10) for a considerably larger range of m than this hypothesis implies.

Young [*3*] has investigated the behaviour of rectangular beams (which are analagous to plates) under the combined action of bending moments and compressive forces for an n-power creep law by taking the position of the neutral axis (which may be outside the section) as a dummy variable, and evaluating expressions for bending moment and rate of change of curvature in terms of this variable. Fig. 6 gives Young's results on a logarithmic plot, with the axes labelled according to the present notation. From the graph it is seen that for small values of $m \, A/H$, Eq. (10) represents very well the behaviour of the section. As m increases, the section at first becomes stiffer in creep, but eventually, as m increases further, less stiff than Eq. (10) indicates. It is the region of enhanced creep stiffness which extends the range of validity of Eq. (10). This increase in creep stiffness, which is not observed for I-section beams or sandwich plates, is due to the participation of the central material of the section in resisting bending. From the figure it is seen that for all values of n up to 5 Eq. (10) is pessimistic with regard to creep resistance to bending for

$$\frac{\dot{C} \, H}{\bar{\dot{\varepsilon}}} \leq 10, \text{ approximately}. \tag{11}$$

From Eq. (10) and the definition of γ [Eq. (9)], we find that Eq. (11) gives

$$\frac{m}{u_0} \leq \frac{5}{6} \, \frac{H}{n \, U_0}. \tag{12}$$

This inequality is used in the Appendix to establish a limit on the validity of the calculations described in the remainder of the paper.

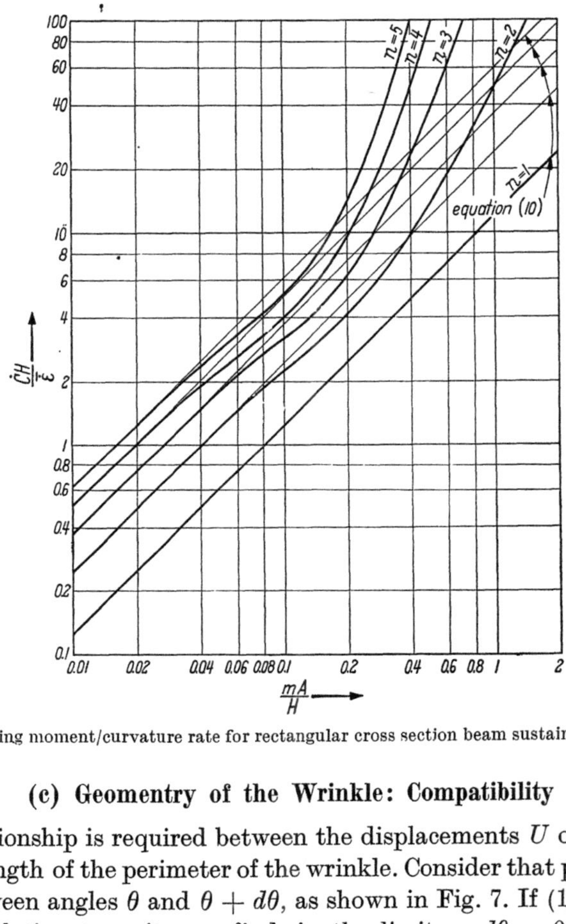

Fig. 6. Bending moment/curvature rate for rectangular cross section beam sustaining axial load

(c) Geomentry of the Wrinkle: Compatibility

A relationship is required between the displacements U of a wrinkle and the length of the perimeter of the wrinkle. Consider that piece of shell lying between angles θ and $\theta + d\theta$, as shown in Fig. 7. If $(1/A)\ (\partial U/\partial \theta)$ is small relative to unity we find, in the limit as $d\theta \to 0$

$$\frac{1}{A}\frac{\partial S}{\partial \theta} = (1+u)\left(1+\frac{1}{2}\left(\frac{\partial u}{\partial \theta}\right)^2\right). \tag{13}$$

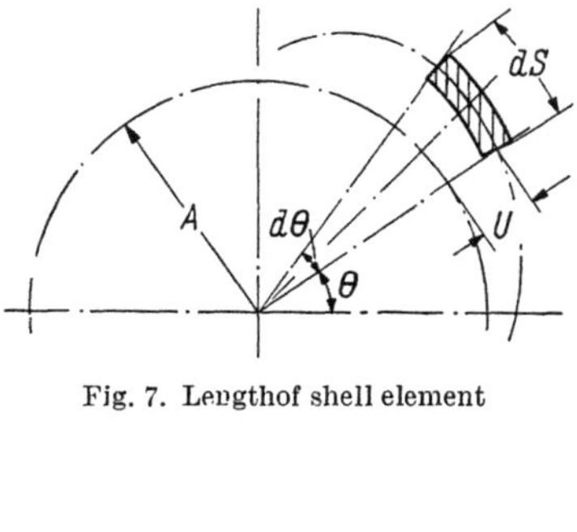

Fig. 7. Lengthof shell element

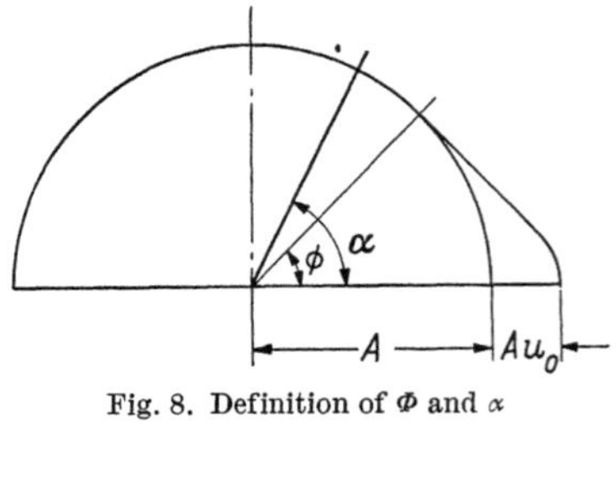

Fig. 8. Definition of Φ and α

Let us examine the relative magnitudes of the various terms in the R. H. S. of Eq. (13). To do this, let us consider a simple wrinkle given by

$$u = \frac{u_0}{2}\left(1 + \cos\frac{\pi\,\theta}{\Phi}\right) \tag{14}$$

that is, a wrinkle of height u_0 and semi-angle Φ (see Fig. 8). Substituting for u in Eq. (13) from Eq. (14) we obtain

$$\frac{1}{A}\frac{\partial S}{\partial \theta} = \left\{1 + \frac{u_0}{2}\left(1 + \cos\frac{\pi\,\theta}{\Phi}\right)\right\}\left\{1 + \frac{u_0^2}{8}\left(\frac{\pi}{\Phi}\right)^2 \sin^2\frac{\pi\,\theta}{\Phi}\right\}. \tag{15}$$

Integrating between $\theta = 0$ and $\theta = \Phi$ we arrive at

$$S = A\,\Phi\left(1 + \frac{u_0}{2}\right)\left(1 + \frac{u_0^2}{16}\left(\frac{\pi}{\Phi}\right)^2\right).$$

As u_0 is small compared to unity by hypothesis, it seems justifiable to neglect the u_0^2 term, and thus to have:

$$S = A\,\Phi\left(1 + \frac{u_0}{2}\right). \tag{16}$$

The accuracy of Eq. (16) depends upon the values of u_0 and Φ: for accuracy better than 1%,

$$\left(\frac{u_0}{4}\cdot\frac{\pi}{\Phi}\right)^2 \leq \frac{1}{100},$$

i. e.

$$u_0 \leq \frac{0\cdot 4\,\Phi}{\pi}. \tag{17}$$

Fig. 9 shows graphically the range of u_0 and Φ for which the accuracy of S, computed from Eq. (16) is better than, or equal to 1%: it also shows relationships between u_0 and Φ for two other limits of accuracy. In general, neglect of the (u_0^2) term is pessimistic from the wrinkling point of view, for the use of Eq. (16) overestimates the height of a wrinkle, given the length of its perimeter and the angle subtended by it.

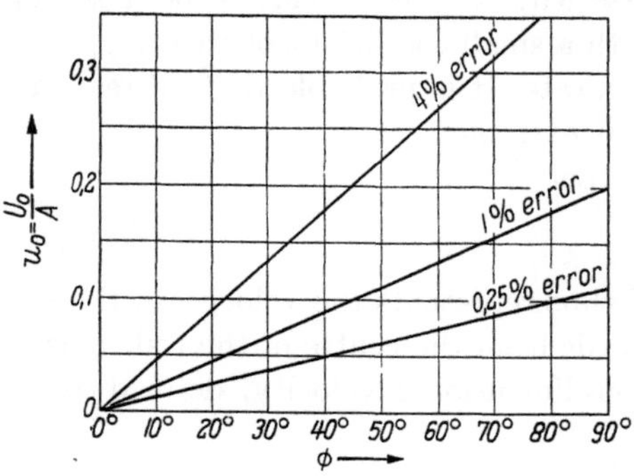

Fig. 9. Accuracy of Eq. (16)

The relationship of Eq. (17) between the values of Φ and u_0 to give a specified accuracy to Eq. (16) is probably approximately correct for most practical wrinkle shapes.

Assuming that the second bracket in Eq. (15) may be replaced by unity, we have

$$\frac{1}{A}\frac{\partial S}{\partial \theta} = 1 + u.$$

Thus the total perimeter of the shell is given by

$$S = 2\pi A + A \int_0^{2\pi} u \, d\theta . \tag{18}$$

Now the length of the perimeter of the shell is changing continuously due to the action of circumferential stress in those parts of the shell which are not fully supported by the rod. Leaving aside for the present the question of how much of the shell is under the full circumferential force PA, and working in terms of an angle α over which full shortening is taking place (see Fig. 8) we have, using Eq. (18)

$$\bar{\dot{\varepsilon}} = -\frac{\partial S}{\partial T} \cdot \frac{1}{A\alpha} = -\frac{1}{A\alpha} \int_0^{\alpha} \left(\frac{\partial U}{\partial T}\right) d\theta . \tag{19}$$

Here, $\bar{\dot{\varepsilon}}$ is the mean (compressive) strain rate over the portion of the shell undergoing compression, and T is time (see Fig. 8). In section 2a it was pointed out that W, the circumferential stress resultant, was approximately equal to PA: it thus seems reasonable to assume that $\bar{\dot{\varepsilon}} = \bar{\dot{\varepsilon}}_1$ for a first approximation. If refinements to the theory are to be made later, α can be adjusted conveniently in compensation. For the present analysis, it seems appropriate to examine the effect of variation of α by considering the limits of all possible variation. It is clear from Fig. 8 that $\Phi \leq \alpha \leq \pi/2$. Thus, as a first attempt it is reasonable to examine both $\alpha = \Phi$ and $\alpha = \pi/2$ for all shell-rod geometries considered. For a real system with a high coefficient of friction between rod and shell, the case with $\alpha = \Phi$ is likely to be most representative, while for a system with a small coefficient of friction, the ($\alpha = \pi/2$) case may well be more accurate. It is probable that all real cases lie somewhere between these two extremes.

Let

$$V = \frac{\partial U}{\partial T} . \tag{20}$$

V thus represents the velocity of the shell, positive when measured outwards from the centre of the rod. It is convenient to work in terms of a non-dimensional velocity, defined as follows:

$$v = \frac{V}{A\,\dot{\varepsilon}} . \tag{21}$$

Using Eqs. (20) and (21) in Eq. (19), we have

$$\alpha = -\int_0^{\pi/2} v \, d\theta . \tag{22}$$

Eq. (22) is the required geometrical relationship.

A further equation is needed, to relate $\dot{C}$, the rate of change of curvature with time, to V, the velocity of the shell. It is readily shown (see, for example, Ref. [4], chapter 4) that, if C represents the curvature of the shell at any point minus the curvature of a shell of radius A

$$C = -\frac{\left(U + \frac{\partial^2 U}{\partial \theta^2}\right)}{A^2}. \tag{23}$$

Differentiating Eq. (23) with respect to time, and substituting from Eqs. (20) and (21), we obtain the non-dimensional relationship

$$\frac{A\,\dot{C}}{\dot{\varepsilon}} = -\left(v + \frac{\partial^2 v}{\partial \theta^2}\right). \tag{24}$$

(d) Governing Equations of the Shell

Having established the equilibrium [(4) and (5)], stress/strain-rate [(7) and (10)] and compatibility [(22) and (24)] relations, we are in a position to attempt a solution to the problem. Combining Eqs. (4), (10) and (24) we have

$$-\gamma\left(v + \frac{\partial^2 v}{\partial \theta^2}\right) = m_0 + u - u_0 \cos\theta - r_0(1 - \cos\theta) - \Sigma\, q_i \sin(\theta - \Phi_i). \tag{25}$$

For the case of the shell shown in Fig. 10, which has axes of symmetry at $\theta = 0$ and $\theta = \pi/2$ we have also in addition to Eqs. (5) and (22) two conditions specifying symmetry of displacement about $\theta = 0$ and $\theta = \pi/2$:

$$\left.\begin{aligned} \frac{\partial v}{\partial \theta} &= 0 \text{ at } \theta = 0, \\ \frac{\partial v}{\partial \theta} &= 0 \text{ at } \theta = \frac{\pi}{2}. \end{aligned}\right\} \tag{26}$$

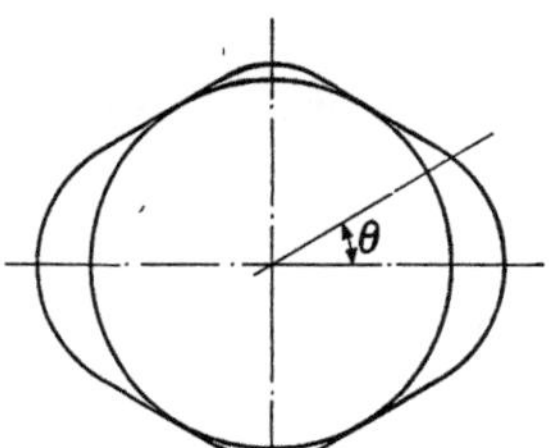

Fig. 10. Symmetrical wrinkle

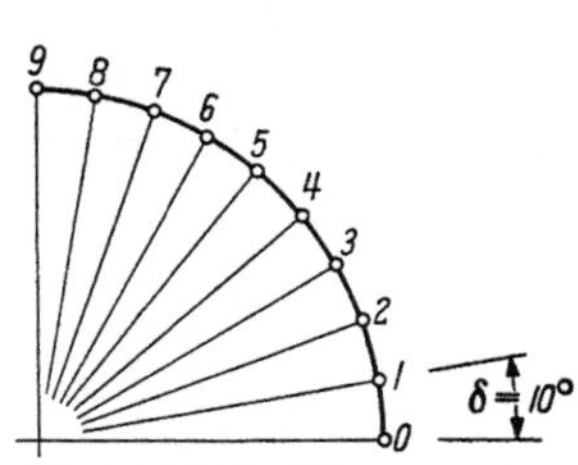

Fig. 11. Finite difference system

Eqs. (25), (5), (22) and (26) may be solved, in principle, to give values of v for any specified set of values of u.

(e) Matrix Formulation of the Governing Equations

In order to obtain an understanding of shell behaviour, it was decided to obtain a numerical solution of the four equations. For this purpose θ was divided into 10° finite-difference intervals, as shown in Fig. 11. The size of finite-difference interval was a compromise between the disadvantages of having to deal with large numbers of equations, and the accuracy of the solution required. The 10° interval served, as will be seen, to lead to a reasonably clear understanding of the creep wrinkling process lacking only, possibly, great accuracy and some detail.

The simplest way of dealing with the reactions q between the rod and the shell is to replace all reactions by a series of concentrated reactions $q_0, q_1, \ldots, q_9$ at the nodal points. The other unknowns are, in general, the 10 nodal velocities $v_0, v_1, \ldots, v_9$, and m_0 and r_0, making, in all, 22 unknowns. The 22 equations in these unknowns are made up as follows:

10 finite-difference formulations of Eq. (25) at the nodal points, incorporating finite-difference equivalents of Eq. (26) at the points 0 and 9.

1 Finite-difference representation of Eq. (22).

1 Equilibrium Eq. (5).

10 Trivial equations of the form q_i or $v_i = 0$; that is, q_i is zero except at nodal points where the shell is considered to touch the rod, which are where $v_i = 0$.

In order to express Eqs. (25) and (22) in finite-difference form, we need finite-difference representations of $\partial^2 v/\partial\theta^2$ and $\int_0^{\pi/2} v\,d\theta$. The simplest finite-difference second derivative formula is:

$$\left.\frac{\partial^2 v}{\partial\theta^2}\right)_i = \frac{v_{i+1} - 2\,v_i + v_{i-1}}{\delta^2}, \qquad i = 1, 2, \ldots, 8, \tag{27}$$

where δ represents (in radian measure) the finite-difference interval of 10°. At points 0 and 9, where the symmetry conditions (26) hold, the equations corresponding to (27) are:

$$\left.\begin{aligned} \left.\frac{\partial^2 v}{\partial\theta^2}\right)_0 &= \frac{2\,v_1 - 2\,v_0}{\delta^2}, \\ \left.\frac{\partial^2 v}{\partial\theta^2}\right)_9 &= \frac{2\,v_8 - 2\,v_9}{\delta^2}. \end{aligned}\right\} \tag{28}$$

The simplest integration formula is the "trapezoidal" rule, which is based upon the assumption that integrand is linear between nodal points. For the system shown in Fig. 11 this gives

$$\int_0^{\pi/2} v\,d\theta = \delta\left(\frac{1}{2}\,v_0 + v_1 + v_2 + \cdots + v_7 + v_8 + \frac{1}{2}\,v_9\right). \tag{29}$$

Now, in order to deal with numbers of convenient magnitude, we normalise the non-dimensional variables to the value of u_0 when the shell first touches the rod; thus

$$\left.\begin{aligned} m' &= \frac{m}{u_0^*}, \\ r' &= \frac{r}{u_0^*}, \\ u' &= \frac{u}{u_0^*}, \\ v' &= \frac{v}{u_0^*}, \\ q' &= \frac{q}{u_0^*}, \end{aligned}\right\} \qquad (30)$$

where u_0^* is the special value of u_0.

In matrix form, the 22 equations in the newly-defined variables are as shown to theright, where

$$c_i = \cos(i\delta)$$

$$c_i' = \cos(i\,\delta) - 1$$

$$k = \frac{\gamma}{\delta^2}$$

and

$$2' = 2 - \delta^2.$$

$$\left|\begin{array}{cc|cccccccccc|cccccccccc}
1 & \cdot & -2' & 2 & \cdot & \cdot & \cdot & \cdot & \cdot & \cdot & \cdot & \cdot & \cdot & \cdot & \cdot & \cdot & \cdot & \cdot & \cdot & \cdot & \cdot & \cdot \\
1 & c'_1 & 1 & -2' & 1 & \cdot & \cdot & \cdot & \cdot & \cdot & \cdot & \cdot & -c_8 & \cdot & \cdot & \cdot & \cdot & \cdot & \cdot & \cdot & \cdot & \cdot \\
1 & c'_2 & \cdot & 1 & -2' & 1 & \cdot & \cdot & \cdot & \cdot & \cdot & \cdot & -c_7 & -c_8 & \cdot & \cdot & \cdot & \cdot & \cdot & \cdot & \cdot & \cdot \\
1 & c'_3 & \cdot & \cdot & 1 & -2' & 1 & \cdot & \cdot & \cdot & \cdot & \cdot & -c_6 & -c_7 & -c_8 & \cdot & \cdot & \cdot & \cdot & \cdot & \cdot & \cdot \\
1 & c'_4 & \cdot & \cdot & \cdot & 1 & -2' & 1 & \cdot & \cdot & \cdot & \cdot & -c_5 & -c_6 & -c_7 & -c_8 & \cdot & \cdot & \cdot & \cdot & \cdot & \cdot \\
1 & c'_5 & \cdot & \cdot & \cdot & \cdot & 1 & -2' & 1 & \cdot & \cdot & \cdot & -c_4 & -c_5 & -c_6 & -c_7 & -c_8 & \cdot & \cdot & \cdot & \cdot & \cdot \\
1 & c'_6 & \cdot & \cdot & \cdot & \cdot & \cdot & 1 & -2' & 1 & \cdot & \cdot & -c_3 & -c_4 & -c_5 & -c_6 & -c_7 & -c_8 & \cdot & \cdot & \cdot & \cdot \\
1 & c'_7 & \cdot & \cdot & \cdot & \cdot & \cdot & \cdot & 1 & -2' & 1 & \cdot & -c_2 & -c_3 & -c_4 & -c_5 & -c_6 & -c_7 & -c_8 & \cdot & \cdot & \cdot \\
1 & c'_8 & \cdot & \cdot & \cdot & \cdot & \cdot & \cdot & \cdot & 1 & -2' & 1 & -c_1 & -c_2 & -c_3 & -c_4 & -c_5 & -c_6 & -c_7 & -c_8 & \cdot & \cdot \\
1 & c'_9 & \cdot & \cdot & \cdot & \cdot & \cdot & \cdot & \cdot & \cdot & 2 & -2' & -c_0 & -c_1 & -c_2 & -c_3 & -c_4 & -c_5 & -c_6 & -c_7 & -c_8 & \cdot \\
\hline
\cdot & \cdot & 1/2 & 1 & 1 & 1 & 1 & 1 & 1 & 1 & 1 & 1/2 & \cdot & \cdot & \cdot & \cdot & \cdot & \cdot & \cdot & \cdot & \cdot & \cdot \\
\cdot & 1 & \cdot & \cdot & \cdot & \cdot & \cdot & \cdot & \cdot & \cdot & \cdot & \cdot & \cdot & c_8 & c_7 & c_6 & c_5 & c_4 & c_3 & c_2 & c_1 & c_0 \\
\hline
\cdot & \cdot & 1/0 & \cdot & \cdot & \cdot & \cdot & \cdot & \cdot & \cdot & \cdot & \cdot & 0/1 & \cdot & \cdot & \cdot & \cdot & \cdot & \cdot & \cdot & \cdot & \cdot \\
\cdot & \cdot & \cdot & 1/0 & \cdot & \cdot & \cdot & \cdot & \cdot & \cdot & \cdot & \cdot & \cdot & 0/1 & \cdot & \cdot & \cdot & \cdot & \cdot & \cdot & \cdot & \cdot \\
\cdot & \cdot & \cdot & \cdot & 1/0 & \cdot & \cdot & \cdot & \cdot & \cdot & \cdot & \cdot & \cdot & \cdot & 0/1 & \cdot & \cdot & \cdot & \cdot & \cdot & \cdot & \cdot \\
\cdot & \cdot & \cdot & \cdot & \cdot & 1/0 & \cdot & \cdot & \cdot & \cdot & \cdot & \cdot & \cdot & \cdot & \cdot & 0/1 & \cdot & \cdot & \cdot & \cdot & \cdot & \cdot \\
\cdot & \cdot & \cdot & \cdot & \cdot & \cdot & 1/0 & \cdot & \cdot & \cdot & \cdot & \cdot & \cdot & \cdot & \cdot & \cdot & 0/1 & \cdot & \cdot & \cdot & \cdot & \cdot \\
\cdot & \cdot & \cdot & \cdot & \cdot & \cdot & \cdot & 1/0 & \cdot & \cdot & \cdot & \cdot & \cdot & \cdot & \cdot & \cdot & \cdot & 0/1 & \cdot & \cdot & \cdot & \cdot \\
\cdot & \cdot & \cdot & \cdot & \cdot & \cdot & \cdot & \cdot & 1/0 & \cdot & \cdot & \cdot & \cdot & \cdot & \cdot & \cdot & \cdot & \cdot & 0/1 & \cdot & \cdot & \cdot \\
\cdot & \cdot & \cdot & \cdot & \cdot & \cdot & \cdot & \cdot & \cdot & 1/0 & \cdot & \cdot & \cdot & \cdot & \cdot & \cdot & \cdot & \cdot & \cdot & 0/1 & \cdot & \cdot \\
\cdot & \cdot & \cdot & \cdot & \cdot & \cdot & \cdot & \cdot & \cdot & \cdot & 1/0 & \cdot & \cdot & \cdot & \cdot & \cdot & \cdot & \cdot & \cdot & \cdot & 0/1 & \cdot \\
\cdot & \cdot & \cdot & \cdot & \cdot & \cdot & \cdot & \cdot & \cdot & \cdot & \cdot & 1/0 & \cdot & \cdot & \cdot & \cdot & \cdot & \cdot & \cdot & \cdot & \cdot & 0/1
\end{array}\right|
\times
\left|\begin{array}{c}
m'_0 \\ r'_0 \\ \hline kv'_0 \\ kv'_1 \\ kv'_2 \\ kv'_3 \\ kv'_4 \\ kv'_5 \\ kv'_6 \\ kv'_7 \\ kv'_8 \\ kv'_9 \\ \hline q'_0 \\ q'_1 \\ q'_2 \\ q'_3 \\ q'_4 \\ q'_5 \\ q'_6 \\ q'_7 \\ q'_8 \\ q'_9
\end{array}\right|
=
\left|\begin{array}{ccccccccccc}
\cdot & \cdot & \cdot & \cdot & \cdot & \cdot & \cdot & \cdot & \cdot & \cdot & \cdot \\
c_1 & -1 & \cdot & \cdot & \cdot & \cdot & \cdot & \cdot & \cdot & \cdot & \cdot \\
c_2 & \cdot & -1 & \cdot & \cdot & \cdot & \cdot & \cdot & \cdot & \cdot & \cdot \\
c_3 & \cdot & \cdot & -1 & \cdot & \cdot & \cdot & \cdot & \cdot & \cdot & \cdot \\
c_4 & \cdot & \cdot & \cdot & -1 & \cdot & \cdot & \cdot & \cdot & \cdot & \cdot \\
c_5 & \cdot & \cdot & \cdot & \cdot & -1 & \cdot & \cdot & \cdot & \cdot & \cdot \\
c_6 & \cdot & \cdot & \cdot & \cdot & \cdot & -1 & \cdot & \cdot & \cdot & \cdot \\
c_7 & \cdot & \cdot & \cdot & \cdot & \cdot & \cdot & -1 & \cdot & \cdot & \cdot \\
c_8 & \cdot & \cdot & \cdot & \cdot & \cdot & \cdot & \cdot & -1 & \cdot & \cdot \\
c_9 & \cdot & \cdot & \cdot & \cdot & \cdot & \cdot & \cdot & \cdot & -1 & \cdot \\
\cdot & \cdot & \cdot & \cdot & \cdot & \cdot & \cdot & \cdot & \cdot & \cdot & 1 \\
1 & \cdot & \cdot & \cdot & \cdot & \cdot & \cdot & \cdot & \cdot & \cdot & \cdot \\
\cdot & \cdot & \cdot & \cdot & \cdot & \cdot & \cdot & \cdot & \cdot & \cdot & \cdot \\
\cdot & \cdot & \cdot & \cdot & \cdot & \cdot & \cdot & \cdot & \cdot & \cdot & \cdot \\
\cdot & \cdot & \cdot & \cdot & \cdot & \cdot & \cdot & \cdot & \cdot & \cdot & \cdot \\
\cdot & \cdot & \cdot & \cdot & \cdot & \cdot & \cdot & \cdot & \cdot & \cdot & \cdot \\
\cdot & \cdot & \cdot & \cdot & \cdot & \cdot & \cdot & \cdot & \cdot & \cdot & \cdot \\
\cdot & \cdot & \cdot & \cdot & \cdot & \cdot & \cdot & \cdot & \cdot & \cdot & \cdot \\
\cdot & \cdot & \cdot & \cdot & \cdot & \cdot & \cdot & \cdot & \cdot & \cdot & \cdot \\
\cdot & \cdot & \cdot & \cdot & \cdot & \cdot & \cdot & \cdot & \cdot & \cdot & \cdot \\
\cdot & \cdot & \cdot & \cdot & \cdot & \cdot & \cdot & \cdot & \cdot & \cdot & \cdot \\
\cdot & \cdot & \cdot & \cdot & \cdot & \cdot & \cdot & \cdot & \cdot & \cdot & \cdot
\end{array}\right|
\times
\left|\begin{array}{c}
u'_0 \\ u'_1 \\ u'_2 \\ u'_3 \\ u'_4 \\ u'_5 \\ u'_6 \\ u'_7 \\ u'_8 \\ u'_9 \\ f
\end{array}\right|
\qquad (31)$$

The meaning of symbols 1/0 and 0/1 is that in the two diagonal sub-matrices, one of the two entries in each row is zero, and the other is unity, depending on whether or not the shell is in contact with the rod at the nodal point in question. Examination of Eq. (31) will reveal the effect of these alternatives on the equations. The quantity f, in the R. H. S. of Eq. (31) characterises the shell. Using Eq. (29), we see that the 11th equation of (31) gives the finite-difference form of

$$\frac{\gamma}{\delta^2 u_0^*} \int_0^{\pi/2} v \, d\theta = f. \qquad (32)$$

Substituting for $\int_0^{\pi/2} v\, d\theta$ from Eq. (22) in Eq. (32) we have

$$f = -\frac{\alpha\,\gamma}{\delta^3\,u_0^*}\,.$$

It is convenient here to define a shell-rod parameter ν, as follows:

$$\nu = \frac{H^2}{n\,A\,\varDelta}\,, \tag{33}$$

where $\varDelta$ represents the mean radial clearance between the shell and the rod at the onset of pressure-loading. Substituting for γ from Eq. (9). we find:

$$f = \frac{-\alpha\,\nu\,\varDelta}{12\,\delta^3\,U_0^*}\,, \tag{34}$$

where U_0^* $(= A\,u_0^*)$ is the displacement of the centre of the wrinkle when the shell first touches the rod.

Now let

$$\frac{U_0^*}{\varDelta} = \beta. \tag{35}$$

The first case to be solved numerically has $\alpha = \pi/2$ and $\delta = \pi/18$, so equation (44) becomes

$$f = -\,24{,}6\,\frac{\nu}{\beta}\,.$$

The parameter β is a measure of what happened to the shell between the moment of application of external pressure, and the moment of first contact with the shell. Discussion of this aspect of the problem is outside the scope of this paper; however, it is possible to impose pessimistic limits on the value of β. In general, wrinkling is more severe the larger the value of U_0^*; hence it is pessimistic from the point of view of wrinkling to assume that between pressurisation and first contact there is no shortening of the perimeter of the tube. Consider a circular shell with original clearance $\varDelta$ between it and the rod, which, under the action of external pressure comes into contact with the rod as shown in Fig. 12. In this Figure, (a) shows the original arrangement, (b) shows a symmetrical tube collpase and (c) shows the most pessimistic case of one-sided collapse. It is easily shown by use of Eq. (18) that if the lobes in (b) and (c) take the form

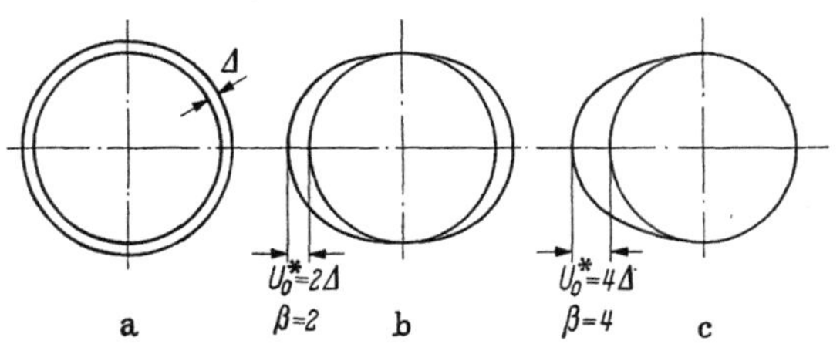

Fig. 12. Examples of β

$$u = \frac{u_0}{2}\,(1 + \cos 2\,\theta)$$

and there is no shortening of the perimeter of the shell, $\beta = 2$ and $\beta = 4$ for cases (b) and (c) respectively.

(f) Results of Matrix Calculations

Starting from a deflected shape given by

$$u' = \frac{1}{2}(1 + \cos 2\,\theta)$$

the progress of wrinkles for a set of different shell/rod geometries was followed by obtaining solutions to Eq. (31) for values of f equal to 0, 0.5, 1, 2, 4 and 8.

At first contact

$$v_9' = q_0' = q_1' = \cdots = q_8' = 0;$$

the Eqs. (31) were reduced to 12 in number by the elimination of these trivial variables, and solved by "DEUCE" computer. When the solution had been made for each value of f, a time-interval was chosen for each case so that, on the assumption of unchanging velocity during the time-interval, the shell just came into contact with the rod at nodal point 8 at the end of the interval. In this way, the continuous rolling of the point of contact was replaced by a series of sudden changes. Although this analysis did not differentiate between the two stages of behaviour between first contact and contact at nodal point 8 (i. e. stage 1; 2 — point contact becomes 4 — point contact: stage 2; points of contact move round to $\Phi = 80°$, as described in the Introduction), it was probably sufficiently accurate for practical purposes.

The new set of values of u' was used in a second calculation, using Eq. (31) with zero variables q_0', q_1', . . ., q_7', v_8' and q_9'. Again a time-interval was chosen for each value of f at the end of which contact had been made at another nodal point. The whole process was repeated several times.

A check was made on each set of results to ensure that there were no inconsistencies between the q'- and v'-variables. If, for example, contact had been at point 7 for one time-interval, at the end of which contact had also occurred at point 6, it was not possible to tell *a priori* whether contact would still be maintained at point 7 during the next time interval, or whether point 7 on the shell would tend to move outwards. The only way to check this was to assume that (say) contact was maintained and to examine the value of q_7' resulting from the solution of Eq. (31). If in the solution q_7' was positive, the assumption about contact was correct: if, however, q_7' was negative, this implied that the shell in fact tended to move outwards, since only positive reactions between the rod and the shell are permissible. Similarly, if it was assumed that there was no contact at point 7, and v_7' turned out to be negative, this would indicate that there was, in fact, contact, and that a re-calculation would have to be made.

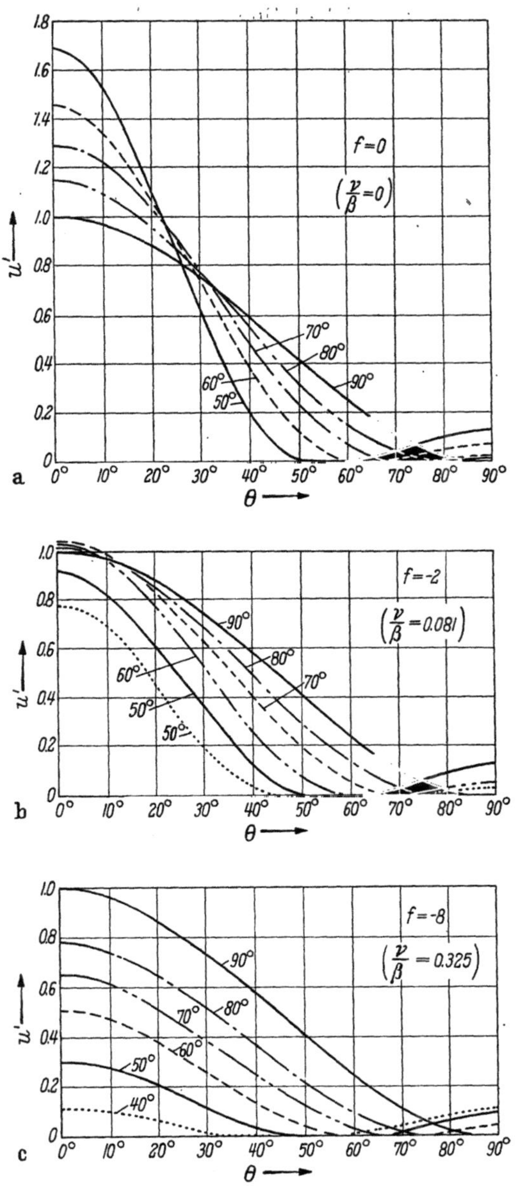

Fig. 13. Progressive deformation of wrinkles

Fig. 13 shows progressive deflection for three different values of f. In these graphs, each curve is labelled with the smallest angle, measured from the centre of the wrinkle, at which contact between shell and rod occurs. Fig. 13a shows that, for $f = 0$, the "wrinkle angle" Φ becomes smaller as time increases. At first, as the point of contact moves round the rod, the part of the shell immediately behind the point of contact lifts up, so that the height of the "back wrinkle" reaches a maximum when Φ is about 50°. At this stage the point of contact splits, so that there is contact at both $\theta = 50°$ and $\theta = 60°$, and the "back wrinkle" begins to flatten. Meanwhile, the height of the main wrinkle increases steadily with time.

For $f = -2$ (Fig. 13b) the behaviour of the shell is similar to that for $f = 0$, except that, apart from a small initial increase in the height of the wrinkle, the wrinkle becomes progressively smaller in height as time passes.

For $f = -8$ (Fig. 13c) the height of the main wrinkle decreases steadily with time, and by the time the semi-wrinkle angle has become about 40°, the height of the back wrinkle is almost equal to that of the main wrinkle.

When 5 time-intervals had been completed for each of the 5 values of f, an attempt was made to correlate the significant features of the wrinkling process for each value of f. It is clear that, if the wrinkle eventually flattens out, its deflection and curvature (strictly, difference in curvature from its curvature in the fully flattened state) both become zero. In particular, the central deflection and curvature both become zero. It would thus be interesting to know what path in deflection/curvature space is taken by the central point of the wrinkle to the origin. Working in non-dimensional terms we have:

$$\left.\begin{aligned} &\text{central deflection } x = u_0', \\ &\text{central curvature } c = C_0\,A = -\left(\frac{\partial^2 u'}{\partial \theta^2} + u'\right)_0. \end{aligned}\right\} \qquad (36)$$

For each of the deflected shapes derived from the numerical work already described, the value of $\partial^2 u/\partial^2\theta)_0$ was computed numerically by the use of Eq. (28) and the plot of Fig. 14 was made. This shows clearly how

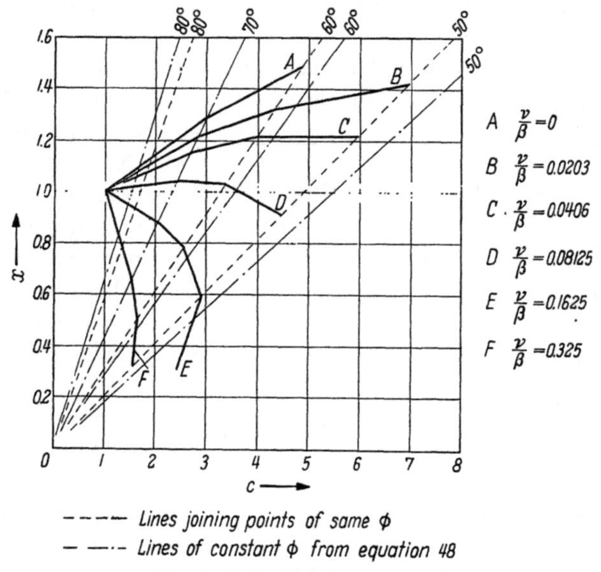

Fig. 14. Deflection/curvature graph

central deflection and curvature are inter-related. For the two largest values of ν/β a maximum curvature was reached, after which both the deflection and curvature decreased. For the smaller values of ν/β, no maximum curvature was reached for $\Phi \geq 50°$, but it seemed plausible that for each small, non-zero value of ν/β, the $c - x$ trajectory would, with decreasing Φ, reach a maximum value of c and eventually pass through the origin.

At this stage, a remarkable feature was noticed in Fig. 14; namely that the points on the different $c - x$ curves corresponding to the same value of Φ lay very nearly on a set of straight lines passing through the origin. This suggested a similarity in shape between the wrinkles, which was confirmed by the following analysis.

Consider a wrinkle defined by

$$u' = \frac{u'_0}{2}\left(1 + \cos\frac{\pi\,\theta}{\Phi}\right), \tag{37}$$

where Φ is the half-wrinkle angle. Use of Eq. (36) shows that the non-dimensional curvature at the centre of the wrinkle ($\theta = 0$) is given by

$$c = x\left(\frac{\pi^2}{2\,\Phi^2} - 1\right). \tag{38}$$

That is, for a given value of Φ, c and x lie on a straight line passing through the origin. Lines of constant Φ according to Eq. (38) are plotted in Fig. 14, and are in reasonably good agreement with those derived from the matrix calculations. Several points illuminated by this discovery merit further discussion.

Firstly, it is apparent that if the matrix calculations are carried on for the smaller values of ν/β, very large steps will be made as Φ becomes smaller than 40° (see Fig. 14). Now the taking of large steps always leads to an over-estimation of the maximum value, c_{max}, of c. This is because (see Fig. 15) the $c - x$ curves are convex away from the $c = 0$ axis, and a "step" is represented by a straight line tangential to a curve closely related to a true $c - x$ curve. For the smaller values of ν/β it is clear that the $c - x$ curves will be sharply convex at the point where c is maximum, and it would thus be particularly unfortunate to have to use large steps in this region.

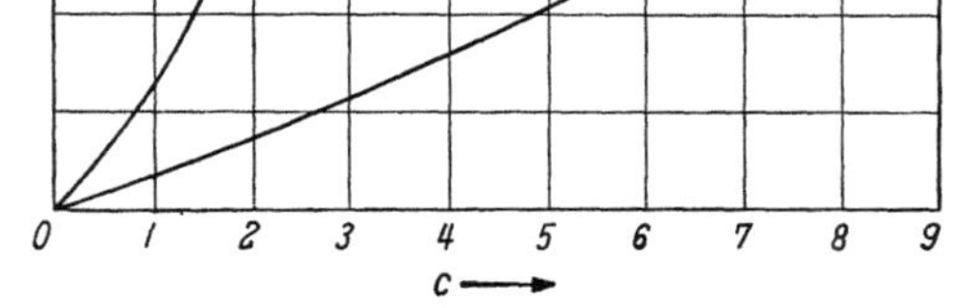

Fig. 15. Construction of c—x curves

Secondly, the fact that all the wrinkles so far charted have conformed closely to Eq. (37) suggests that, if some way of dealing with the subsidiary wrinkles can be found, it may be possible to build up $c - x$ curves by using a set of 'standard' solutions (one for each value of Φ) based on the deflected shape given by Eq. (37). The $c - x$ trajectory would then be made by successive integration of m_0' and v_0', thus eliminating the necessity for enquiring closely about conditions at the extremities of the wrinkle.

The desirability of taking step-sizes in Φ of less than 10° suggests two possible courses of action.

i to use a smaller finite-difference angle.

ii somehow to interpolate between angles Φ spaced by 10°.

Of these alternatives, the second is more attractive, when used in conjunction with the "standard solutions" concept discussed above.

A study of the wrinkles and sub-wrinkles in Fig. 13 reveals that the maximum height of the sub-wrinkles is generally not great relative to the height of the associated main wrinkle. Thus it may be possible to use as a basis for the "standard solution" at each stage a deflected shape given by

$$\left.\begin{aligned} u' &= \frac{u_0'}{2}\left(1+\cos\frac{\pi\,\theta}{\Phi}\right) \qquad 0\le\theta\le\Phi, \\ u' &= 0 \qquad\qquad\qquad\qquad \Phi\le\theta\le\pi/2. \end{aligned}\right\} \tag{39}$$

It is convenient for the "standard solutions" to normalise all variables by dividing by u_0'; thus f is replaced by f/u_0', $\gamma\, v_0$ by $\gamma\, v_0'/u_0'$ (or, equivalently, $\gamma\, v_0/u_0$) and m_0' by m_0'/u_0' (or, equivalently, m_0/u_0).

In order to test the hypothesis of the validity of the use of Eq. (39) for the standard cases, a set of matrix solutions was obtained for each of $\Phi = 80°$, 70° and 60°. When these solutions (which were made for $u_0' = 1$) had been suitably scaled, it was found that they were in remarkably good agreement with the previous solutions both as regards v_0' and m_0'. For each of these calculations it was surmised that there would only be one point of contact with the shell, and that the "back wrinkle" would always tend to lift away from the rod. This was in fact the case except that for $\Phi = 60°$ and $f = 0$, v_9' was small and negative. For $\Phi = 50°$, 40° and 30°, it was found that the assumption about single-point bearing gave results in which the back-wrinkle tended to move towards the centre of the rod, which was clearly contrary to hypothesis. In order to find the correct mode of support in each case, a set of solutions was made, each with a different arrangement of support. It was found that the correct mode of support for any value of Φ depended upon the value of f/u_0'; the results are summarised in Table 1.

In general the length of back-wrinkle lifted away from the rod increases as f/u_0' decreases.

A remarkable feature of the results of this set of calculations was the insensitivity of m_0/u_0 and $\gamma\, v_0/u_0$ to the made of support assumed. Fig. 16 shows, for example, the relationship between m_0/u_0, $\gamma\, v_0/u_0$ and f/u_0' for $\Phi = 50°$; it is seen that the different modes of support have little effect

Table 1

Φ	range of $\frac{f}{u'_0}$	points of support of shell						
		30°	40°	50°	60°	70°	80°	90°
50°	$-0.11 \leq f/u'_0 \leq 0$			//		//	//	//
	$-3.5 \leq f/u'_0 \leq -0.11$			//			//	//
	$-24 \leq f/u'_0 \leq -3.5$			//				//
40°	$-1.9 \leq f/u'_0 \leq 0$		//		//	//	//	//
	$-13 \leq f/u'_0 \leq -1.9$		//			//	//	//
	$f/u'_0 \leq -13$		//				//	//
30°	$-0.72 \leq f/u'_0 \leq 0$	//	//	//	//	//	//	//
	$-11 \leq f/u'_0 \leq -0.72$	//		//	//	//	//	//
	$f/u'_0 \leq -11$	//			//	//	//	//

upon the results, as the three segments corresponding to the three regions of f/u'_0 are very nearly in line with each other. Thus, for practical pur-

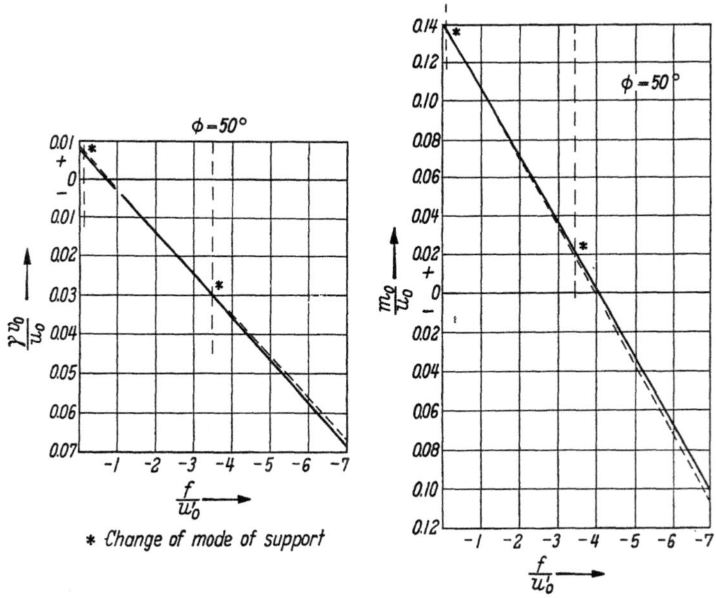

Fig. 16. Typical $\gamma v_0/u_0$, f/u'_0 and m_0/u_0, f/u'_0 graphs

poses, each "standard solution" may be represented by two straight lines on the $\gamma\, v_0/u_0$, f/u'_0 and m_0/u_0, f/u'_0 diagrams respectively. This confirms the use of the simplified deflected shape given by Eq. (39), for the length of back-wrinkle not in contact with the rod will, in reality, depend to some extent upon the previous history of the shell. A study of the numerical results for $\Phi = 50°$, $40°$ and $30°$ showed that the greatest velocities in the back-wrinkle were only about 5% of those at the centre of the main wrinkle, indicating that the back-wrinkle deflections would be small. Fig. 17 summaries the results further by showing values of m_0/u_0

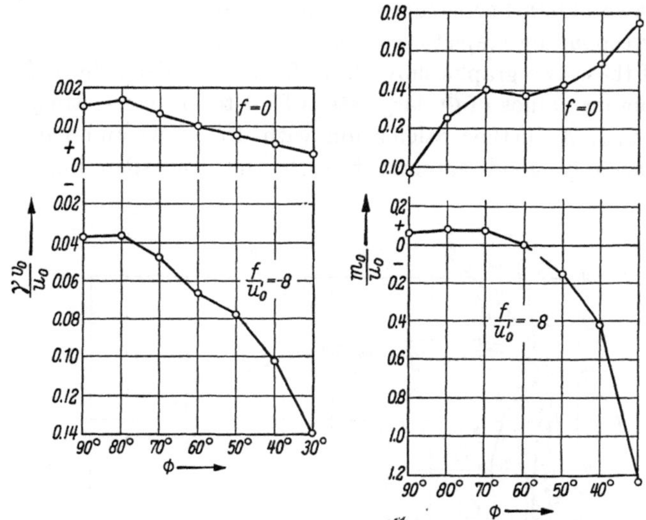

Fig. 17. Relevant information from "standard case" calculations

and $\gamma\, v_0/u_0$ for two different values of f/u_0' at different values of Φ. From this figure, the $\gamma\, v_0/u_0'$, f/u_0' and m_0/u_0, f/u_0' diagrams may be constructed for any value of Φ between 90° and 30° by graphical interpolation.

The method used for step-by-step construction of $c - x$ trajectories is described as follows. Assuming that one (c, x) point is known, the next point is found by the use of the following procedure.

The value of ν/β for the system is used to calculate f/u_0' from the equation

$$\frac{f}{u_0'} = \frac{2.75}{x}\left(\frac{\alpha}{\delta}\right)\left(\frac{\nu}{\beta}\right). \tag{40}$$

This equation is a re-arrangement of Eq. (34) for $\delta = 10°$. The value of f/u_0' is used to obtain values of $\beta\, v_0/u_0$ from the graph like Fig. 17 for the current value of Φ. "Increments" Δx and Δc are computed by use of the formulae

$$\Delta x = \left(\frac{k}{\gamma}\right)\gamma\,\frac{v_0}{u_0},$$

$$\Delta c = \left(\frac{k}{\gamma}\right)\frac{m_0}{u_0},$$

where k/γ is a constant chosen only to give convenient numerical values to Δx and Δc. The *direction* of the $c - x$ trajectory is found by constructing, on the $c - x$ graph, the line joining (c, x) to $(c + \Delta c, x + \Delta x)$. The point at which the line intersects the line of constant Φ corresponding to the end of the 'step' is the new (c, x) point.

Fig. 18a shows $c - x$ trajectories for values of ν/β equal to 0.512, 0.256, 0.128 and 0.064 constructed by taking 10° steps in Φ, and Fig. 18b shows corresponding trajectories constructed in 5° steps. A visual comparison of these two graphs shows how, for the smaller values of ν/β, and for the smaller values of Φ, large steps lead to an over-estimate of the value of c_{max}. A further calculation with 10° steps and $\nu/\beta = 0.162$ gave remarkably good agreement with the corresponding curve in Fig. 14.

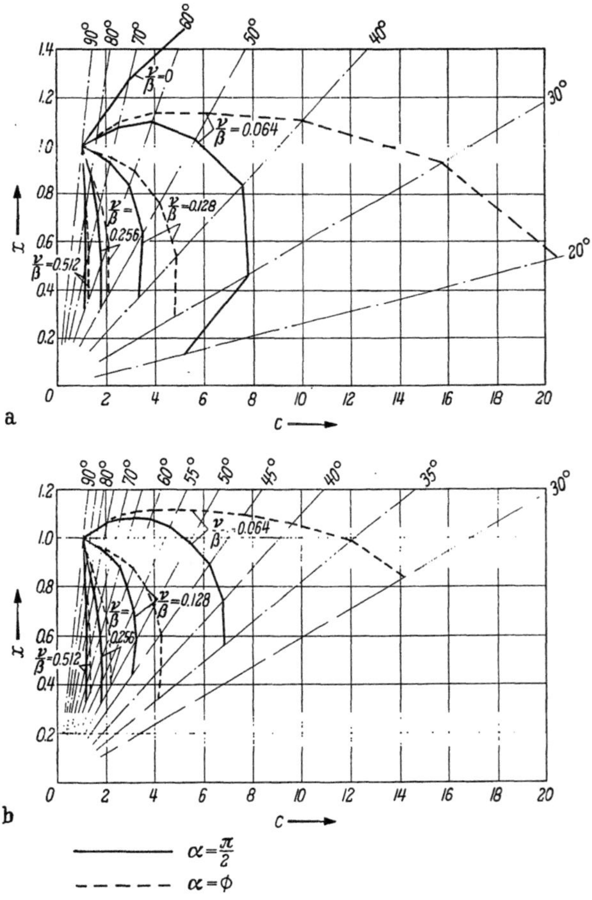

Fig. 18. c—x curves (a) 10° steps (b) 5° steps

In Fig. 18, $c - x$ trajectories are plotted for both $\alpha = \pi/2$ and $\alpha = \Phi$. For the larger values of ν/β, c_{max} is not much larger for $\alpha = \Phi$ than for

$\alpha = \pi/2$, whereas for the smaller values of ν/β (in particular for $\nu/\beta = 0.064$), there is a wide divergence between the two cases.

(g) Interpretation of Results

It is difficult to know how to judge the severity of a wrinkle from the point of view of the material. Assuming that strain is a measure of damage to the material during wrinkling, it seems proper to define a "strain intensification factor" I as follows:

$$I = \frac{\text{Maximum strain due to bending of shell}}{\text{Strain resulting from uniform collapse of shell}}$$

where "strain" means circumferential strain. It is a matter of simple geometry to show that

$$\text{Bending strain} = \frac{H\, U_0^*\, c}{2\, A^2} \tag{41}$$

and

$$I = \frac{c}{2} \cdot \beta \cdot \frac{H}{A}\,. \tag{42}$$

Therefore, as we have some values (from Fig. 18b) of $c_{\max}$ as a function of ν/β, it is appropriate to plot $(I/\beta)\,(A/H)$ as a function of ν/β. This is done in Fig. 19. The two curves in this figure represent the cases $\alpha = \Phi$

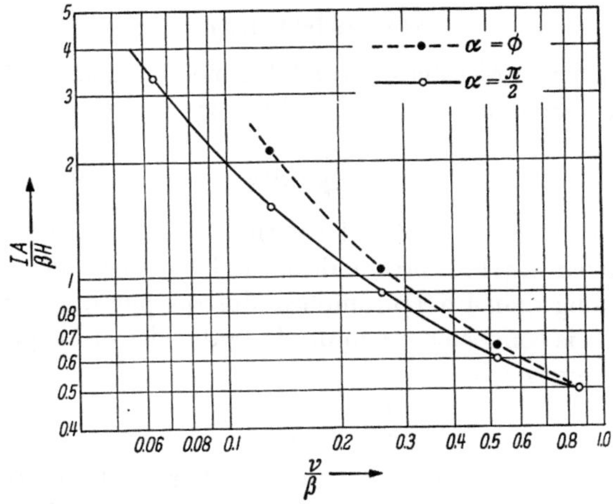

Fig. 19. Strain intensification factor I as a function of ν, β, A and H.

and $\alpha = \pi/2$ respectively; they intersect at the point where c is a maximum at $\Phi = 90°$, i. e. the point where $m_0 = 0$ at $\Phi = 90°$, for which the value of ν/β is easily found from the m_0/u_0, f/u_0' graph for $\Phi = 90°$.

The range of ν/β covered by Fig. 19 should be adequate for the study of many fuel-can designs. If smaller values of ν/β are of interest, it will be necessary to consider the behaviour of wrinkles of semi-angle less than 30°.

(h) Conclusions and Recommendations for further Study

The work described in this paper leads to what is believed to be a satisfactory understanding of the principles involved in the creep pressurisation of oversize tubes onto circular rods.

The linearisation of the equilibrium, geometry, and bending moment/curvature rate equations does not detract from the validity of the analysis for sufficiently shallow wrinkles, but it would be worthwhile in the future to have a more thorough estimate of the limits on deflection imposed by these simplifications than has been provided in this paper.

It would also be of interest to investigate the effect of the use of a large finite-difference angle and simple finite-difference expressions on the solution, and to ascertain the effects of ignoring the circumferential variation of r.

A problem largely untouched by the work of this paper is the effect of friction between the shell and the rod. A realistic solution of this problem presents many difficulties which may not be resolved by purely theoretical means.

Finally, the Author hopes that this paper may be of use to those concerned with the analysis of nuclear reactor fuel element behaviour.

Acknowledgement

The author acknowledges his indebtedness to the English Electric Co. Ltd. for permission to publish this paper.

Appendix

In section 2b it was pointed out that when m/u_0 exceeds a certain value, the linear Eq. (10) breaks down and the shell becomes less resistant to bending than Eq. (10) implies. For the purpose of investigating limits on m, it is convenient to multiply both sides of inequality (12) by U_0/U_0^*. This gives:

$$\frac{m}{u_0^*} \leq \frac{5}{6} \frac{H}{n\, U_0^*}. \tag{43}$$

An analysis of the bending moments found from the matrix solutions showed that, for the values of ν/β considered, the bending moments in the shell were generally greatest at the point of contact with the rod at $\theta = \Phi$, but in no case were greater than

$$\frac{m}{u_0^*} = 0{,}5.$$

Combining this with inequality (43) we have, approximately,

$$\frac{n\, U_0^*}{H} \leq 1{,}5. \tag{44}$$

Inequality (44) gives a conservative limit on the initial height of the wrinkle for $\nu/\beta \geq 0.064$ for which the linear analysis is valid.

Another way of testing the validity of the solutions is to note that inequality (11) specifies that the rate of incease of bending strain with time must not be more than 5 times greater than the mean (compressive) strain rate. Therefore the "strain intensification factor" I has an upper bound of 5 if the limit of inequality (11) is not to be exceeded. This, however, can only be used as a rough check; if I is nearly equal to 5, a check on m_0/u_0^* should also be made.

References

[1] Hoff, N. J., W. E. Jahsman and W. Nachbar: A study of creep collapse a of long circular cylindrical shell under uniform external pressure. J. Aero/Space Sciences Oct. 1959.

[2] Timoshenko, S. P.: Strength of Materials, Part II, 3rd Edition. D. van. Nostrand Company, Inc.

[3] Young, A. G.: Private communication.

[4] Timoshenko, S. P.: Theory of Elastic Stability. McGraw-Hill Book Company, Inc.

On the Mechanics of Column Creep

By

R. L. Carlson and **W. W. Breindel**

Battelle Memorial Institute, Columbus, Ohio, USA

1. Introduction

In recent years, a considerable amount of research has been devoted to the phenomenon described as creep buckling. Although some studies have dealt with plates and shells, the greater part of the effort has been on the creep buckling of columns. Since the investigation to be described in this paper dealt with column behavior, the discussion will be confined to creep buckling as it is manifested in columns.

In terms of the perspective to be developed here, buckling — due to creep — is an event that occurs at the critical time or what might be described as the end of the column lifetime. The properties of the column prior to buckling are also to be of interest in this development, however, so more specifically, we are concerned not just with creep buckling, but with the mechanics of a column whose deflections are increasing with time due to creep.

One method of deducing the properties of a creeping column is to perform an inspection at the time of interest. The present paper begins, therefore, with a consideration of procedures for inspecting a creeping column for stability. The studies of Rabotnov and Shesterikov [*1*] and those of Fraeijs de Veubeke [*2*] and Hoff [*3*] are interpreted as examples of two essentially different possible procedures. An inspection procedure similar, to but more general than, that of Fraeijs de Veubeke is proposed, and an analysis of the mechanics involved is performed. The equations developed are then used to examine the Hoff [*3*] column-creep-buckling solution, and to illustrate the essential differences in column action between purely viscous materials, viscoelastic materials, and materials which possess both time-dependent and nonlinear time-independent components of deformation. Finally, the results of column-creep experiments in which inspections were performed are described, and a discussion of the correlations with the analysis is presented.

2. General Discussion

(a) Inspection Procedures

Analyses of the column creep problem have been based either on initially imperfect columns or on initially perfect (straight, axially loaded) columns. In many respects, an analysis based on imperfect columns is more attractive, since real columns are imperfect. Also, in the limit, it may be possible to approach perfection analytically as closely as desired in the imperfect-column analysis. The perfect-column analysis, in contrast cannot be analytically extended to describe the imperfect column.

In a sense, the above reasoning may be misleading. It tends to suggest, perhaps, that perfect-column analysis is of little value. A review of static (no time effects) column theory indicates that this is not the case; rather, there is much to be gained from considering the perfect column in static buckling.

The value of the perfect column in creep-buckling analysis is, however, less obvious. Several creep-buckling theories [*1*, *4*, *5*] use a perfect column as a basis for analysis. Hoff [*6*], however, has demonstrated that these theories do not provide equivalent results. It is possible that the differences in results could arise from differences in the procedure used to inspect for stability. One must then, however, examine the acceptability of the different inspection procedures.

A comparison of the details of inspection procedures is difficult. Only the theory of Rabotnov and Shesterikov [*1*] appears to be explicit in this respect. Their inspection procedure is stated in detail. The theories of Shanley [*4*] and Gerard [*5*] involve implicit assumptions regarding column action. To a large extent, the use of these assumptions has been supported on the basis of providing satisfactory correlations with experimental data. Although, this may be encouraging it is not sufficient to dispel the reservations that exist regarding development of the analyses.

The theories developed by Shanley [*4*] and Gerard [*5*] postulate that a perfect column that is initially stable in the straight form may become susceptible to buckling after a certain amount of creep has occurred. This would be possible if the effective bending stiffness decreased with increasing time. In these developments, the concept of a time-dependent modulus is then introduced to reflect the decrease in bending stiffness. The formulae for short-time buckling conditions then become available for creep buckling predictions. Following the outlined procedure, Shanley proposed a time-dependent modulus based on the use of isochronous stress-strain curves, and Gerard proposed what has been described as a time-dependent secant modulus.

Although the selection of the appropriate time-dependent modulus apparently can be arbitrary, the basic approach involved is not unsound. The inspection procedure associated with this approach is essentially "static" in nature; i. e., inertia effects are neglected. For it, one is concerned with the column response (at the time of inspection) *as* a disturbance is administered. By way of contrast, the dynamic inspection procedure of RABOTNOV and SHESTERIKOV is concerned with the column response (at the time of inspection) *after* a disturbance is administered.

For the static procedure, one should be concerned with the response developed during an instantaneous inspection. Permitting the inspection to take place noninstantaneously may occur as a possibility. One must then be concerned with inspection deflection rates, however, and this does not lead to a definitive procedure.

The use of a static inspection procedure was discussed in 1953 by CARLSON [7], who considered the stability of a perfect column subject to creep. It was concluded that "a perfect column that is loaded to a value of load which is less than its tangent-modulus load will become unstable in time only if the static stress-strain properties are affected by either the creep that has occurred or by metallurgical change". This statement focused attention on the importance of the short-time stress-strain properties, whereas in the application of the theories of GERARD and SHANLEY, the amount of creep that has occurred governs the results.

Recently, B. FRAEIJS DE VEUBEKE [2] applied a static inspection procedure to imperfect creeping columns and produced, as the imperfection became vanishingly small, the same result predicted by CARLSON [7]. FRAEIJS DE VEUBEKE's analysis also includes a description of column behavior for loads greater than the tangent-modulus load. It appears, therefore, that a theory does exist that can describe both imperfect- and perfect-column creep-buckling behavior.[1]

The analysis of FRAEIJS DE VEUBEKE [2] is an extension of earlier work by HOFF [3]. HOFF observed that, for a material possessing properties found in metals and alloys (time-independent elastic and inelastic strain, time-dependent inelastic strain or creep), the critical deflection depended only on the time-independent properties. Knowing the critical deflection, the critical time could be computed by the use of the basic equation governing the problem.

FRAEIJS DE VEUBEKE showed that the critical deflection could be obtained by "inspecting" the column for stability at a deflection, w, by the consideration of a neighboring configuration, $w + dw$. Although

[1] The creep-buckling behavior predicted applies for materials with a nonlinear static stress-strain curve. Viscous and viscoelastic materials without nonlinear time-independent properties behave differently. This point is discussed in greater detail in another section.

the final result that he obtained was available from HOFF's work, the new treatment helped to focus attention on the details of the creep-buckling process, and thereby gave rise to subsequent simplified procedures of solution by HOFF and his co-workers [8].

As noted above, the inspection procedure of FRAEIJS DE VEUBEKE involves an examination during an instantaneous increment in deflection. In the present study, a more general inspection procedure was utilized. This involved an increase in the column load, which, of course, also produces a corresponding deflection increase. This procedure is simpler to achieve experimentally than deflection increases under constant load.

(b) A New Inspection Procedure[1]

As a basis for this discussion, consider an imperfect column made of a material possessing the mechanical properties common to structural metals and alloys. Under a constant load or average stress, and after a certain period of time during which the column deflection is increasing steadily because of creep, collapse will occur; i. e., the column will no longer be able to support its load. To illustrate how creep buckling is in essence similar to static or short-time column buckling, reference will be made to Fig. 1, which is a plot of column average stress versus column deflection. The solid curve, starting at the origin, describes the shorttime (no creep) variation of average stress with deflection. For a column load producing an average stress such as σ_1, stability can be examined by an infinitesimal increase in stress, $d\sigma$. If $\frac{d\sigma}{dy}$ is greater than zero, the column is considered stable under σ_1. When the column average stress is increased, the level designated as σ_m will ultimately be reached. Here,

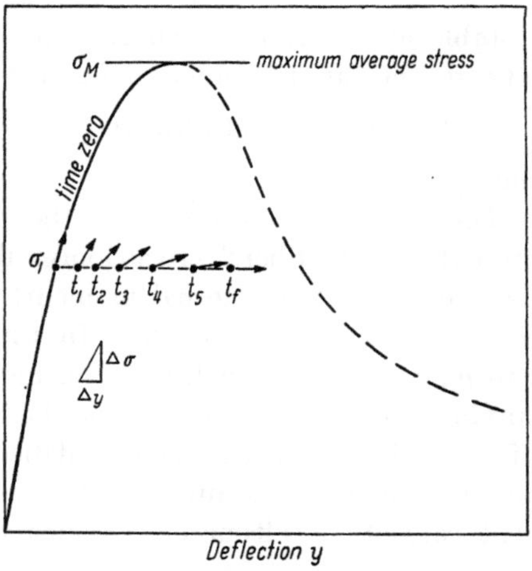

Fig. 1. Column average stress versus deflection

$$\frac{d\sigma}{dy} = 0,$$

[1] This procedure was originally proposed by one of the authors in Ref. [9]. Since this reference is not readily available the details are repeated here.

and the column is considered unstable; i. e., collapse will occur and deflections greater than that corresponding to the maximum average stress can be realized only for average stresses less than σ_m (see the dashed curve). This interpretation of stability has been presented by Bleich [*10*] and by Hoff [*11*].

The stability of a column whose deflection is steadily increasing due to creep (under a constant load) can be examined in essentially the same manner. Initially, the column is loaded to an average stress of σ_1. At time zero, the column is stable in the same sense as previously. If no creep occurs, the stress-deflection point remains on the solid curve. If creep can occur, however, and the load remains constant, the stress-deflection point will move horizontally and to the right with increasing time. A sequence of times $t_1, t_2, \ldots$ can be depicted, then, as in Fig. 1. To examine the stability of the loaded column after a time t_1 has elapsed, the inspection described earlier can be utilized. An infinitesimal increase in stress, $d\sigma$, can be applied instantaneously (since creep can occur, the stress increase should not take any time).

If, as is indicated at t_1, the value $\frac{d\sigma}{dy}$ is greater than zero, the column is stable at the time t_1. Subsequent examinations at $t_2, t_3, \ldots$ would indicate that, as the column deflection increases with time, the value of $\frac{d\sigma}{dy}$ would decrease, and ultimately become zero at the critical or failure time t_f.

The physical significance of the collapse phenomenon is that the internal column "fibers" are no longer able to provide an internal moment that can resist the external moment, i. e., their capacity to resist the external moment is exceeded. In the static or short-time case, this limiting moment is reached by an increasing column load and accompanying deflection increases. For the creep-buckling case, the moment is increased to a limiting value by deflection increases alone, as the external load remains constant. In both instances, however, the internal moment capacity is ultimately exhausted.

It should be noted that in the preceding discussion, the conclusions regarding the state of the column at any given time is based on an externally observable response. The internal stress distribution prevailing is not explicitly specified. In an actual column, the internal history may be too complex (due to relaxation, creep recovery, strain hardening due to creep, etc.) to permit an exact specification of the stress distribution. This, however, would be a limitation of available analytical methods, not of the concepts involved in the inspection procedure. The inspection procedure is, therefore, valid regardless of whether or not we can "solve" for the prevailing stress distribution.

The problem from this point on, then, becomes one of developing a model that can simulate the actual column behavior in an acceptable manner. As will be seen in the subsequent discussion, the model used in the solution proposed by HOFF [3] appears to possess the properties required to provide a good description of metal and alloy column behavior.

(c) Analysis

In order to facilitate subsequent comparisons, the notation used will be essentially that given by HOFF [3]. The definitions can be obtained from the list below or by reference to Fig. 2.

A	Total cross-sectional area of column
L	Length of column
h	Distance between flanges of column
$x = \frac{x'}{L}$	Nondimensional axial coordinate
$w_0 = \frac{2}{h} y_0$	Initial, unloaded deviation from straightness (imperfection)
$w = \frac{2}{h} y$	Nondimensional deflection due to load
P	Axial force (negative when compressive)
E	Young's modulus
m	Exponent in creep law of Eq. (1)
μ	Material constant in creep law of Eq. (1)
F	A creep strain-rate function
ε_c	Strain in concave flange (negative when compressive)
ε_t	Strain in convex flange (negative when compressive)
σ	Average stress, P/A
σ_c	Stress in concave flange
σ_t	Stress in convex flange

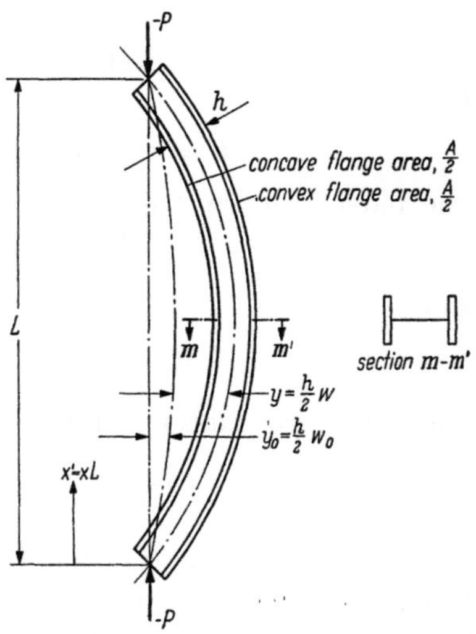

Fig. 2. Idealized H-section column

The creep law used as a basis for analysis is that first proposed by ODQVIST [12] and used by HOFF [3] in his analysis.

$$\dot{\varepsilon} = \frac{\dot{\sigma}}{E} + k_1 \left(\frac{\sigma}{\mu}\right)^m \frac{\dot{\sigma}}{\mu} + F. \tag{1}$$

The constant k_1 assumes values necessary to describe the manner of loading. Although unloading will be considered, it will be assumed that, for the case considered, the stresses are compressive only, and will not reverse sign.[1]

$$\text{When } \sigma < 0,\ \dot{\sigma} < 0,\ k_1 = (-1)^m,$$
$$\text{When } \sigma < 0,\ \dot{\sigma} > 0,\ k_1 = 0.$$

For the conditions considered, F, the creep strain-rate function, will be negative, and the constant m will be considered an odd integer (ODQVIST [12] and HOFF [3] specified F as a power term in σ).

As the column creeps, the strain rates in the flanges will be

$$\dot{\varepsilon}_c = \frac{\dot{\sigma}_c}{E} - \left(\frac{\sigma_c}{\mu}\right)^m \frac{\dot{\sigma}_c}{\mu} + F_c, \tag{2}$$

and

$$\dot{\varepsilon}_t = \frac{\dot{\sigma}_t}{E} + F_t. \tag{3}$$

The moment, M, developed at any section of the column by the load P is

$$M = -P(y + y_0) = -\frac{Ph}{2}(w + w_0). \tag{4}$$

The flange stresses are

$$\sigma_c = \frac{P}{A} - \frac{2M}{hA} = \sigma(1 + w + w_0), \tag{5}$$

and

$$\sigma_t = \frac{P}{A} + \frac{2M}{hA} = \sigma(1 - w - w_0). \tag{6}$$

The curvature relation for the problem is

$$\varepsilon_t - \varepsilon_c = -\frac{1}{2}\left(\frac{h}{L}\right)^2 \left(\frac{d^2w}{dx^2}\right). \tag{7}$$

Eqs. (2), (3), (5), (6), and (7) form the basis for analysis. These equations will now be operated on to convert them into a form convenient for an analysis of the inspection procedure.

From Eqs. (5) and (6), the changes in stress accompanying an increase in the column load can be written

$$d\sigma_c = (1 + w + w_0)\, d\sigma + \sigma\, dw, \tag{5'}$$

and

$$d\sigma_t = (1 - w - w_0)\, d\sigma - \sigma\, dw. \tag{6'}$$

Since an instantaneous inspection is desired, $dt = 0$, and

$$d\varepsilon_c = \frac{d\sigma_c}{E} - \left(\frac{\sigma_c}{\mu}\right)^m \frac{d\sigma_c}{\mu}. \tag{2'}$$

$$d\varepsilon_t = \frac{d\sigma_t}{E} - \left(\frac{\sigma_t}{\mu}\right)^m \frac{d\sigma_t}{\mu} \text{ for } |\sigma\, dw| < |(1 - w - w_0)\, d\sigma|, \tag{3'}$$

[1] Other cases can be treated as shown by HOFF [3].

and

$$d\varepsilon_t = \frac{d\sigma_t}{E} \text{ for } |\sigma\, dw| > |(1 - w - w_0)\, d\sigma|. \tag{3''}$$

The curvature relation, [Eq. (7)], can be written as

$$d\varepsilon_t - d\varepsilon_c = -\frac{1}{2}\left(\frac{h}{L}\right)^2 d\left(\frac{d^2 w}{dx^2}\right). \tag{7'}$$

When the creeping column is inspected by an increase, $d\sigma$, in the column average stress, the states of loading in the column flanges will be functions of the column deflection, w. The stress change in the concave flange will occur as a loading for all increase in $d\sigma$. The change of stress in the convex face, $d\sigma_t$, is given by Eq. (6′). An inspection of this equation reveals, that if

$$|\sigma\, dw| > |(1 - w - w_0)\, d\sigma|,$$

$d\sigma_t$ will be decreasing, or "unloading". This will be designated as Case II. If

$$|\sigma\, dw| < |(1 - w - w_0)\, d\sigma|,$$

$d\sigma_t$ will be increasing, or "loading". This will be designated as Case I.

Case I (Loading on Convex Flange). For Case I, the conditions require that the differential strain law for $d\varepsilon_t$ describe loading. Hence, the relation

$$d\varepsilon_t = \frac{d\sigma_t}{E} - \left(\frac{\sigma_t}{\mu}\right)^m \frac{d\sigma_t}{\mu} \tag{3'}$$

must be used. Through the use of Eqs. (2′), (3′), (5′), (6′), and (7′), the equation governing the "inspection" is obtained:

$$\begin{aligned} &\frac{1}{2} E\left(\frac{h}{L}\right)^2 d\left(\frac{d^2 w}{dx^2}\right) \\ &= \left\{2\,\sigma - E\left(\frac{\sigma}{\mu}\right)^{m+1} [(1 + w + w_0)^m + (1 - w - w_0)^m]\right\} dw \\ &+ \left\{2(w + w_0) - \frac{E}{\mu}\left(\frac{\sigma}{\mu}\right)^m [(1 + w + w_0)^{m+1} - (1 - w - w_0)^{m+1}]\right\} d\sigma. \end{aligned} \tag{8}$$

Let the initial imperfection be

$$w_0 = a_0 \sin \pi x;$$

assume

$$\begin{aligned} w &= a \sin \pi x, \\ dw &\doteq da \sin \pi x; \end{aligned} \tag{9}$$

and note that

$$\varepsilon_E = -\frac{\pi^2}{4}\left(\frac{h}{L}\right)^2.$$

Making use of these relations, the following result is obtained from Eq. (8):

$$\frac{d\sigma}{dw}=\frac{-2\,E\,\varepsilon_E+2\,\sigma-E\left(\frac{\sigma}{\mu}\right)^{m+1}\{[1+(a+a_0)\sin\pi x]^m+[1-(a+a_0)\sin\pi x]^m\}}{\frac{E}{\mu}\left(\frac{\sigma}{\mu}\right)^m\{[1+(a+a_0)\sin\pi x]^{m+1}-[1-(a+a_0)\sin\pi x]^{m+1}\}-2\,(a+a_0)\sin\pi x}. \qquad (10)$$

Eq. (10) provides a method for computing the slopes of loading inspection curves of the type shown in Fig. 1. This completes the analysis for inspections in which the compressive loading in both the convex and concave flanges increases.

Case II (Unloading on Convex Flange). This case deals with the problem when the stress in the convex flange is decreasing, or "unloading". Hence,

$$|\sigma\,dw| > |(1-w-w_0)\,d\sigma|^*,$$

and the following relation must be used:

$$d\varepsilon_t=\frac{d\sigma_t}{E}. \qquad (3'')$$

Through the use of Eqs. (2′), (3′), (5′), (6′), and (7′), the equation governing the "inspection" is obtained:

$$-\frac{1}{2}\left(\frac{h}{L}\right)^2 d\left(\frac{d^2w}{dx^2}\right)=\left[-2\frac{\sigma}{E}+\left(\frac{\sigma}{\mu}\right)^{m+1}(1+w+w_0)^m\right]dw$$
$$+\left[E\left(\frac{\sigma}{\mu}\right)^m(1+w+w_0)^{m+1}-2\,\mu\,(w+w_0)\right]\frac{d\sigma}{E\,\mu}. \qquad (11)$$

Again, using the relations under (9), the following result is obtained from Eq. (11):

$$\frac{d\sigma}{dw}=\frac{2\,\sigma-2\,E\,\varepsilon_E-E\left(\frac{\sigma}{\mu}\right)^{m+1}[1+(a+a_0)\sin\pi x]^m}{\frac{E}{\mu}\left(\frac{\sigma}{\mu}\right)^m[1+(a+a_0)\sin\pi x]^{m+1}-2\,(a+a_0)\sin\pi x}. \qquad (12)$$

Eq. (12) provides a method for computing the slopes of the loading inspection curves of the type shown in Fig. 1.

As noted in the preceding section, the critical time occurs when $\frac{d\sigma}{dw}=0$.[2] Setting Eq. (12) equal to zero yields

$$+2\,\sigma-2\,E\,\varepsilon_E-E\left(\frac{\sigma}{\mu}\right)^{m+1}[1+(a+a_0)\sin\pi x]^m=0. \qquad (13)$$

[1] In this development it is assumed that the given inequality holds along the entire convex flange. Actually, the region of unloading begins at the column midpoint and grows toward the column ends. It will be seen, however, that for the condition of primary interest, unloading occurs all along the convex flange.

[2] The nondimensional deflection, w, corresponds to the deflection, y, of Fig. 1; i. e., the conditions $\frac{d\sigma}{dw}=0$ and $\frac{d\sigma}{dy}=0$ are equivalent.

As a sample calculation, let $m = 1$ and consider a one-point collocation that satisfies the equation at $x = \frac{1}{2}$ (the column midpoint). (The following result is then obtained for the total critical deflection:

$$(a + a_0)_{\text{crit.}} = 2\left(\frac{\mu}{\sigma}\right)^2 \left[\frac{\sigma}{E} - \varepsilon_E - \frac{1}{2}\left(\frac{\sigma}{\mu}\right)^2\right]. \tag{14}$$

For the conditions specified, this corresponds to the value of critical deflection that would be obtained from HOFF's solution [3] with the following qualifications:

1. HOFF uses w as the total deflection, whereas in this analysis $(w + w_0)$ is the total deflection. Setting $a_0 = 0$, therefore, makes the results equivalent.

2. HOFF defines $\varepsilon_E = \frac{\pi^2}{4}\left(\frac{h}{L}\right)^2$, whereas we have used $\varepsilon_E = -\frac{\pi^2}{4}\left(\frac{h}{L}\right)^2$. Substituting the defined quantities back into the corresponding equations for critical deflection would produce equivalent results.

(d) Significance of Analysis

The equivalence of results noted above for two different procedures should have been expected, since, at the critical time (see Fig. 1), the inspection procedure discussed here coincides with that of FRAEIJS DE VEUBEKE, i. e., w increases, but σ remains constant.

From an experimental point of view, however, times less than the critical are easier to examine, and the present analysis provides a method for accomplishing the examination. The examination, in turn, provides a relatively direct check on the validity of the concept that the initial stress-strain curve governs instantaneous, incremental inspections. By increasing the column load quickly at a time less than the critical time, experimental values can be obtained for $\frac{d\sigma}{dw}$. Through use of the basic material properties, a computed value can be obtained from Eq. (12). These values can then be compared.

Previous checks [8] on the validity of the concepts involved have been less direct. They involve a comparison of experimental and computed curves of column load versus critical time. This type of check is less direct in that it introduces the computation of the critical time. In effect, this results in an accumulation of additional assumptions, which, in turn, should be checked.

From a practical point of view, it may also be noted that disturbances can be expected to be transmitted through the ends of the column as load increases. One would be interested, therefore, in knowing the "stiffness" remaining after creep deflections had occurred and, in fact, may wish to specify a limiting deflection based on an allowable reserve of stiffness.

Calculation of Critical Time. The analysis conducted to this point has produced only a part of the solution of the problem. One must proceed further for a complete analysis in which the critical time is required. It is worth noting, however, at this point that the "stiffness" and the critical deflection were computed without specifying the creep strain-rate function, F. This could have been a power term in σ, a strain-hardening relation involving the stress and the creep strain, or a more general relationship. This part of the solution will not be considered in this paper.

Viscous and Viscoelastic Materials. If the nonlinear time-independent term had not appeared in Eq. (1) and subsequently in (2′) and (3′), the inspection criterion would not have evolved as it did. A column of a viscous material [only a function F would appear on the right side of Eq. (1)], for example, would be infinitely rigid to an instantaneous inspection.

A column of a viscoelastic material would merely adjust, with the necessary elastic stress changes, to an instantaneous inspection. There would, moreover, be no maximum condition to the average stress. This can be deduced from Eq. (12) by letting $\mu \to \infty$, and noting that only as $a \to \infty$ would $\frac{d\sigma}{dw}$ at $x = \frac{1}{2}$ approach zero.[1]

For the type of behavior characterized by the preceding analysis, the nonlinear time-independent term is, therefore, essential. It is of interest to note, then that the, details of failure or collapse for the different materials cited are fundamentally different from the another. one

Examination of Deformation Law Used. In the inspection procedure used in the analysis presented here, and that by FRAEIJS DE VEUBEKE, it is tacitly assumed that

$$d\varepsilon = f(\sigma)\, d\sigma,$$

or

$$\frac{d\sigma}{d\varepsilon} = \frac{1}{f(\sigma)},$$

regardless of the time or the creep strain that has accrued. Excluding metallurgical changes, which would obviously invalidate this assumption, it is possible that this assumption is, at best, an approximation even for so-called metallurgically stable materials. It is possible, for example, that a relation of the type $\frac{d\sigma}{d\varepsilon} = g(\sigma, \varepsilon_c)$ where ε_c is the creep strain, may

[1] Since the curvature relations (7) and (7′) are approximate, permitting "a" to become very large in Eq. (12) is not actually an acceptable procedure. It does, nevertheless, provide added insight into the problem.

be effective. This would not invalidate the inspection procedure. It would, however, require that the modified value of $\frac{d\sigma}{d\varepsilon}$ at the moment of inspection be known. Even though we are concerned only with metallurgically stable materials, it is still proper to examine this assumption more closely.

Two references were found in the literature that, although they do not duplicate the conditions of interest completely, possess value in examining the validity of the assumption involved. The studies described by the references involve experiments in which tensile specimens were crept various amounts at elevated temperatures. They were subsequently tested at room temperature to determine the effect on short-time stress-strain behavior of the accrued creep strain. Although these studies were conducted under either a constant stress or a constant load, and the effect on the room-temperature rather than "at temperature" stress-strain properties was found, the trends observed are translatable in terms of the present problem.

In the first study to be cited, FLOYD, HAZLETT, and PARKER [*13*] obtained results on commercial "A" nickel that had been precrept at 700 C. These results are shown in Fig. 3. The effect of precreep has been to raise the stress-strain curve, and the greater the precreep, the higher the stress-strain curve. Calculations presented previously [*8*] indicate that the creep strains indicated in Fig. 3 may be considerably greater than those experienced in creep buckling. For this reason, curves of flow stress for various tensile strains versus creep strain were prepared from Fig. 3, and from them, the flow stresses for a precreep of 0.5 per cent were picked off. The dashed curve of Fig. 3 was obtained, then, from the interpolated values.

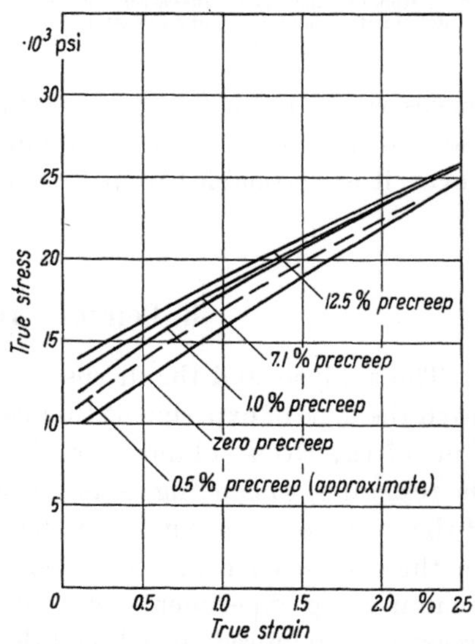

Fig. 3. Room temperature stress strain curves for commercial "A" nickel precrept at 700 C from Ref. [*13*]

The shift in the stress-strain curve is significant for 0.5 per cent precreep. It should be noted, however, that, for a given level of σ, the difference between the slope of the dashed curve and the slope of the zero-precreep curve appears to be relatively small. The latter difference, rather than the difference in the level of the stress-strain curves, is, of course, the difference of interest here.

In the second study of interest, SHERBY, GOLDBERG, and DORN [14] obtained extensive results on high-purity aluminium precrept at several elevated temperatures. Typical results obtained for precreep at 300 F are presented in Fig. 4. Again, since the values of precreep were quite large, an interpolated curve for 0.5 per cent has been constructed. Here again, the differences in slope for a given σ appear to be small.

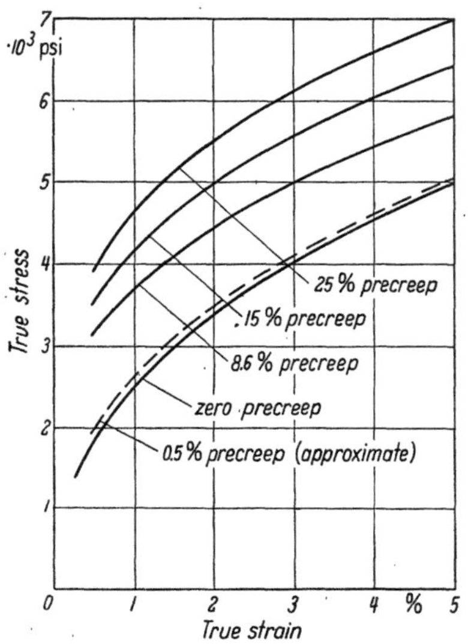

Fig. 4. Room temperature stress strain curves for high purity aluminium precrept at 300 F from Ref. [14]

For the results reviewed, it would appear that $\frac{d\sigma}{d\varepsilon}$ probably is a function, not only of the stress, but also of the creep strain. This indicates that a solution for these two materials based on the assumption that $\frac{d\sigma}{d\varepsilon}$ is a function only of stress would be approximate. It should be noted, also, that both the precreep strains and the smallest strains shown for the stress-strain data are greater than those of interest to this study. The details of the manner in which the curves would begin to separate[1] in the region of interest cannot be deduced, therefore, from Fig. 3 and 4. On the basis of these data, however, it is reasonable to expect that some difference in slopes would exist.

3. Column Creep Experiments

The analysis and the discussion in the preceding sections have indicated the type of experiment that can be utilized to examine the assumptions of the HOFF column-creep-buckling solution. Since the change in the resistance to bending — as measured by the instantaneous derivative of the average stress with respect to the deflection — serves as the basis for the examination procedure developed in this paper, it follows that column creep experiments in which the average stress is suddenly increased are required. A detailed description of experiments designed to provide the required data is presented in Ref. [15], so only the results will be given here.

[1] The initial parts of all the curves would coincide, since YOUNG's modulus would not be changed by prior creep.

(a) Experimental Results

The experiments being considered can be illustrated by reference to Fig. 5. The graphs of Fig. 5 describe the history of an interrupted column-creep-buckling experiment in terms of column load, lateral deflection, and time. First, the column is loaded quickly[1] to point 0, which represents the time $t = 0$ condition. With the load remaining constant, the column was free to deflect in time as indicated in the lower graph. At a time corresponding to the point designated as A, the column load was increased to the magnitude indicated by B. The transition from A to B, which ideally occurs instantaneously, results in the deflection increase shown. The slopes of the curve at points along A—B are the measures of bending stiffness referred to previously. It can be noted that, beyond B, the load has been indicated as being constant. Since there is nothing in the basic analysis of the problem that indicates that additional interruptions beyond B could not be made, a second interruption is indicated between C and D. It may also be noted that, since the equation for the derivative of the average stress with respect to the deflection is a function of the variables, average stress and deflection, it also should be possible to compute the slope at any point along A to B or C to D. Of course, the proper values of stress and deflection must be used.

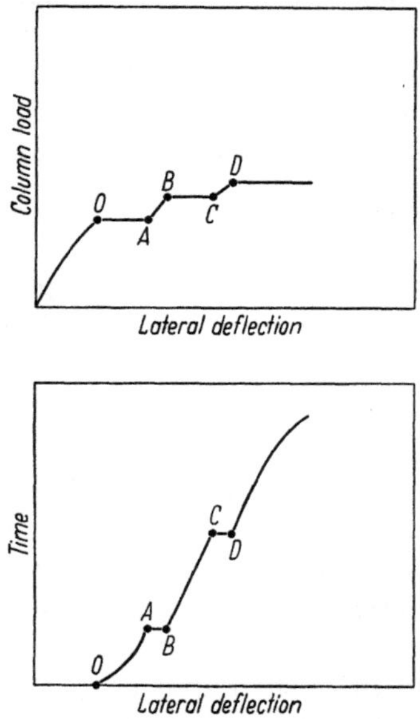

Fig. 5. History of interrupted column-creep-buckling experiment

The material used for the experimental investigation was the aluminium alloy 7075-0. All experiments were conducted at 325 F. The parameters derived from the short-time stress-strain curve were as follows:

$$\mu = 43{,}1 \cdot 10^3 \text{ psi},$$
$$E = 9{,}1 \cdot 10^6 \text{ psi},$$
$$m = 5.$$

Two column experiments were conducted. Both of the column specimens had rectangular cross sections 0.400 inch by 0.500 inch and lengths

[1] In the experiments conducted, the total load was applied in increments in about 50 seconds.

of 7.000 inches (slenderness ratio of 60.6). During the column experiments, a continuous record of the lateral deflection of the column midpoint was obtained. Strain records at the column midpoint and at quarter points for the concave and convex faces were also obtained.

For the experiments conducted, the imperfections were controlled by adjustments prior to testing. This was accomplished by applying loads at room temperature, observing the measured strains and deflections, and then adjusting the and eccentricities to achieve the desired imperfections. The effective imperfections were estimated by the use of LUNDQUIST plots of the room-temperature load-deflection data. The effective imperfections, $y_0\left(\frac{L}{2}\right)$, are tabulated below:

Specimen Number	$y_0\left(\frac{L}{2}\right)$ 10^{-3} inch
75—11	0,1
75—12	2

To perform computations, it was necessary to convert the solid rectangular cross-section to an equivalent two-flange cross section. It was assumed, as by CHAPMAN, et al [8], that the solid rectangular cross section could be replaced by a two-flange cross section in which one half of the area is concentrated in each flange, and the flanges were separated by twice the radius of gyration of the rectangular cross section. If the depth of the rectangular cross section is denoted by d, then it follows that

$$h = \frac{\sqrt{3}}{3} d .$$

For the column conditions considered, it was found that the condition

$$\left|(1 - w - w_0)\, d\sigma\right| > \left|\sigma\, dw\right|$$

held for all interruptions (finite increments were substituted for $d\sigma$ and dw). Eqs. (10) was used, therefore, to determine $\frac{d\sigma}{dw}$ at $x = \frac{1}{2}$.

The conditions for which computations were made and the results obtained are presented in Table 1. A review of the results presented in Table 1 reveals that the computed value of $\frac{d\sigma}{dy}$ at $x' = \frac{L}{2}$ is in each case less than the experimental value.

Considering the fact that the analysis contained several simplifying assumptions, and that in each case two independent[1], and fairly complex experiments were involved, the general agreement is encouraging. The

[1] The short-time compression test and the column-creep experiment.

fact that the calculated values for all interruptions were less than the measured values is felt to be significant. To provide some insight into the observed trend, a more detailed consideration of the factors involved must be undertaken.

Table 1. *Summary of Interrupted Creep Buckling Experimental Results*

Specimen	Interruption	Average Stress, psi	Midpoint Deflection $y\left(\frac{L}{2}\right)$, 10^{-3} inch	Calculated $\frac{d\sigma}{dy}$ at $x' = \frac{L}{2}$, 10^3 psi inch^{-1}	Experimental $\frac{d\sigma}{dy}$ at $x' = \frac{L}{2}$, 10^3 psi inch^{-1}	Difference[1], per cent
75—11	First	—9110[2]	11.5[2]	—810	—950	15
75—11	Second	—9430[2]	21.5[2]	—370	—410	10
75—12	First	—9010	11.4	—720	—810	11
75—12	Second	—9480	30.0	—230	—320	28

(b) Discussion of Results

One of the assumptions introduced in the analysis involved the stipulation of the form or configuration of the column during the interruption. Data from the strain measurements and from shape measurements after the experiments [*15*] indicate that the half-sine-wave assumption is good for the times the interruptions were made. It is of interest to note that this assumption was not a good one during approximately the first half of the experiments. The tendency toward the half-sine wave developed as the experiments proceeded. These results suggest that the assumption regarding the shape is not the source of the trend observed in Table 1 between the experimental and calculated values.

Another possible source of error in the analysis involves the assumption that the rate of strain hardening of the material depends only on the stress; i. e., it is not influenced by prior creep. This effect has been discussed in a previous section of this paper. Data were presented that indicated that prior creep can increase the rate of strain hardening. As creep occurs during the column experiment, it is probable that the instantaneous rate of strain hardening at stress levels above the elastic limit would be increased above that indicated by the short-time compression stress-strain curve. Since the latter curve forms the basis for the determination of the inelastic constants used in the calculations, it follows that the calculations would predict, from this effect, a lower bending stiffness than actually exists. This conforms with the trend observed for the results presented in Table 1. From a practical viewpoint, it may

[1] Based on experimental value.

[2] Values corresponding to the beginning of the second increment in interruption.

be noted that this effect produces, in the computation, a conservative effect[1].

An additional possible source of error arises from the tacit assumption that the strain of the short-time compression stress-strain curve has no time-dependent component. In order to minimize this effect, the stress-strain curve was recorded autographically to permit the use of a relatively high strain rate (0.002 inch per inch per minute). This effect could contribute to the trend indicated in Table 1, since it would result in a "lowering" of the stress-strain curve from which the inelastic constants were obtained. The probable relative effect of this assumption will be discussed in conjunction with another assumption below.

Two additional assumptions that should be considered are as follows:

1. The assumption that the loading interruption occurs instantaneously.

2. The assumed equivalence between the actual cross section and the two-flange cross section.

The first assumption listed above would tend to produce a trend opposite to that observed in Table 1. Practically, the loading interruption cannot occur instantaneously; time must elapse, and, consequently, time-dependent deformation can occur.

It will be noted that the latter effect, although due to rate of loading, would produce a result opposite to that associated with the stress-strain test. To compare the relative significance of these two "rate" effects, the strain rates will be compared. For the compressive stress-strain test, the strain rate was 0.002 inch per inch per minute. The strain rate during the interruptions in the column experiments must be estimated. Considering a half-sine-wave incremental increase in deflection, the strain rate on the concave face at the column midpoint (due to the increase in direct loading and bending) is estimated to be of the order of 0.0004 inch per inch per minute. A comparison of these values suggests that the rate of loading of the columns during the interruptions would tend to be more critical in terms of producing trends in the results in Table 1. For these two effects only, then, the experimental values should tend to be lower than the calculated values. Since this is opposite to the trend observed, it appears probable that the dominant effect is not associated with rate of loading.

The second assumption, associated with the effective flange separation, is valid if the stress distribution acting on the actual rectangular cross section is linear. Then, as noted previously,

$$h = \frac{\sqrt{3}}{3} d,$$

[1] For a metallurgically nonstable material, a much more detailed review of possible effects would be necessary, of course.

where h is the effective flange separation and d is the depth of the rectangular cross section. When the stress distribution on the actual cross section becomes nonlinear, however, the above relation no longer holds, and

$$h < \frac{\sqrt{3}}{3} d .$$

As creep occurs, the stress distribution can be expected to become nonlinear. It follows, then, that the value of h used in the computations is too large. The computed values of Table 1 should tend, from this effect, to be larger than the observed values. It should be emphasized that the nonlinearity referred to here results not only from stress-strain-curve nonlinearity, but also from the tendency for elements under higher stresses to relax. A quantitative evaluation of this effect, therefore, should incorporate the effects of creep on the actual stress distribution.

From the general considerations above, it can be concluded that several effects could produce differences between the calculated and the experimental values of the derivative of the average stress with respect to the deflection. In view of the difference trend observed in Table 1, and the discussion that has been presented, it appears that the dominant effect is that associated with the increase in the rate of strain hardening that can accompany the occurrence of creep.

To assess the magnitude of individual effects more completely than is possible with the data available, future studies with the following objectives are desirable.

1. The acquisition of additional data of establish more definitely the probable error of the approximations introduced in the analysis and experiments.

2. The development of a method for estimating the decrease in the effective flange separation with increasing nonlinearity in the stress distribution on the actual cross section.

3. Acquisition of data similar to those obtained by Lloyd, et al. [*13*], and Sherby, et al. [*14*], but for ranges of creep strain and subsequent static of interest to the problem of creep buckling.

The results of the preceding presentation are considered as support for the Hoff-Fraeijs de Veubeke approach to the column-creep-buckling problem. Although confirmation is not complete, the agreement between the analytical and experimental results is considered good enough to indicate that the method is basically sound, and that a continued evaluation is warranted.

Acknowledgement

The authors would like to acknowledge the encouragement of GEORGE MANNING and the interest and helpful comments of Professor N. J. HOFF. The studies were conducted under the sponsorship of the Aeronautical Research Laboratory, Directorate of Research, Wright Air Development Division, Contract No. AF 33(616)—6301.

References

[1] RABOTNOV, Y. N., and S. A. SHESTERIKOV: Creep Stability of Columns and Plates, Journal of the Mechanics and Physics of Solids, **6**, 1 (1957).
[2] FRAEIJS DE VEUBEKE, B.: Creep Buckling, Chapter 13 of High Temperature Effects in Aircraft Structures, edited by N. J. HOFF and published for AGARD by Pergamon Press (1958).
[3] HOFF, N. J.: Creep Buckling, Aeronautical Quarterly, **7**, Part 1 (1956).
[4] SHANLEY, F. R.: Weight-Strength Analysis of Aircraft Structures, New York: McGraw-Hill Book Company, Inc. (1952), Chapter 19.
[5] GERARD, G.: A Creep Buckling Hypothesis, Journal of the Aeronautical Sciences, **23**, 9 (1956).
[6] HOFF, N. J.: A Survey of the Theories of Creep Buckling, Proceedings of the Third US National Congress of Applied Mechanics, New York: The American Society of Mechanical Engineers (1959).
[7] CARLSON, R. L.: The Behavior of Perfect Columns at Elevated Temperatures, Proceedings of the First Midwestern Conference on Solid Mechanics, University of Illinois (April, 1953).
[8] CHAPMAN, J. C., B. ERICKSON and N. J. HOFF: A Theoretical and Experimental Investigation of Creep Buckling, PIBAL Report No. 406, Polytechnic Institute of Brooklyn, October ,1957 and International Journal of Mechanical Sciences, Vol. 1, Nos. 2/3, p. 145, April 1960.
[9] CARLSON, R. L.: A Summary of Compressive-Creep Characteristics of Metal Columns at Elevated Temperatures, WADC Technical Report 57—96 (1957).
[10] BLEICH, F.: Buckling Strength of Metal Structures, New York: McGraw-Hill Book Company, Inc., (1952).
[11] HOFF, N. J.: Buckling and Stability, Journal of the Royal Aeronautical Society, **58** (517) 1 (1954).
[12] ODQVIST, F. K. G.: Influence of Primary Creep on Column Buckling, Journal of Applied Mechanics, **21** (3), 295 (1954).
[13] LLOYD, K. B., T. H. HAZLETT and E. R. PARKER: Change of State of Nickel During Secondary Creep at High Temperatures, University of California Institute of Engineering Rssearch, Seventh Technical Report (May, 1950).
[14] SHERBY, O. D., A. GOLDBERG and J. E. DORN: Effect of Prestrain Histories on the Creep and Tensile Properties of Aluminium, Transactions of the American Society for Metals, **46**, 681 (1954).
[15] CARLSON, R. L., W. W. BREINDEL and G. K. MANNING: An Investigation of Column Action During Creep Buckling, WADD TR 60-7 (1960).

Comparison of Ranges of Applicability of Predictions of Creep Buckling Time[1]

By

Brian M. Lempriere

Stanford University, Stanford, California, USA

Summary

The most rigorous theory of the creep buckling of initially imperfect columns (due to HOFF and FRAEIJS DE VEUBEKE) is examined for possible simplifications. It is found that the influence of plasticity on the life of a column can be ignored under certain conditions characterized by the proportional limit stress, a creep parameter, the applied stress, and the initial curvature. These conditions are conveniently presented in the form of graphs for steady creep as represented by the power and the exponential laws.

The simplification possible in the regions so defined permits the application of the early theories such as those due to KEMPNER, LIBOVE, etc. For steady creep these lead to simple formulae which consist of the life determined according to GERARD's suggestion multiplied by a correction factor. Some published data are examined in terms of the present ideas and give encouraging agreement with the theoretical results, for values of the correction factor ranging from 0.0074 to 1.25.

Information is given on preliminary results of an experimental investigation now in progress at Stanford University.

Notation

A	creep rate coefficient in exponential law
B	creep rate exponent in exponential law
E	Young's modulus of instantaneous deformation
I	deflection integral defining time for power law
J	deflection integral defining time for exponential law
L	length of column
a	non-dimensional lateral deflection amplitude
b	symbol quoted from Ref. [*11*] identical to h

[1] The work here presented was supported in whole by the United States Navy under Contract Nonr 225 (47), monitored by the Mechanics Branch of the Office of Naval Research.

h	depth of I-section (or twice radius of gyration of rectangular section)
n	creep rate index in power law
t	time
w	lateral deflection
x	axial position along column
α	applied load ratio $= \bar{\sigma}/\sigma_E$
ε	strain
λ	creep rate coefficient in power law
$\varkappa$	curvature
σ	flange stress
σ_E	EULER stress $= \pi^2 E\, h^2/4\, L^2$

Subscripts, etc.

c	creep
cr	critical (at instability)
E	EULER
0	initial condition
pl	proportional limit
red	reduced (VON KÁRMÁN'S)
S. T.	Simplified Theory
t	tangent
1, 2	concave and convex flange, respectively

The dot above a symbol indicates a time derivative.

1. Introduction

In the recent technical literature considerable attention has been given to the instability of columns in the presence of creep, and the basic phenomenon causing a finite life is now well understood. A reasonable set of assumptions has been introduced and examined to permit the logical derivation of a theory. The most general theory of initially slightly curved columns is due to HOFF [1] and FRAEIJS DE VEUBEKE [2], but this has the disadvantages that calculations based on it are rather lengthy and require detailed information on material properties. Under certain conditions, to be investigated in detail in this paper, the theory can be simplified and reduced to an earlier one due to KEMPNER [3] or to a form which can be deduced from the theory of LIBOVE [4].

The purpose of the present paper is to define the region in which the HOFF-FRAEIJS DE VEUBEKE theory can be simplified, and to develop for that region a convenient method of estimating column life in terms of simple parameters. The formula so obtained can be split into two factors. The first one is the column life time corresponding to GERARD'S conjecture [5], which is not based on consistent physical principles. The second can be considered a correction factor to the GERARD life.

2. Summary of Theoretical Considerations

(a) Theory of Initially Slightly Curved Columns

The simple case of an idealized I-section has been adopted widely as a foundation for the application of buckling theories as it exhibits all the important phenomena and introduces much simplification. Further, it has been shown by PATEL and WAGLE [6] that equivalent I and rectangular sections have virtually the same life.

The assumption of a sinusoidal deflection pattern simplifies the problem by reducing it to the solution of an ordinary differential equation. The effect of this assumption was examined by HOFF [7].

Finally, the deformations of the material are taken as composed of an instantaneous and of a time-dependent part given in the form of an incremental relation.

The usual engineering assumptions of plane sections, etc. are also made.

The physically necessary relations for the deformation geometry, for the equilibrium of forces and moments, and for the material behavior give the following basic equations for quantities at the mid-section of the column:

Deflection:

$$w(x) = -(a\,h/2)\sin(\pi\,x/L), \tag{1}$$

$$w(L/2) = -(a\,h/2).$$

Curvature:

$$\varkappa = (\varepsilon_1 - \varepsilon_2)/h = 2\,\varepsilon_E\,a/h. \tag{2}$$

Flange Stresses:

$$\sigma_{1,2} = \bar{\sigma}(1 \pm a). \tag{3}$$

Deformation Law:

$$\dot{\varepsilon} = [\dot{\sigma}/E_t(\sigma)] + \dot{\varepsilon}_c(\sigma, \ldots), \tag{4}$$

where all quantities are defined in the Notation; the first part of the deformation law includes elastic and plastic deformations, the second part represents the creep deformation.

The expressions for the stresses can be substituted from Eq. (3) into Eq. (4) to give the strains in terms of the lateral displacement, which then may be put into Eq. (2) to obtain the desired differential equation for the lateral displacement:

$$\dot{a} = (\dot{\varepsilon}_{c_1} - \dot{\varepsilon}_{c_2})/[2\,\varepsilon_E\,\{1 - (\bar{\sigma}\,E/\sigma_E\,E_{\mathrm{red}})\}], \tag{5}$$

where $E_{\mathrm{red}} = 2\,E\,E_t/(E + E_t)$ is von KÁRMÁN's reduced modulus.

The theory of HOFF and FRAEIJS DE VEUBEKE states that this equation has no meaning and the column is neutrally stable when the deformation reaches the point at which the effective stiffness of the column is reduced to the extent that the denominator of the right side of Eq. (5) vanishes, that is, when

$$E_{\text{red}}/E = \bar{\sigma}/\sigma_E = \alpha . \tag{6}$$

This equation can be solved for the critical deflection using Eq. (3) and the complete stress-strain curve for the material.

The life of the column is then found by integration of Eq. (5) in the form

$$t_{cr} = \int_{a_0}^{a_{cr}} da/\dot{a}(a) . \tag{7}$$

This generally requires numerical evaluation, even if a linearization of the relation between tangent modulus and stress is feasible. The calculations would obviously be tedious.

(b) Gerard's Conjecture

Based on Stowell's [8] equilibrium analysis of structural elements, Gerard stipulated that a straight column would become unstable when the compressive stress is

$$\sigma_E = \varepsilon_E E_t . \tag{8}$$

He then conjectured that in creep buckling a reasonable approximation might be to set the tangent modulus equal to the total secant modulus (including creep):

$$E_t = E_{s_{\text{total}}} . \tag{9}$$

This approximation is obviously inaccurate except for extremely small creep strains. The total strain at buckling under the mean stress is then, by Eq. (8), equal to the Euler strain:

$$\varepsilon_{\text{total}} = \sigma/E_{s_{\text{total}}} = \varepsilon_E . \tag{10}$$

The elastic strain due to the mean stress may be considered separately as in Eq. (4), and is given by:

$$\varepsilon_{el} = \sigma/E = \alpha \varepsilon_E . \tag{11}$$

The difference between Eqs. (10) and (11) is the creep strain at buckling:

$$\varepsilon_{cr} = \varepsilon_E (1 - \alpha) . \tag{12}$$

The time to attain this value of creep strain can be read directly off curves obtained from constant-stress creep experiments. In the case of steady creep Eq. (12) can be written explicitly in terms of the life:

$$t_G = \varepsilon_E (1 - \alpha)/\dot{\varepsilon}_0(\bar{\sigma}) . \tag{13}$$

3. Insignificance of Plastic Deformation

Many sample calculations based on the rigorous theory outlined above (even that presented by CHAPMAN, et al. [9] in verification of the theory) have shown that the critical deflection is not always significant in determining the life. The increments of time corresponding to fixed incre-

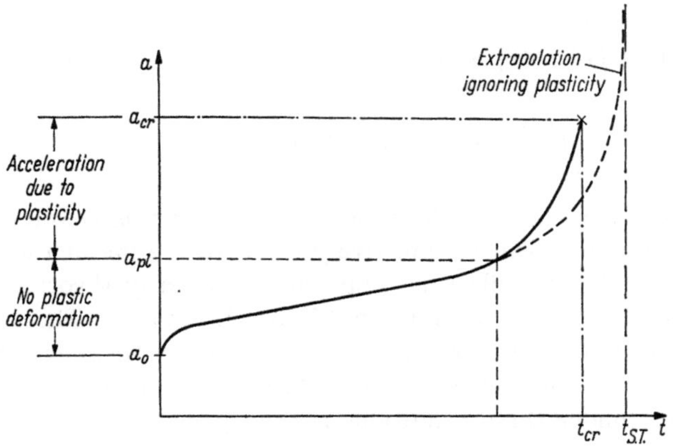

Fig. 1. Typical column deflection curve

ments of deflection can become negligibly small while the stresses remain below the proportional limit. This suggests that the time integral of Eq. (7) should be divided into two parts which can be considered separately (see Fig. 1):

$$t_{cr} = \int_{a_0}^{a_{pl}} da/\dot{a} + \int_{a_{pl}}^{a_{cr}} da/\dot{a}. \tag{14}$$

The first part is the time to reach a deflection at which the stress in the concave flange is raised to the proportional limit. This defines a deflection by Eq. (3) as

$$a_{pl} = (\sigma_{pl}/\bar{\sigma}) - 1 \tag{15}$$

below which the stress dependent tangent modulus E_t in Eq. (5) becomes the constant modulus of elasticity E.

The second part is the time to further increase the deflection to the critical value which is implied by Eq. (6). This part is more complicated in that it involves plastic as well as creep deformations. Now the effect of plasticity is to accelerate the deformation, so that the time taken for this part of the column deflection is less than would be obtained by ignoring plasticity:

$$\int_{a_{pl}}^{a_{cr}} da/\dot{a} \leq \int_{a_{pl}}^{a_{cr}} da/\dot{a}\,(E_{\text{red}} = E). \tag{16}$$

This second integral is an extrapolation of the first part of Eq. (14), and increases monotonically with deflection because it is a measure of time. Obviously its value for the finite critical deflection is less than the value corresponding to infinite deflection:

$$\int_{a_{pl}}^{a_{cr}} da/\dot{a}\,(E_{\text{red}} = E) \leq \int_{a_{pl}}^{\infty} da/\dot{a}\,(E_{\text{red}} = E). \tag{17}$$

Using Eqs. (14), (16) and (17), limits on the life-time can be established:

$$\int_{a_0}^{a_{pl}} da/\dot{a} \leq t_{cr} \leq \int_{a_0}^{\infty} da/\dot{a}\,(E_{\text{red}} = E). \tag{18}$$

The ideas expressed in the first paragraph of this chapter suggest that the lower limit can be effectively equal to the life of the column. The upper limit is the life time as given by the early theories of creep buckling, such as those of KEMPNER [3] and LIBOVE [4]. These theories can therefore be expected to give good results when the above limits are close to the life-time and so to each other. The simplest expression for the life-time under these conditions is the upper limit above:

$$t_{cr_{\text{S. T.}}} = \int_{a_0}^{\infty} da/\dot{a}\,(E_{\text{red}} = E). \tag{19}$$

4. Applicability of Simplified Theory

For engineering purposes the column life will be given with sufficient accuracy by Eq. (19) when it differs by less than 5% from the lower limit of Eq. (18). Thus the simplified theory which neglects plastic deformations is useful when

$$\int_{a_0}^{a_{pl}} da/\dot{a} \geq 0.95 \int_{a_0}^{\infty} da/\dot{a}\,(E_{\text{red}} = E). \tag{20}$$

By the additive property of integrals this reduces to

$$\int_{a_{pl}}^{\infty} da/\dot{a}\,(E_{\text{red}} = E) \leq 0.05 \int_{a_0}^{\infty} da/\dot{a}\,(E_{\text{red}} = E). \tag{21}$$

Physically, this inequality represents conditions under which plastic deformation does not develop in the first 95 per cent of the life.
empirical creep laws

To derive practical results from these considerations, the empirical laws of creep must be introduced into the expression of Eq. (5) for the lateral velocity.

The two most widely used laws for steady creep are the power law:

$$\dot{\varepsilon} = (\sigma/\lambda)^n \tag{22}$$

and the exponential law:

$$\dot{\varepsilon} = A \exp(B\,\sigma). \tag{23}$$

These laws are usually derived at high stress levels and do not take account of the sign of the stress. Thus they do not fulfill the physically necessary condition of representing the creep rate as an odd function of stress. Since the stress in the convex flange of the column changes sign when $a = 1$, this is a matter that should not be taken lightly. The power law can be converted into an odd function for all values of n by writing it as

$$\dot{\varepsilon} = (\operatorname{sgn} \sigma)\,(|\sigma|/\lambda)^n. \tag{24}$$

The exponential law is not applicable for low stresses since it gives a finite creep rate at zero stress. For the present problem where a wide range of stress is encountered, the hyperbolic sine law is a useful extension of the exponential law as it has neither of the above shortcomings, and is identical with the exponential law at high stresses. Thus the exponential law is taken in the form:

$$\dot{\varepsilon} = 2\,A \sinh(B\,\sigma). \tag{25}$$

With these expressions and with $E_{\text{red}} = E$, Eqs. (3) and (5) give the following formulae for $\dot{a}$:

$$\dot{a} = \dot{\varepsilon}_0\{(1 + a)^n - \operatorname{sgn}(1 - a)\,(|1 - a|)^n\}/2\,\varepsilon_E(1 - \alpha) \tag{26}$$

and

$$\begin{aligned}\dot{a} &= 2\,A\{\sinh[B\,\bar{\sigma}(1 + a)] - \sinh[B\,\bar{\sigma}(1 - a)]\}/2\,\varepsilon_E(1 - \alpha)\\ &= \dot{\varepsilon}_0\{\coth(B\,\bar{\sigma})\sinh(B\,\bar{\sigma}\,a)\}/\varepsilon_E(1 - \alpha),\end{aligned} \tag{27}$$

where the creep rates corresponding to the mean applied stress are used as abbreviations for the expressions

$$\dot{\varepsilon}_0 = (\bar{\sigma}/\lambda)^n \text{ or } 2\,A \sinh(B\,\bar{\sigma}). \tag{28}$$

The integrals required for Eqs. (19) and (21) are therefore of the following forms:

$$I(a, n) = 2\int_a^\infty da/\{(1 + a)^n - \operatorname{sgn}(1 - a)\,(|1 - a|)^n\} \tag{29}$$

which was evaluated by Kempner [3] using a notation related to the one used in this paper through the equation

$$I(a, n) = 2^{2-m}\,\tau_{cr}(f_{T_0} = (a/2), m); \tag{30}$$

and

$$J(a, B\,\bar{\sigma}) = \tanh(B\,\bar{\sigma}) \int\limits_a^\infty da/\sinh(B\,\bar{\sigma}\,a)$$
$$= -(1/B\,\sigma)\tanh(B\,\bar{\sigma})\ln\tanh(B\,\bar{\sigma}\,a/2). \qquad (31)$$

The limiting condition of Eq. (21) can now be expressed in terms of these integrals as

$$I(a_{pl}, n) \leq 0.05\, I(a_0, n) \qquad (32)$$

and

$$J(a_{pl}, B\,\bar{\sigma}) \leq 0.05\, J(a_0, B\,\bar{\sigma}). \qquad (33)$$

If the definition of a_{pl} given in Eq. (15) is used, these are seen to define relations between the following convenient parameters: applied stress, initial curvature, the proportional limit stress and a creep constant. Curves defining the regions in which the simplified theory is applicable are given in Fig. 2 for the power law. These are derived from the results of Ref. [*3*]. For a column whose parameters are represented by a point lying below the line which characterizes its material the simplified theory is applicable; otherwise the HOFF-FRAEIJS DE VEUBEKE theory is required.

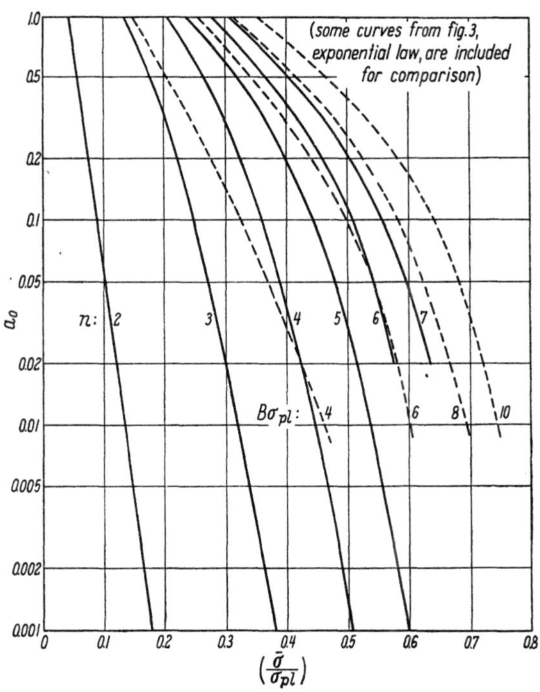

Fig. 2a. Limitations of simplified theory (power law)

Since the exponential law permits explicit evaluation of the integral, as in Eq. (31), the condition of Eq. (33) can be expressed as:

$$-\ln\tanh(B\,\bar{\sigma}\,a_{pl}/2) \leq$$
$$-0.05\ln\tanh(B\,\bar{\sigma}\,a_0/2). \qquad (34)$$

Solutions of this equation are plotted in Fig. 3 for various values of $B\,\sigma_{pl}$ and are compared to the results for the power law in Fig. 2. Note that a convenient approximation to this equation can be derived as follows: it is assumed that the arguments of both sides are reasonably large, so that the the hyperbolic tangents approach unity. They can then be expressed in terms of exponentials:

$$\tanh u = \{\exp u - \exp(-u)\}/\{\exp u + \exp(-u)\} = 1 - 2\exp(-2u), \quad (35)$$

where terms of higher order are neglected. The series expansion of the natural logarithm can be used to give

$$\ln \tanh u = \ln\{1 - 2\exp(-2u)\} = -2\exp(-2u). \quad (36)$$

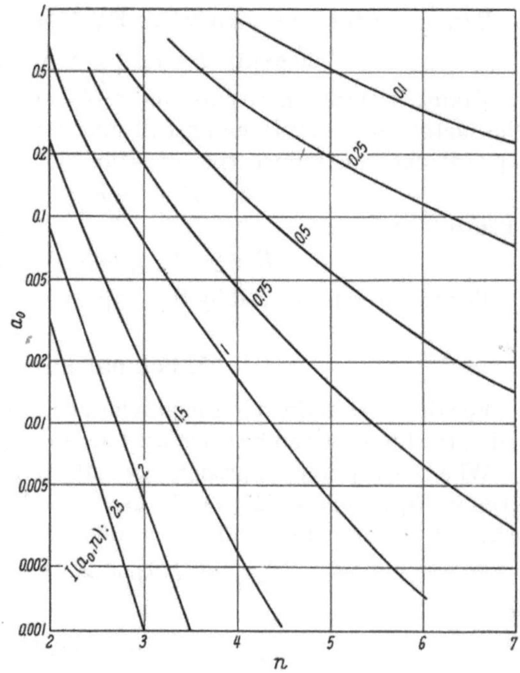

Fig. 2b. Chart of correction factors (power law).

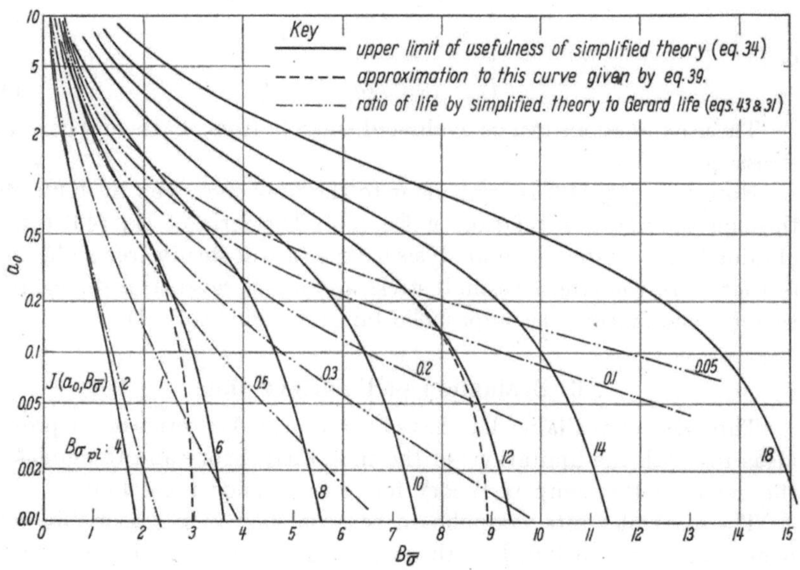

Fig. 3. Limitations of simplified theory and chart of correction factors (exponential creep law)

Because of this approximation, Eq. (34) can be written as

$$20 \exp(-B \cdot \bar{\sigma}\, a_{pl}) \leq \exp(-B\, \bar{\sigma}\, a_0). \tag{37}$$

Taking natural logarithms, and replacing ln 20 by 3 for convenience (the factor 20 is purely empirical) and using the form of a_{pl} given by Eq. (15) one can reduce this equation to

$$3 - B(\sigma_{pl} - \bar{\sigma}) \leq - B\, \bar{\sigma}\, a_0, \tag{38}$$

whence

$$B\, \bar{\sigma} \leq (B\, \sigma_{pl} - 3)/(1 + a_0). \tag{39}$$

The curves represented by this approximation are also given in Fig. 3.

5. Column Life

For the cases defined in the previous chapter and illustrated in Figs. 2 and 3, the life-time can be calculated according to the simplified Eq. (19).

When steady creep is predominant, the expressions for lateral velocity given in Eqs. (26) and (27) can be used. The life-time can be expressed in terms of the integrals defined in Eqs. (29) and (31):

$$t_{cr_{\text{S. T.}}} = I(a_0, n)\, \varepsilon_E\, (1 - \alpha)/\dot{\varepsilon}_0 \tag{40}$$

or

$$t_{cr_{\text{S. T.}}} = J(a_0, B\, \bar{\sigma})\, \varepsilon_E (1 - \alpha)/\dot{\varepsilon}_0. \tag{41}$$

The combination of terms excluding the integrals can be identified with the GERARD life-time given by Eq. (13) for steady creep. The simplified theory then leads to the final expression for the column life:

$$t_{cr_{\text{S. T.}}} = I(a_0, n)\, t_G \tag{42}$$

or

$$t_{cr_{\text{S. T.}}} = J(a_0, B\, \bar{\sigma})\, t_G. \tag{43}$$

These expressions can be evaluated entirely from charts and constant stress creep data.

Note that Eq. (42) is equivalent to the life-time given by KEMPNER through the notation relation of Eq. (30). The form of Eq. (43) can be obtained from LIBOVE's [4] analysis for the case of steady creep, although in that work the effect of small stress and stress reversal is disregarded thereby eliminating the hyperbolic tangent from Eq. (31).

6. Evaluation of Published Data

Three sets of available data have been analyzed in terms of the present ideas to check the limitations of the simplified theory and to emphasize the need for the complex theory for cases outside these limits.

The particular sets were chosen because they were reasonably complete and they included both stress-strain curves and creep curves from which the necessary parameters could be evaluated.

CHAPMAN's Data [9] — Al. Alloy 2024-T 4 at 500°F. The creep data in this report were given in the form of curves and so the creep law had to be evaluated: it was found to be given with reasonable accuracy by the exponential law, with a value of $B = 3.5 \times 10^{-4}$ per psi.

Since for each test the applied stress and column size were given, the GERARD time was easily evaluated according to Eq. (12). The initial unloaded non-dimensional eccentricity was quoted as the constant value 0.10. With the aid of the standard Southwell equation, the initial eccentricity of the loaded column (i. e. axis offset plus elastic displacement) was computed from

$$a_0 = 0.1/(1-\alpha). \tag{44}$$

The values of $B\,\sigma$ and of the ratio $(t_{\exp}/t_G)$ were calculated and the results plotted in Fig. 4. This is a chart of the form of Fig. 3, and includes the limit line corresponding to the proportional limit stress of 33,000 psi, which is evident from the tangent modulus curve presented in the report. All these quantities are collected in Table 1.

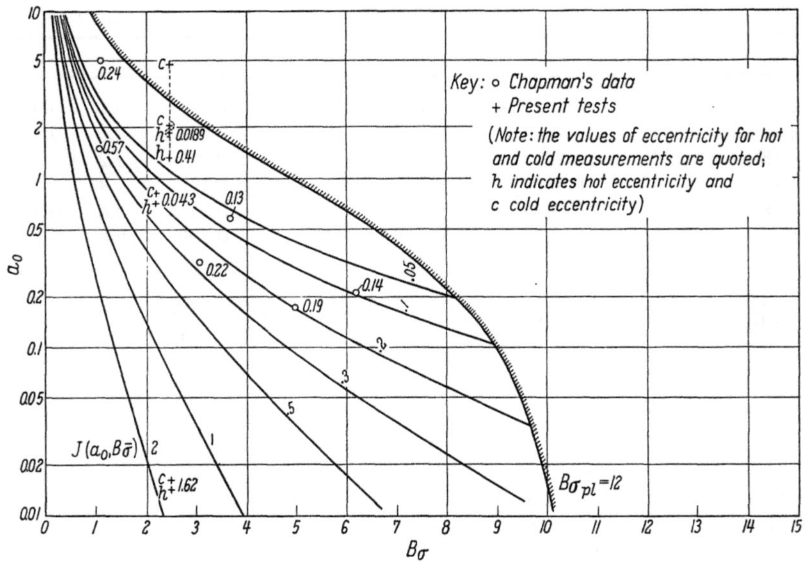

Fig. 4. Comparison of CHAPMAN's data and of present results with theory

Table 1. *Evaluation of Chapman's Data (9)* — $B\,\sigma_{pl} \sim 12$

σ(psi)	$\varepsilon_E \times 10^4$	$(1-\alpha)$	$\dot{\varepsilon}_0$(/min)	t_G(mins)	$(t_{\exp}/t_G)$	$B\bar{\sigma}$	a_0
2880	3.56	0.047	9.5×10^{-8}	176	0.57	1.11	1.5
2960	3.56	0.020	9.6×10^{-8}	74	0.24	1.14	5.0
8000	13.62	0.309	5.9×10^{-7}	714	0.22	3.09	0.32
9600	13.62	0.171	1.0×10^{-6}	224	0.13	3.70	0.59
12800	35.8	0.575	3.2×10^{-6}	64	0.19	4.94	0.17
16000	35.8	0.470	1.0×10^{-5}	168	0.14	6.18	0.21

Data of JACKSON, et al. [*10*] — 7075-T 6 Alloy at 600°F. Although here the creep data were presented in the form of an exponential law, this cannot be used directly as a function of time was included in the formula; consequently the law did not represent steady creep. The creep curves of Fig. 17.3 in the reference were therefore approximated by straight lines whose slopes were plotted against stress on semi-logarithmic paper. From this graph the value of B was found to be 2.4×10^{-3} per psi. The unloaded eccentricity was given as a dimension a, from which the loaded non-dimensional eccentricity was derived according to the formula

$$a_0 = (\sqrt{12}a)/h\ (1-\alpha). \tag{45}$$

The proportional limit stress was taken from Fig. 17.7 of the reference to be 5,500 psi so that $B\,\sigma_{pl} \sim 13$. The remaining parameters were evaluated directly. These results are quoted in Table 2 and are plotted in Fig. 5 for comparison with the theory.

Data of MATHAUSER and BROOKS [*11*] 7075-T 6 Alloy at 600°F. Only the high temperature results were examined since they represent

Table 2. *Evaluation of Data of Jackson, et. al.* (*10*) — $B\,\sigma_{pl} \sim 13$

σ(psi)	$\varepsilon_E \times 10^4$	$(1-\alpha)$	$\dot{\varepsilon}_0$(/min)	t_G(mins)	(t_{exp}/t_G)	$B\bar{\sigma}$	a_0
2500	8.05	0.402	6×10^{-8}	5400	0.37	6.0	0.07
2500	8.05	0.402	6×10^{-8}	5400	0.0076	6.0	0.24
3000	8.05	0.282	2×10^{-7}	1140	0.0074	7,2	0.15
3500	8.05	0.163	6.5×10^{-7}	202	0.11	8.4	0.05
4000	8.05	0.044	2.1×10^{-6}	17	0.37	9.6	0.32
4000	14.3	0.462	2.1×10^{-6}	314	0.029	9.6	0.10

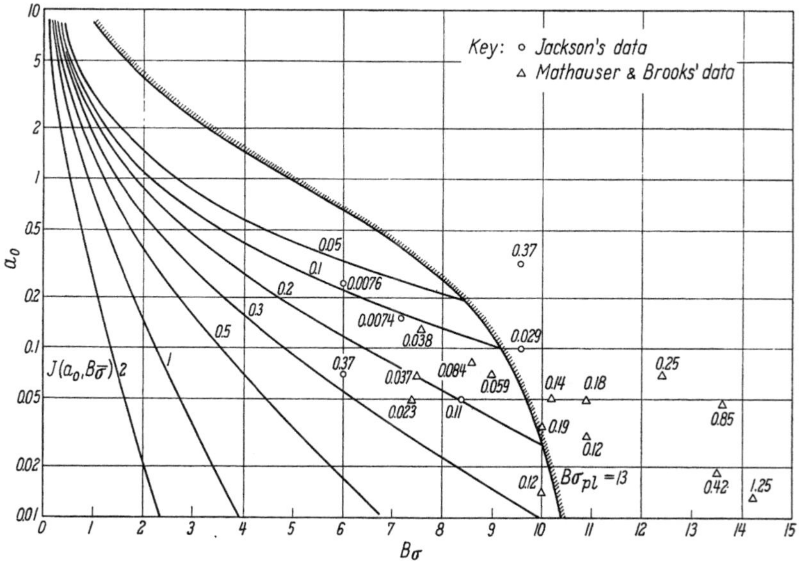

Fig. 5. Comparison of data of JACKSON et al. and of MATHAUSER and BROOKS with theory

the same material and temperature as do the data of the previous section. The same creep law (with the same constants, though independently obtained) were quoted, and hence the creep results for that section were used, i. e. the values of $B = 2.4\times10^{-3}$ and of $B\,\sigma_{pl} \sim 13$ were taken. The same graph was used to obtain the steady creep rates. The unloaded eccentricity was quoted relative to the section depth, and so a factor was introduced to convert it to the radius of gyration. Then the non-dimensional loaded eccentricity was calculated from

$$a_0 = \sqrt{12}\,(d_0/b)/(1-\alpha)\,, \tag{46}$$

where b in the notation of the reference is identical to h in the present notation.

The other parameters were computed directly, and plotted on Fig. 5. They are quoted in Table 3.

Table 3. *Evaluation of Mathauser and Brook's Data (11)* — $B\,\sigma_{pl} \sim 13$

σ (psi)	$\varepsilon_E \times 10^3$	$(1-\alpha)$	$\dot{\varepsilon}_0$ (/min)	t_G (mins)	$(t_{\exp}/t_G)$	$B\bar{\sigma}$	a_0
5930	11.4	0.90	2.17×10^{-4}	47	1.25	14.2	0.013
5660	12.3	0.90	1.11×10^{-4}	100	.85	13.6	0.046
5160	12.3	0.91	3.3×10^{-5}	340	.25	12.4	0.069
5610	5.94	0.82	9.9×10^{-5}	48	.42	13.5	0.018
4540	5.87	0.85	7.7×10^{-5}	650	.18	10.9	0.049
4180	3.91	0.82	3.3×10^{-6}	970	.12	10.0	0.014
4540	2.87	0.70	7.7×10^{-6}	260	.12	10.9	0.030
4270	2.85	0.71	4.0×10^{-6}	510	.14	10.2	0.050
3570	2.87	0.76	7.7×10^{-7}	2800	.084	8.6	0.082
4150	1.84	0.56	3.0×10^{-6}	340	.19	10.0	0.034
3750	1.90	0.62	1.18×10^{-6}	1000	.057	9.0	0.070
3130	1.04	0.40	2.7×10^{-7}	1500	.037	7.5	0.068
3160	0.99	0.38	2.9×10^{-7}	1300	.038	7.6	0.13
3080	0.81	0.27	2.4×10^{-7}	910	.023	7.4	0.049

Remarks on the Published Data. It is important to note that for short time tests the experimental lifetime may be quite inaccurate. The loading time may amount to a significant fraction of the life and experimental errors are more serious. Poor correlation is to be expected for tests with large eccentricities or with high stresses.

The use of steady creep alone in the present analysis, and the corresponding approximation to the creep data inevitably reduces the value of the correlation for the experiments discussed in sections (b) and (c) in which primary creep was not unimportant.

With these remarks in mind, the results obtained in this paper seem gratifying.

7. Experimental Project

An experimental project for studying column buckling was undertaken to complete the evaluation of the HOFF-FRAEIJS DE VEUBEKE theory as started by CHAPMAN, et al. [9]. The method of life estimation proposed by GERARD was to be examined for its usefulness. The foregoing theoretical study was used as a guide in setting up a program of research. In order to implement this, a machine having refined construction and instrumentation was designed and built as shown in the photograph of Fig. 6. It was based on the one used by CHAPMAN, et al. The instrumen-

Fig. 6. Overall view of buckling machine

tation of the CHAPMAN set-up was improved to permit the recording of the load while it was applied. Thus data could be obtained for the evaluation of the eccentricity during the actual test. The lateral deflection was measured by a caliper pivoted between the column ends and provided with differential transformers.

The machine was of the double lever type having a lever ratio of 50:1. The lever pivots were made of Stellite tool pieces and they bore on oil-hardened pads in which was ground a shallow V.

The test column was heated in a resistance-element oven which was controlled by a Leeds and Northrup Micromax recorder and Series 60 Duration Adjusting controller.

The determination of material properties has not yet been started, but is under consideration. The same material was used as in CHAPMAN's buckling tests, so that the material properties were tentatively assumed to be as given in CHAPMAN's report.

A preliminary series of seven columns has been tested. Three of these columns did not fail in a reasonable time according to the estimates based on CHAPMAN's data and the present methods (i. e. for an expected life of less than one hour, there was only a slow steady creep displacement after three hours). This was found to be due to a misalignment of the loading knife edges. The main lever of the buckling machine had been accidentally moved sideways on its knife-edge bearings, so that the column was made to rest at an angle in the plane of the knife-edges, with contact at opposite corners of the end pieces. On loading the column, moments were induced about the major axis of the section as well as the intended moments about the minor axis. The resultant stress distribution caused apparent changes of eccentricity (as calculated from the recorded data) and invalidated the tests. It is probable that the misalignment was partially present during the second and third tests also, but was eliminated before the last. Thus only the first and last tests should be regarded as complete. Their results and the operation of the instrumentations were satisfactory, justifying the start of the final program of tests.

Table 4.
Results of Present Tests — $B\,\sigma_{pl} \sim 12$
All tests on columns of the same size at the same load:

$\sigma = 6500$ psi, $\varepsilon_E = 0.958 \times 10^{-3}$ | $B\,\bar{\sigma} = 2.5$, $(1-\alpha) = 0.2$
$t_G = 53$ mins

a_0(cold)	a_0(hot)	t_{exp}/t_G
0.016	0.0145	1.62
2.02	1.31	0.41
4.79	1.85	0.02
0.11	.075	—
0.075	~.005	—
0.82	0.76	—
0.87	0.64	0.04

Acknowledgement

The author wishes to express his appreciation for the help given him by Professor N. J. HOFF in the presentation of his ideas, and for the assistance of Mr. MARTIN LANFRANCO in performing the calculations and experimentation.

References

[1] HOFF, N. J.: Creep Buckling, The Aeronautical Quarterly, Vol. 7, No. 1, (Feb 1956) p. 1.

[2] FRAEIJS DE VEUBEKE, B.: Creep Buckling, Chapter 13 of High Temperature Effects in Aircraft Structures, edited by N. J. HOFF, published for Agard by Pergamon Press, London: 1958.

[3] KEMPNER, J.: Creep Bending and Buckling of Non-Linearly Viscoelastic Columns, NACA Technical Note 3137, January 1954.

[4] LIBOVE, C.: Creep Buckling of Columns, Journal of the Aeronautical Sciences, **19**, No. 7, 459 (1952).

[5] GERARD, G.: A Creep Buckling Hypothesis, Journal of the Aeronautical Sciences, **23**, No. 9, 879 (1956).

[6] PATEL, S. A., and V. D. WAGLE: Creep Buckling of Rectangular Section Columns, PIBAL Report No. 358, Brooklyn, N. Y.: Polytechnic Institute of Brooklyn, June 1957.

[7] HOFF, N. J.: Buckling and Stability, Forty-First Wilbur Wright Memorial Lecture, Journal of the Royal Aeronautical Society, **58**, No. 517, 30 (1954).

[8] STOWELL, E. Z.: A Unified Theory of Plastic Buckling of Columns and Plates, NACA Report No. 898, 1948.

[9] CHAPMAN, J. C., B. ERICKSON and N. J. HOFF: A Theoretical and Experimental Investigation of Creep Buckling, PIBAL Report No. 406, Polytechnic Institute of Brooklyn, N. Y., October 1957; also International Journal of Mechanical Sciences, **1**, Nos. 2/3, 145 (1960).

[10] JACKSON, L. R., A. D. SCHWOPE and F. R. SHOBER: Stress-Strain-Time Properties of Some Aircraft Materials, Chapter 17 of Weight-Strength Analysis of Aircraft Structures, by F. R. SHANLEY, New York N. Y.: McGraw-Hill Book Co., 1952.

[11] MATHAUSER, E. E., and W. A. BROOKS, jr.: An Investigation of the Creep Lifetime of 75 S-T 6 Aluminum-Alloy Columns, NACA Technical Note 3204, July 1954.

Geometrically Non-Linear Creep Buckling of Bars

By

Michał Życzkowski

Technical University and Polish Academy of Sciences, Kraków, Poland

1. Introductory Remarks

There exist two fundamental approaches to the theory of creep buckling. One, commonly used, consists of analyzing the deflections as a function of time (technical buckling); the bar is initially curved or eccentrically loaded and no problem of stability is formulated (N. J. HOFF [3], J. A. H. HULT [5], A. R. RZHANITSYN [14, 15] and others). The other approach, proposed by G. N. RABOTNOV and S. A. SHESTERIKOV [12], uses a dynamical criterion of stability of a straight bar. Recently S. A. SHESTERIKOV [16] formulated a new stability criterion for a curved bar, connected with the sign of acceleration. This criterion may be used for unsteady creep; the case of steady creep is always that of instability, according to SHESTERIKOV's criterion.

The present paper is connected with the first approach to creep buckling — the analysis of deflections of an initially curved bar. In many papers using this approach an important qualitative difference was found between the phenomenon which takes place if the material of the bar is subjected to a linear creep law and the phenomenon corresponding to a non-linear creep law (power- or exponential-law). For a power creep law, which, with elastic strains taken into account, has the form (F. H. NORTON [8])

$$\dot{\varepsilon} = \frac{\dot{\sigma}}{E} + \frac{\sigma^n}{\lambda} \tag{1.1}$$

(where ε denotes longitudinal strain, σ — normal stress, E — YOUNG's modulus, λ — creep modulus and dots — differentiation with respect to time t) the existence of a critical time was stated. This critical time was understood to be the time after which deflections of the bar tend to infinity. The statement was found to be true for $n > 1$, n odd; it is also valid for n even, but then the basic law (1.1) must be written in a modified form

$$\dot{\varepsilon} = \frac{\dot{\sigma}}{E} + \frac{\sigma|\sigma|^{n-1}}{\lambda} \tag{1.2}$$

because of the conditions of symmetry (ε is assumed to be an odd function of both $\dot{\sigma}$ and σ). Setting $n = 1$ in (1.1) we obtain the linear law (MAXWELL's body):

$$\dot{\varepsilon} = \frac{\dot{\sigma}}{E} + \frac{\sigma}{\lambda}\,; \tag{1.3}$$

for such a body the critical time tends to infinity [6].

These conclusions are in doubt, however, because the small-deflection theory (geometrical linearity) is assumed to hold, and at the same time the deflections are allowed to tend to infinity. We propose to clarify the problem from the standpoint of geometrical non-linearity.

2. Finite deflections of a physically linear column

Consider a bar with one end clamped and the other free (Fig. 1). At the moment $t = 0$ we subject it to the compressive force P. First the elastic deflections should be calculated — they appear immediately when the load is applied. The problem is to find the deflection δ_+ after loading ($t = +0$) as a function of the initial deflection δ_- ($t = -0$) and the force P. This problem was solved and extensively analyzed by the author [18] for circular initial curvature and the results of the computations will be treated here as known.

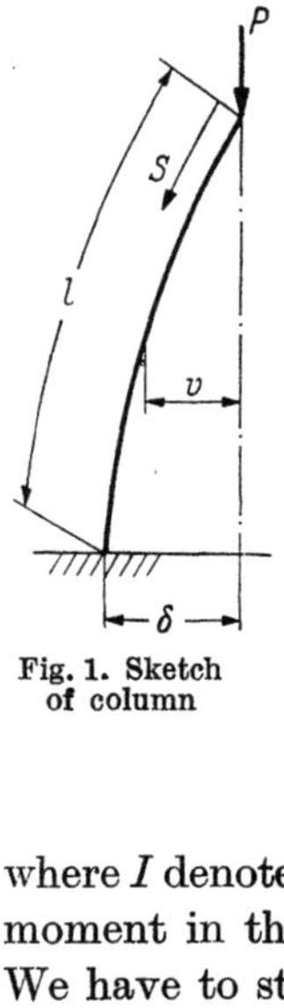

Fig. 1. Sketch of column

We start therefore with the creep deflections, first in the physically linear case [MAXWELL's law (1.3)]. Taking BERNOULLI's assumption of plane cross sections into account we obtain for any shape of cross section the following equation of creep bending and creep buckling [15, 19]

$$\frac{\dot{M}}{E} + \frac{M}{\lambda} = \dot{\varkappa}\, I\,, \tag{2.1}$$

where I denotes the moment of inertia of the cross section, M the bending moment in this section, and $\varkappa$ the curvature of the axis of the beam. We have to stress that the initial curvature $\varkappa_-$ does not appear in this equation; it is hidden in the initial conditions only. To obtain a further simplification we introduce the new, dimensionless time

$$\tau = \frac{E}{\lambda}\, t \tag{2.2}$$

obtaining finally

$$\dot{M} + M = \dot{\varkappa}\, E\, I\,; \tag{2.3}$$

dots denote here differentiation with respect to τ.

Assume non-cartesian co-ordinates $s - v$ (s measured along the curved axis of the bar, Fig. 1), and introduce the dimensionless variables x, y, defined by the formulae

$$x = \frac{\pi}{2\,l}\,s, \quad y = \frac{\pi}{2\,l}\,v, \tag{2.4}$$

l being the length of the bar. Then the exact expression for the curvature takes the form [18],

$$\varkappa = -\frac{\pi}{2\,l}\,\frac{y'}{\sqrt{1-y'^2}}, \tag{2.5}$$

where dashes denote differentiation with respect to x. We substitute this expression in (2.3), and differentiate with respect to the dimensionless time τ. The bending moment is given by:

$$M = P\,v = \frac{2\,l}{\pi}\,P\,y. \tag{2.6}$$

If the dimensionless force is introduced through the formula

$$m = \frac{P}{P_E} = \frac{4\,P\,l^2}{\pi^2\,E\,I}, \tag{2.7}$$

we get the following non-linear partial differential equation of the third order:

$$\dot{y}''(1-y'^2) + y'\,y''\,\dot{y}' + m(y+\dot{y})\,(1-y'^2)^{3/2} = 0. \tag{2.8}$$

To this equation should be added the following boundary conditions:

$$\left.\begin{aligned} x &= 0 \\ y &= 0 \end{aligned}\right\} \qquad \left.\begin{aligned} x &= \frac{\pi}{2} \\ y' &= 0, \end{aligned}\right\} \tag{2.9}$$

and one initial condition

$$y = y_+(x) \text{ for } \tau = 0. \tag{2.10}$$

Eq. (2.8) probably cannot be integrated exatly; even many of the approximate methods are difficult to apply to it because of the non-linearity and because of the order of the equation. We use, therefore, a very simple method, namely the collocation method, proposed by R. A. Fraser, W. P. Jones and S. W. Skan [2]. This method gives good results in such a case only if we can forecast the solution relatively accurately. However, we know in this case that the shape of the deflection curve, which changes with time during creep buckling, can differ only slightly from the shape of the elastic deflection curve, at all values of the force m. The exact expression for the elastic deflections of an initially circularly curved bar — Eq. (3.8) in Ref. [18] — seems to be too complicated for our purpose; thus we use the equation for an initially straight bar, namely

$$y = f\sqrt{1 - \frac{m f^2}{4}}\,\frac{\operatorname{sn}(x\sqrt{m})}{\operatorname{dn}(x\sqrt{m})} \tag{2.11}$$

hoping that the initial curvature influences only slightly the shape of the curve (if the maximum deflection $f = \pi\,\delta/2\,l$ is fixed). In (2.11) the functions $\operatorname{sn}(x\sqrt{m})$ and $\operatorname{dn}(x\sqrt{m})$ are Jacobian elliptic functions with the modulus $k = f\sqrt{m}/2$. Now we replace the parameter m in (2.11) by a function of time, denoting it, for convenience, by $4\,\alpha^2$, and we assume the solution of (2.8) in the form

$$y(x,\tau) = f\sqrt{1 - \alpha^2 f^2}\,\frac{\operatorname{sn}(2\,\alpha\,x)}{\operatorname{dn}(2\,\alpha\,x)}, \tag{2.12}$$

where $\alpha = \alpha(\tau)$ and $f = f(\tau)$ are two unknown functions of time; the modulus of the elliptic integrals is here equal to

$$k = \alpha\,f. \tag{2.13}$$

Eq. (2.12) fulfills exactly the boundary conditions (2.9), but the initial condition (2.10) is fulfilled exactly only if bar is initially straight (buckling is here possible in the supercritical range only); for an initially curved bar the condition (2.10) will be fulfilled only approximately, in so far as the maximum deflection $f = f_+$ is exact at $\tau = 0$.

The functions $\alpha = \alpha(\tau)$ and $f = f(\tau)$ are not independent because for $x = \pi/2$ we must have $y = f$ (dimensionless deflection):

$$f = f\sqrt{1 - \alpha^2 f^2}\,\frac{\operatorname{sn}(\alpha\,\pi)}{\operatorname{dn}(\alpha\,\pi)} \tag{2.14}$$

whence $sn(\alpha\,\pi) = 1$ and we obtain

$$\alpha\,\pi = K(\alpha\,f) \tag{2.15}$$

as the relation connecting α with f at every moment τ.

Now we turn to the collocation method. As we shall determine only one parameter, either α or f, as a function of the dimensionless time τ, we can require the agreement of (2.12) with (2.8) along one line, not perpendicular to the τ — axis, in the plane $\tau - x$. We choose a line parallel to this axis, that is a line $x =$ const. The most important value of x is $x = \pi/2$ (fixed end), because at this point both the bending moment and the curvature have their maximal values. Substituting in (2.8) the values $x = \pi/2$, $y' = 0$, we obtain the condition that must be fulfilled by (2.12)

$$\dot{y}'' + m(y + \dot{y}) = 0. \tag{2.16}$$

The second derivative of (2.12) with respect to x is equal to

$$y'' = 4\,\alpha^2 f\sqrt{1 - \alpha^2 f^2}\,[k^2\operatorname{cn}^2(2\,\alpha\,x) - 1 + k^2]\,\frac{\operatorname{sn}(2\,\alpha\,x)}{\operatorname{dn}^3(2\,\alpha\,x)}, \tag{2.17}$$

and for $x = \pi/2, y = f,\ \operatorname{sn}(\alpha\,\pi) = 1,\ \operatorname{cn}(\alpha\,\pi) = 0, \operatorname{dn}(\alpha\,\pi) = \sqrt{1 - \alpha^2 f^2}$, we get

$$y''\left(\frac{\pi}{2}, \tau\right) = -4\,\alpha^2 f. \tag{2.18}$$

Differentiating this equation with respect to the dimensionless time τ we obtain

$$y''\left(\frac{\pi}{2}, \tau\right) = -8\,\alpha\,\dot{\alpha}\,f - 4\,\alpha^2\,\dot{f}. \tag{2.19}$$

Substitution of (2.19) in (2.16) gives — together with Eq. (2.15) — a system of two equations, one being transcendental and the other an ordinary differential equation of the first order. They determine the functions $f = f(\tau)$ and $\alpha = \alpha(\tau)$:

$$\left.\begin{aligned} \alpha\,\pi &= K(\alpha\,f) \\ -8\,\alpha\,\dot{\alpha}\,f - 4\,\alpha^2\,\dot{f} + m(f + \dot{f}) &= 0. \end{aligned}\right\} \tag{2.20}$$

The only initial condition for this set is given by $f = f_+$ for $\tau = 0$; numerical values of the function $f_+ = f_+(f_-, m)$ are collected in Table 7 of Ref. [*18*].

To solve the system (2.20) we introduce the new dependent variable (function of time)

$$u = \alpha\,f; \tag{2.21}$$

then the first of Eq. (2.20) yields

$$\alpha = \frac{K(u)}{\pi}, \tag{2.22}$$

$$f = \frac{\pi\,u}{K(u)}. \tag{2.23}$$

Dividing the second Eq. (2.20) by f and substituting the above values into it we obtain

$$-\frac{8}{\pi^2} K(u)\,K'(u)\,\dot{u} - \frac{4}{\pi^2} K^2(u)\left[\frac{\dot{u}}{u} - \frac{K'(u)\,\dot{u}}{K(u)}\right] + m\left[1 + \frac{\dot{u}}{u} - \frac{K'(u)\,\dot{u}}{K(u)}\right] = 0. \tag{2.24}$$

This equation can be easily integrated except for one term:

$$-\frac{2}{\pi^2} K^2(u) - \frac{4}{\pi^2}\int \frac{K^2(u)}{u}\,du + m\,\tau + m\ln u - m\ln K(u) = m\,C \tag{2.25}$$

whence

$$\tau = \ln\frac{K(u)}{u} + \frac{2}{\pi^2\,m}\left[K^2(u) + 2\int\frac{K^2(u)}{u}\,du\right] + C. \tag{2.26}$$

Eq. (2.26) together with (2.23) determines in parametric form the function $f = f(\tau)$ sought, u being a parameter. The constant C can be found from the initial condition $f = f_+$ for $\tau = 0$ as (2.23) determines the initial value of u.

The solution obtained contains the integral

$$\overline{K}_{-1}(k) = \int \frac{K^2(k)}{k}\, dk. \tag{2.27}$$

This integral is a new non-elementary function which cannot be expressed by means of elementary functions or elliptic integrals. Recurrence formulae for the family of such functions, namely for the functions (the argument k is omitted in the notation)

$$\left.\begin{aligned} \overline{K}_n &= \int k^n K^2(k)\, dk, \\ \overline{E}_n &= \int k^n E^2(k)\, dk, \\ \overline{M}_n &= \int k^n E(k)\, K(k)\, dk, \end{aligned}\right\} \tag{2.28}$$

were given earlier by the author [*20*]. All functions of this kind can be reduced to four independent new non-elementary functions, which were analyzed and tabulated in Ref. [*20*]. The function $\overline{K}_{-1}(k)$, (2.27), can be expressed by means of $\overline{E}_{-1}(k)$ (which is more convenient for tabulation) and by elliptic integrals:

$$\overline{K}_{-1} = \frac{1}{2} k^2 K^2 - \frac{1}{2} K^2 - \frac{1}{2} E^2 + \overline{E}_{-1}. \tag{2.29}$$

Substitution of (2.29) in (2.26) yields

$$\tau = \ln \frac{K(u)}{u} + \frac{2}{\pi^2 m} \left[u^2 K^2(u) - E^2(u) + 2\overline{E}_{-1}(u)\right] + C. \tag{2.30}$$

Using the table of $\overline{E}_{-1}(k)$ and A. FLETCHER's table [*1*] we can determine the numerical value of C and find the numerical values of the two functions $\tau = \tau(u)$ and $f = f(u)$ for any value of the parameter u.

A more convenient form of the solution may be obtained, however, if we introduce the quantity $\bar{\tau}$,

$$\bar{\tau} = \tau - C; \tag{2.31}$$

this quantity will be called the "translative dimensionless time". The relation between $\bar{\tau}$ and u or $\bar{\tau}$ and f does not depend on the initial condition; it can be represented for example in a graphical form in a completely general way for a given force m; and subsequently the origin of the axis τ may be selected so as to fulfill the required initial condition. We should clearly stress here that this possibility is due to the use of the one-point collocation method which in effect reduced the system with an infinite number of degrees of freedom to a system with one degree of freedom. Our solution will be now written concisely as

$$\left.\begin{aligned} \bar{\tau} &= \omega_1(u) + \frac{1}{m}\, \omega_2(u), \\ f &= \frac{\pi u}{K(u)}, \end{aligned}\right\} \tag{2.32}$$

where the functions $\omega_1(u)$ and $\omega_2(u)$ are determined by the formulae

$$\left.\begin{aligned}\omega_1(u) &= \ln\frac{K(u)}{\pi u}\,,\\ \omega_2(u) &= \frac{2}{\pi^2}\left[u^2 K^2(u) - E^2(u) + 2\,\overline{E}_{-1}(u)\right].\end{aligned}\right\} \tag{2.33}$$

Numerical values of the functions $\omega_1(u)$, $\omega_2(u)$ and $f(u)$ are collected in Table 1.

Table 1. *Exact values of auxiliary functions $\omega_1(u)$ and $\omega_2(u)$, deflections $f(u)$ and vertical displacements $w(u)$ for linear (Maxwellian) creep $n = 1$*

u	ω_1	ω_2	f	w
0	$+\infty$	$-\infty$	0	0
0.001	6.2146	−7.4078	0.0020	0.0000
0.002	5.5215	−6.7146	0.0040	0.0000
0.005	4.6052	−5.7983	0.0100	0.0000
0.01	3.9121	−5.1051	0.0200	0.0002
0.02	3.2190	−4.4118	0.0400	0.0006
0.03	2.8136	−4.0061	0.0600	0.0014
0.04	2.5261	−3.7181	0.8000	0.0025
0.05	2.3032	−3.4945	0.0999	0.0039
0.10	1.6119	−2.7976	0.1995	0.0157
0.15	1.2097	−2.3857	0.2983	0.0355
0.20	0.9265	−2.0890	0.3960	0.0632
0.25	0.7092	−1.8540	0.4920	0.0990
0.30	0.5343	−1.6567	0.5861	0.1430
0.35	0.3891	−1.4843	0.6777	0.1955
0.40	0.2663	−1.3289	0.7662	0.2568
0.45	0.1611	−1.1849	0.8512	0.3271
0.50	+0.0706	−1.0486	0.9318	0.4068
0.55	−0.0073	−0.9166	1.0073	0.4965
0.60	−0.0739	−0.7859	1.0767	0.5969
0.65	−0.1298	−0.6535	1.1386	0.7090
0.70	−0.1752	−0.5156	1.1915	0.8341
0.75	−0.2094	−0.3674	1.2330	0.9741
0.80	−0.2308	−0.2011	1.2596	1.1320
0.85	−0.2356	−0.0026	1.2656	1.3130
0.90	−0.2150	+0.2601	1.2398	1.5276
0.95	−0.1418	0.5633	1.1523	1.8040
0.99	+0.0762	1.7555	0.9266	2.1790
1.00	$+\infty$	$+\infty$	0	3.1416

Using Table 1 we can determine the relation between f and $\bar{\tau}$ graphically or numerically and then fix the origin of time so as to obtain $f = f_+$ for $\tau = 0$. The graphical method is especially convenient here. Since f_+ is, for a given force m, a function of f_- (the values of which are given in Table 7 of Ref. [*18*]) we may add an auxiliary graph of f_-, corresponding to f_+, and find the origin of time immediately for a given f_- without a separate evaluation of f_+. Such a diagram, very convenient for practical applications, has an auxiliary character only, because the line $f_- = f_-(\bar{\tau})$ is, of course, without any physical meaning.

In Figs. 2, 3 and 4 there are shown the graphs $f = f(\tau)$ for $m = 0.8$ (subcritical range), $m = 1$ (EULER force) and $m = 1.1$ (supercritical

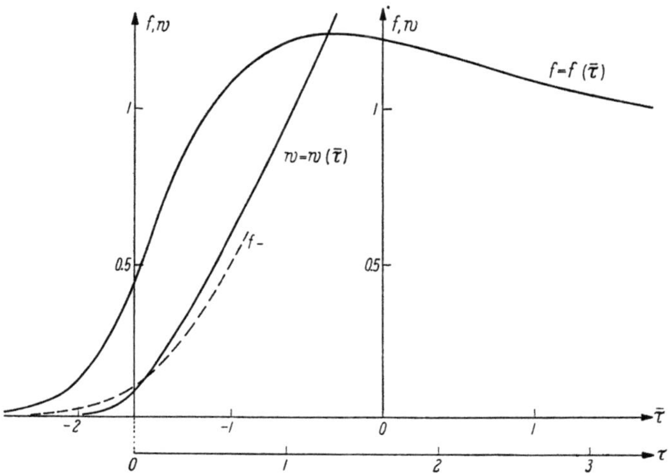

Fig. 2. Development of buckling for $m = 0.8$ (subcritical range) and $n = 1$ (MAXWELL's law). The origin of time is fixed for $f_- = 0.1$

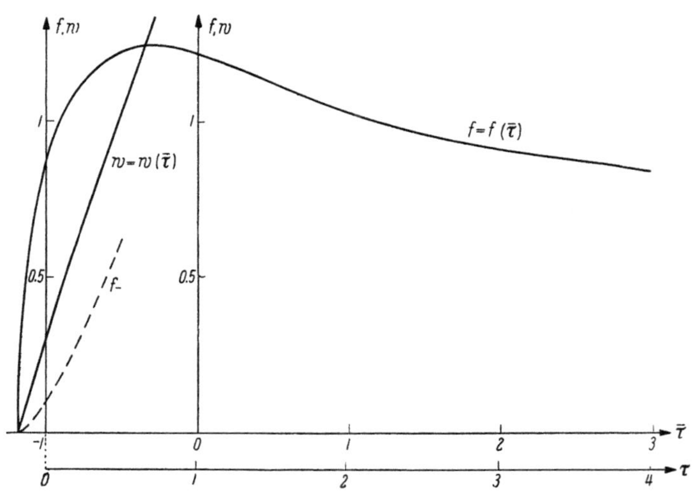

Fig. 3. Development of buckling for $m = 1$ (EULER force) and $n = 1$ (MAXWELL's law). The origin of time is fixed for $f_- = 0.1$

range) with the auxiliary graphs for f_-. In all the three cases the origin of time τ is fixed for $f_- = 0.1$. For a different value of f_- the phenomenon will take place in a similar manner but with a changed origin of time ("translation" of time). In the subcritical range this translation may be very great (up to infinity), and in the supercritical range it is very small

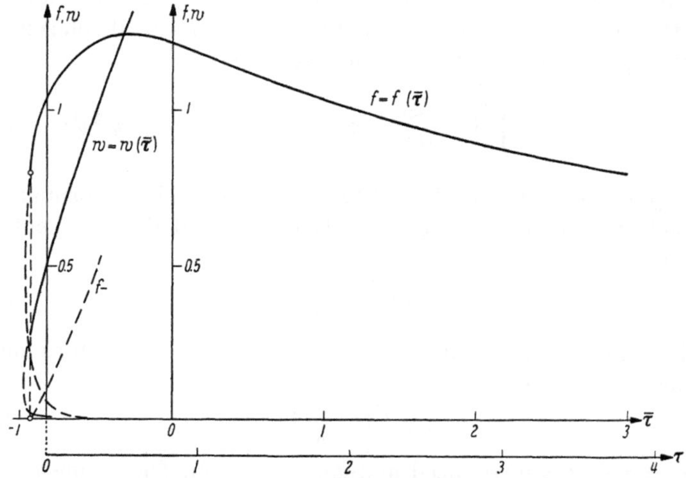

Fig. 4. Development of buckling for $m = 1.1$ (supercritical range) and $n = 1$ (MAXWELL's law). The origin of time is fixed for $f_- = 0.1$

as a rule and finite even for $f_- = 0$ (mathematical buckling) — in this range the initial deflection influences very slightly the relatively quickly occurring phenomenon. In the supercritical range only part of line $f = f(\bar{\tau})$ may have a physical interpretation, namely the part beginning with the deflection f_+ corresponding to mathematical buckling ($f_- = 0$). The remaining part of the curve $f = f(\bar{\tau})$ is marked by a dotted line in Fig. 4.

The parameter u used in our calculations has a very simple geometric meaning. The derivative of the function (2.12) with respect to x is

$$y' = 2\,\alpha\, f\,\sqrt{1 - \alpha^2 f^2}\;\frac{\operatorname{cn}(2\,\alpha\,x)}{\operatorname{dn}^2(2\,\alpha\,x)}\,. \qquad (2.34)$$

In our system of co-ordinates $x - y$, where x is measured along the axis of the deformed bar, this derivative represents the sine of the angle φ subtended by the tangent to the axis of the bar and the direction of the force. At the free end, for $x = 0$, $\operatorname{cn} 0 = \operatorname{dn} 0 = 1$ and

$$\sin\varphi_0 = 2\,\alpha\, f\sqrt{1 - \alpha^2 f^2} = 2\,u\,\sqrt{1 - u^2}\,. \qquad (2.35)$$

Using the formula for the sine of the double argument we find at once

$$u = \sin\frac{\varphi_0}{2}\,. \qquad (2.36)$$

Thus the parameter u represents the sine of the half angle between the tangent at $x = 0$ and the direction of the force P, and its values are contained in the interval $0 \leq u < 1$.

Now it is very easy to calculate the second important quantity determining the progress of buckling, namely the vertical displacement ξ (the lowering of the free end of the bar). Since

$$\xi = l - \int_0^l \sqrt{1 - \left(\frac{\partial v}{\partial s}\right)^2} ds = l - \frac{2l}{\pi} \int_0^{\pi/2} \sqrt{1 - y'^2}\, dx \tag{2.37}$$

and $y = y(x)$ as shown in Eq. (2.12), we find easily a formula that corresponds to the formula for the elastic vertical displacement of a buckled bar:

$$\xi = 2l\left[1 - \frac{E(u)}{K(u)}\right], \tag{2.38}$$

where the parameter u is still defined by (2.21). Writing this formula in a dimensionless form we get

$$w = \pi\left[1 - \frac{E(u)}{K(u)}\right], \tag{2.39}$$

where $w = \pi\,\xi/2\,l$ is introduced in analogy to (2.4). The values of w are added to Table 1 and the diagrams of the function $w = w(\bar{\tau})$ are shown in Figs. 2, 3 and 4 together with the function $f = f(\bar{\tau})$.

Although by introduction and tabulation of the new non-elementary function $\bar{E}_{-1}$ we obtained the exact solutions (2.32) and (2.33) of the system (2.20), it will be very convenient to introduce M. G. PUWEIN's approximation for the complete elliptic integral of the first kind, used previously in the theory of buckling in Refs. [*17*] and [*18*]. Making use of this approximation we express the solution by means of elementary functions, which are better known and which have been investigated more thoroughly. In addition, in comparing the two solutions we shall find the error to be very small; this statement will be very important in connection with the calculations dealing with non-linear creep where only PUWEIN's approximation will be used.

PUWEIN's approximation [*11*] consists in replacing $K(k)$ by the function

$$K(k) \approx \frac{\pi}{2\sqrt[4]{1-k^2}}. \tag{2.40}$$

Using (2.40) we write the approximate relations

$$\int \frac{K^2(u)}{u}\, du = \frac{\pi^2}{4} \int \frac{du}{u\sqrt{1-u^2}} = -\frac{\pi^2}{4} \ln \frac{1 + \sqrt{1-u^2}}{u} = -\frac{\pi^2}{4} \operatorname{Arch} \frac{1}{u}, \tag{2.41}$$

where $\operatorname{Arch} u = \cosh^{-1} u$ denotes the inverse hyperbolic cosine. Using PUWEIN's approximation in the remaining terms of (2.26) and in (2.23), we get

$$\bar{\tau} = \omega_1(u) + \frac{1}{m}\,\omega_2(u), \qquad f = 2\,u\sqrt[4]{1-u^2}, \tag{2.42}$$

where

$$\omega_1(u) = -\frac{1}{4}\ln(1-u^2) - \ln u - \ln 2, \qquad \omega_2(u) = \frac{1}{2\sqrt{1-u^2}} - \operatorname{Arch}\frac{1}{u} - (1-\ln 2). \tag{2.43}$$

The differences between the exact formulae (2.33) and the approximate ones (2.43) vanish for $u \to 0$ (because of suitably chosen constants) and the two formulae can now be compared. Approximate values of the functions $\omega_1(u)$, $\omega_2(u)$ and $f(u)$, computed from (2.43) and (2.42) are collected in Table 2.

Table 2. *Approximate values of auxiliary functions $\omega_1(u)$ and $\omega_2(u)$ and deflections $f(u)$ for linear (Maxwellian) creep $n = 1$*

u	ω_1	ω_2	f
0	$+\infty$	$-\infty$	0
0.001	6.2146	−7.4078	0.0020
0.002	5.5215	−6.7146	0.0040
0.005	4.6052	−5.7983	0.0100
0.01	3.9121	−5.1051	0.0200
0.02	3.2190	−4.4118	0.0400
0.03	2.8136	−4.0061	0.0600
0.04	2.5261	−3.7181	0.0800
0.05	2.3032	−3.4945	0.0999
0.10	1.6119	−2.7976	0.1995
0.15	1.2097	−2.3857	0.2983
0.20	0.9265	−2.0890	0.3960
0.25	0.7093	−1.8539	0.4920
0.30	0.5344	−1.6565	0.5860
0.35	0.3893	−1.4839	0.6775
0.40	0.2667	−1.3281	0.7659
0.45	0.1619	−1.1837	0.8505
0.50	+0.0719	−1.0465	0.9306
0.55	−0.0053	−0.9131	1.0053
0.60	−0.0708	−0.7805	1.0733
0.65	−0.1251	−0.6450	1.1333
0.70	−0.1681	−0.5023	1.1831
0.75	−0.1988	−0.3463	1.2199
0.80	−0.2146	−0.1667	1.2394
0.85	−0.2102	+0.0566	1.2339
0.90	−0.1726	+0.3731	1.1884
0.95	−0.0599	0.9714	1.0617
0.99	+0.2962	3.0956	0.7413
1.00	$+\infty$	$+\infty$	0

Comparing the numerical values in Tables 1 and 2 we observe the high accuracy of PUWEIN's approximation: up to $u = 0.2$ no differences appear even in the fourth decimal place. With increasing u the differences increase, but they influence the curve $\bar{\tau} - f$ only slightly: for large u we get higher values of $\bar{\tau}$ and lower values of f and the point passes along the curve changing its shape only slightly, since the function $f = f(\bar{\tau})$ decreases in this range. For all practical applications the accuracy of (2.42) and (2.43) is therefore completely satisfactory.

The analysis of the phenomenon here presented is very simple because it reduces — for a constant force m — all cases of the initial deflection to one common case with changing origin of time only (the scale of time may be stretched by variation of E and λ, formula (2.2)). This conclusion is connected with the collocation method used and with the assumption (2.12). In reality, the cases of different initial deflections will differ not only in the origin of time, but in the shape of the curve $\bar{\tau} - f$ as well; we estimate these differences to be small because the deflection curve (2.12) was chosen very carefully and no doubt it expressed with a good approximation the real behaviour of the bar.

3. Finite deflections of a physically non-linear bar

We pass now to the general case of NORTON's law (1.1), limiting ourselves to the idealized H-section to avoid the integration of the stresses over the area of the cross-section which would introduce essential complications. Furthermore we assume n to be odd to avoid the inconvenient notation of (1.2).

BERNOULLI's assumption of plane cross-sections yields

$$\frac{\dot{\sigma}}{E} + \frac{\sigma^n}{\lambda} = \dot{\varkappa}\, z. \tag{3.1}$$

Substituting $z = h/2$ and $z = -h/2$ (h is the depth of the cross-section) and denoting the corresponding stresses by σ_r (tension) and σ_c (compression) we get

$$\begin{aligned} \frac{\dot{\sigma}_r}{E} + \frac{\sigma_r^n}{\lambda} &= \dot{\varkappa}\, \frac{h}{2}, \\ \frac{\dot{\sigma}_c}{E} + \frac{\sigma_c^n}{\lambda} &= -\dot{\varkappa}\, \frac{h}{2}. \end{aligned} \tag{3.2}$$

Subtraction of these equations yields

$$\frac{\dot{\sigma}_r - \dot{\sigma}_c}{E} + \frac{\sigma_r^n - \sigma_c^n}{\lambda} = \dot{\varkappa}\, h. \tag{3.3}$$

Expressing the stresses σ_r and σ_c by the bending moment M and the normal force N we have

$$\begin{aligned} \sigma_r &= \frac{N}{F} + \frac{2M}{Fh}, \\ \sigma_c &= \frac{N}{F} - \frac{2M}{Fh}, \end{aligned} \tag{3.4}$$

where F denotes the cross-sectional area of the bar; substitution of (3.4) in (3.3) yields finally

$$\frac{4\dot{M}}{E F h} + \frac{1}{\lambda F^n}\left[\left(N + \frac{2M}{h}\right)^n - \left(N - \frac{2M}{h}\right)^n\right] = \dot{\varkappa}\, h. \qquad (3.5)$$

We have obtained the basic equation of bending and buckling for the non-linear creep law (1.1). The essential difference between (3.5) and (2.1) is the appearance of the normal force N although a constant length of the bar was assumed; substitution of $n = 1$ leads, of course, to a vanishing of N. We obtain then (2.1) if we set $F h^2/4 = I$ (moment of inertia for ideal H section).

Returning now to the column of Fig. 1 we substitute $M = P v$ and $N = -P\cos\varphi$ in (3.5) and obtain

$$\frac{4P\dot{v}}{E F h^2} + \frac{P^n}{\lambda F^n h}\left[\left(\frac{2v}{h} + \cos\varphi\right)^n + \left(\frac{2v}{h} - \cos\varphi\right)^n\right] - \dot{\varkappa} = 0. \qquad (3.6)$$

Next we introduce dimensionless quantities. The dimensionless variables x and y are defined by (2.4), whence

$$\cos\varphi = \sqrt{1 - y'^2}. \qquad (3.7)$$

Denote by η the reciprocal of the slenderness of the bar:

$$\eta = \frac{\pi h}{4l}. \qquad (3.8)$$

The dimensionless force will be defined, as before, by

$$m = \frac{P}{P_E} = \frac{16 P l^2}{\pi^2 E F h^2}: \qquad (3.9)$$

Finally, substituting (2.5) we get

$$\frac{\partial}{\partial t}\left[\frac{y''}{\sqrt{1-y'^2}}\right] + m\dot{y} + m^n \frac{2\pi^{n-1} h^{n-1} E^n}{4^n l^{n-1}\lambda}\left[(y + \eta\sqrt{1-y'^2})^n + \right.$$
$$\left. + (y - \eta\sqrt{1-y'^2})^n\right] = 0. \qquad (3.10)$$

We introduce the dimensionless time τ,

$$\tau = \frac{\pi^{n-1} h^{n-1} E^n}{4^{n-1} l^{n-1}\lambda}\, t = \frac{\eta^{n-1} E^n}{\lambda}\, t. \qquad (3.11)$$

Thus the number of parameters is now reduced to two, which are m and η:

$$y''(1 - y'^2) + y' y'' \dot{y}' + m\dot{y}(1 - y'^2)^{3/2} +$$
$$+ \frac{m^n}{2}\left[(y + \eta\sqrt{1-y'^2})^n + (y - \eta\sqrt{1-y'^2})^n\right](1 - y'^2)^{3/2} = 0. \qquad (3.12)$$

This equation, which is in a certain sense a generalization of (2.8) (more general creep law, but less general form of the cross-section) is again a non-linear partial differential equation of the third order. We seek the solution of the equation with the boundary conditions (2.9) and the initial condition (2.10).

As in the case of linear creep we use the one-point collocation method, assuming the equation of the deflection curve in the form (2.12). The condition (2.15), connecting the functions α and f, still holds. The second condition will be obtained by collocation at $x = \pi/2$. Substituting $x = \pi/2$, $y' = 0$ in (3.12) we get

$$\dot{y}'' + m\,\dot{y} + \frac{m^n}{2}\left[(f+\eta)^n + (f-\eta)^n\right] = 0, \tag{3.13}$$

and making use of (2.12) we obtain the following system of equations which is a generalization of (2.20):

$$\begin{gathered} \alpha\,\pi = K(\alpha\,f) \\ -8\,\alpha\,\dot{\alpha}\,f - 4\,\alpha^2\,\dot{f} + m\,\dot{f} + \frac{m^n}{2}\left[(f+\eta)^n + (f-\eta)^n\right] = 0\,. \end{gathered} \tag{3.14}$$

The initial condition accompanying this set has the form $f = f_+$ for $\tau = 0$; numerical values of $f_+ = f_+(f_-, m)$ are collected in Table 7 of Ref. [*18*].

Writing the second of Eq. (3.14) in the differential form

$$d\tau = -\frac{2}{m^{n-1}}\,\frac{df}{(f+\eta)^n + (f-\eta)^n} + \frac{8}{m^n}\,\frac{d(\alpha^2 f)}{(f+\eta)^n + (f-\eta)^n} \tag{3.15}$$

we observe that the first term can be integrated directly, without introducing the parameter u. We have to introduce it, however, in the second term; since

$$d(\alpha^2\,f) = d(\alpha\,u) = \frac{1}{\pi}\left[K(u) + u\,K'(u)\right] du \tag{3.16}$$

we obtain finally

$$\begin{gathered} \tau = -\frac{2}{m^{n-1}}\int\frac{df}{(f+\eta)^n + (f-\eta)^n} + \frac{8}{\pi\,m^n}\int\frac{K(u) + u\,K'(u)}{\left(\frac{\pi\,u}{K(u)} + \eta\right)^n + \left(\frac{\pi\,u}{K(u)} - \eta\right)^n}\,du\,, \\ f = \frac{\pi\,u}{K(u)}\,. \end{gathered} \tag{3.17}$$

These equations constitute the exact solution of the system (3.14), written in a "partially-parametrical" form. The advantage of this notation is the ease of integration of the first term and we see that it is only the second term that cannot be expressed by means of elementary functions. Since this integral has a complicated form, omitting an exact analysis we pass to PUWEIN's approximation which was found to be sufficiently accurate for our purposes. Introducing (2.40) for $K(u)$ and the corresponding expression for $K'(u)$ we obtain in the place of (3.17) the following expressions:

$$\tau = -\frac{2}{m^{n-1}} \int \frac{df}{(f+\eta)^n + (f-\eta)^n}$$

$$+ \frac{2}{m^n} \int \frac{(2-u^2)\,du}{[(2u\sqrt[4]{1-u^2}+\eta)^n + (2u\sqrt[4]{1-u^2}-\eta)^n](1-u^2)^{5/4}}, \tag{3.18}$$

$$f = 2u\sqrt[4]{1-u^2}.$$

The second term in the first formula is not an elliptic integral, as it might be thought when observing the expression $\sqrt[4]{1-u^2}$. Inside the square bracket we get only odd powers of the expression, which multiplied by $(1-u^2)^{5/4}$ always give an expression of the type $(1-u^2)^{j/2}$, where j is an integer; thus the integral can be expressed by elementary functions.

Assume, for example, $n = 3$. At first we get

$$\tau = -\frac{1}{m^2} \int \frac{df}{f^3 + 3\eta^2 f} + \frac{1}{2m^3} \int \frac{(2-u^2)\,du}{4u^3(1-u^2)^2 + 3\eta^2 u(1-u^2)^{3/2}}. \tag{3.19}$$

The form of the integrals shows clearly that the integrals can be expressed by elementary functions; in the general case, however, the calculations are very cumbersome. We present here the results for a limiting case only, namely for $\eta = 0$. This case corresponds to a very great slenderness of the bar, as can be seen from formula (3.8). Introducing the "translative" time $\bar{\tau}$ and omitting the constants of integration we obtain

$$\bar{\tau} = \frac{1}{2m^2 f^2} + \frac{1}{16m^3}\left[6\ln u - 3\ln(1-u^2) - \frac{2}{u^2} + \frac{1}{1-u^2}\right]. \tag{3.20}$$

Now we change the "partially-parametrical" notation into the ordinary parametrical one, substituting for f the corresponding formula (3.18.2); the result will be written concisely in the form

$$\bar{\tau} = \frac{1}{m^2}\,\omega_1(u) + \frac{1}{m^3}\,\omega_2(u),$$
$$f = 2u\sqrt[4]{1-u^2}, \tag{3.21}$$

where

$$\omega_1(u) = \frac{1}{8u^2\sqrt{1-u^2}},$$
$$\omega_2(u) = \frac{1}{16}\left[6\ln u - 3\ln(1-u^2) - \frac{2}{u^2} + \frac{1}{1-u^2}\right]. \tag{3.22}$$

Numerical values of the functions $\omega_1(u)$, $\omega_2(u)$ and $f(u)$ are collected in Table 3. Corresponding values of $w(u)$ (vertical displacements) may be taken from Table 1.

Table 3. *Approximate values of auxiliary functions $\omega_1(u)$ and $\omega_2(u)$ and deflections $f(u)$ for non-linear (Nortonian) creep $n = 3$*

u	ω_1	ω_2	f
0	$+\infty$	$-\infty$	0
0.01	1250.0625	−1251.6644	0.0200
0.02	312.5625	−313.9044	0.0400
0.03	138.9527	−140.1411	0.0600
0.04	78.1875	−79.2692	0.0800
0.05	50.0630	−51.0603	0.0999
0.10	12.5630	−13.2985	0.1995
0.15	5.6191	−6.1988	0.2983
0.20	3.1894	−3.6558	0.3960
0.25	2.0656	−2.4411	0.4920
0.30	1.4559	−1.7540	0.5860
0.35	1.0893	−1.3184	0.6775
0.40	0.8524	−1.0178	0.7659
0.45	0.6912	−0.7959	0.8505
0.50	0.5774	−0.6227	0.9306
0.55	0.4948	−0.4803	1.0053
0.60	0.4340	−0.3574	1.0733
0.65	0.3893	−0.2462	1.1333
0.70	0.3572	−0.1401	1.1831
0.75	0.3360	−0 0322	1.2199
0.80	0.3255	+0.0862	1.2394
0.85	0.3284	+0.2316	1.2339
0.90	0.3540	+0.4465	1.1884
0.95	0.4436	0.9198	1.0617
0.99	0.9041	3.7438	0.7413
1.00	$+\infty$	$+\infty$	0

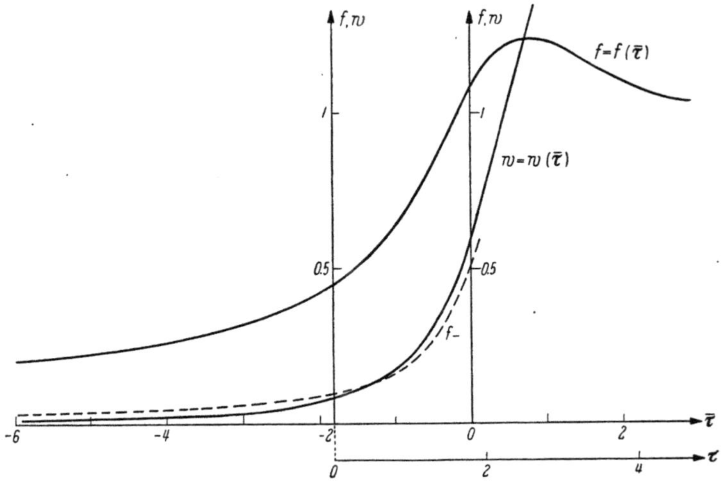

Fig. 5. Development of buckling for $m = 0.8$ (subcritical range) and $n = 3$ (NORTON's law). The origin of time is fixed for $f_- = 0.1$

The diagrams of Figs. 5 and 6 show the functions $f = f(\bar{\tau})$ and $w = w(\bar{\tau})$ for $m = 0.8$ (subcritical range) and $m = 1.1$ (supercritical

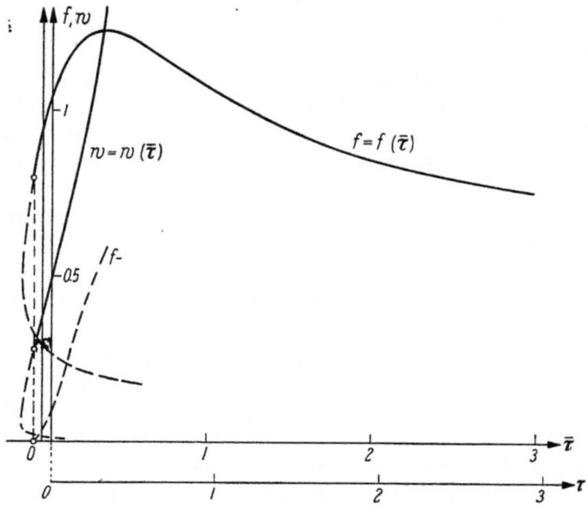

Fig. 6. Development of buckling for $m = 1.1$ (supercritical range) and $n = 3$ (NORTON's law). The origin of time is fixed for $f_- = 0.1$

range); auxiliary graphs of f_- are also added. In the same manner as in Figs. 2 to 4, the origin of time τ was fixed for $f_- = 0.1$, that is for $f_+ = 0.4505$ and $f_+ = 1.0161$ (Table 7 of Ref. [*18*]).

4. Conclusions

The finite-deflection-theory gives no significant differences between the linear creep law ($n = 1$) and the non-linear creep law ($n = 3$), except for a more rapid development of the buckling in the latter case. The "critical time" τ_k does not exist from the standpoint of finite deflections. The general criterion of failure of a structure subjected to creep buckling should be chosen in a different way, for instance as a criterion of limited deflections or a criterion of limited strains. If strongly non-linear materials are considered (high values of n), admittedly, these criteria and the "critical time" — criterion may give similar results. However, the quantity τ_k should be used with great caution. In physically linear cases a deformation-criterion only can give correct results and should be used for design purposes.

References

[*1*] FLETCHER, A.: A table of complete elliptic integrals, Phil. Mag. **30**, 516—519 (1940), Seventh Series.

[*2*] FRAZER, R. A., W. P. JONES and SYLVIA W. SKAN: Approximation to functions and to solutions of differential equations, Rep. and Memor. nr. 1799, Aero. Res. Com., 1937.

[3] HOFF, N. J.: Creep buckling, Aeron. Quart. **7**, 1—20 (1956).
[4] HOFF, N. J.: A survey of the theories of creep buckling, Proc. Third US Nat. Congr. Appl. Mech., Brown Univ. 1958, publ. Pergamon Press 1958.
[5] HULT, J. A.: Creep buckling, Instn. Hallfasthetslära Kungl. Tekniska Högskolan, Publ. nr. 111, Stockholm 1955.
[6] KEMPNER, J., and F. V. POHLE: On the nonexistence of finite critical time for linear viscoelastic columns, J. Aero. Sci. **20**, 572—573 (1953).
[7] LIBOVE, CH.: Creep buckling of columns, J. Aero. Sci. **19**, 459 (1952).
[8] NORTON, F. H.: The creep of steel at high temperatures, New York: McGraw-Hill, 1929.
[9] ODQUIST, F. K. G.: Primärkrypningens inverkan vid knäckning av strävor, Tekn. Tidskr. **84**, nr. 29, (1954). 654. English summary: J. Appl. Mech. **21** (1954).
[10] PATEL, S. A.: Buckling of columns in the presence of creep, Aero. Quart. **7**, 125—134. (1956)
[11] PUWEIN, M. G.: Die mutierte Pendellänge, Öster. Ing.-Archiv **8**, 54—55 (1954).
[12] RABOTNOV, G. N., and S. A. SHESTERIKOV: Creep stability of columns and plates, J. Mech. Phys. Solids **6**, 27—34 (1957). Russian original: Prikl. Mat. Miech. **21**, 406—412 (1957).
[13] ROZENBLUM, W. I.: Ustoichivost szhatogo stierzhnya v sostoyanii polzuchesti (Stability of compressed bar in the presence of creep), Inzhen. Sbornik **18**, 99—104 (1954).
[14] RZHANITSYN, A. R.: Processy dieformirovanya konstrukcij iz uprugovyazkikh elementov, (Deformation of structures with viscoelastic elements), Dokl. Akad. Nauk SSSR, **52**, 25 (1946).
[15] RZHANITSYN, A. R.: Nekotorye voprosy mekhaniki sistiem deformiruyushchikhsya vo vremeni (Some problems of mechanics of systems deforming in time) Gostekhizdat, Moscow—Leningrad 1949.
[16] SHESTERIKOV, S. A.: O kriterii ustoichivosti pri polzuchesti (On the criterion of stability in the presence of creep), Prikl. Mat. Mekh. **23**, 1101—1106 (1959).
[17] WNUK, M., and M. ŻYCZKOWSKI: Application of the PUWEIN formula to the theory of buckling, Arch. Mech. Stos. **9**, 293—300 (1957).
[18] ŻYCZKOWSKI, M.: Skończone ugięcia mimośrodowo ściskanych prętów o krzywiźnie pierwotnej (Finite deflections of eccentrically compressed bars with initial curvature), Księga Jubil. Prof. W. WIERZBICKIEGO, Warszawa 1959, PWN, 479—518. English Summary: Bull. Acad. Polon., Serie Sc. Techn. **7**, 1 (1959).
[19] ŻYCZKOWSKI, M.: Some problems of creep buckling of homogeneous and non-homogeneous bars, Proc. Intern. Symposium on Non-Homogeneity in Elasticity and Plasticity, London: Pergamon Press 1959.
[20] ŻYCZKOWSKI, M.: Całki względem modułu z kwadratów pełnych całek eliptycznych (Some integrals involving complete elliptic integrals squared), Zastosowania Matematyki **6**, 67—80 (1961).

Discussion

JAN HULT, Royal Institute of Technology, Stockholm: The changes in shape taking place in a column subjected to creep buckling were studied in some detail by HULT in 1955, see author's reference 5.

The deflections of a column (99.25% Al, annealed, extruded and milled to 8×15 mm H-section) were measured at the three quarter points in the course of a creep buckling test (77.5 °C). The magnitude of the third harmonic in a Fourier series describing the deflection curve was calculated from these data. As long as the total deflection was less than the half depth of the column, the magnitude of this third harmonic was found to be in close accordance with the predictions of a theoretical analysis based on geometric linearity. This implies that the third harmonic is fully negligible.

This result, however, does not contradict the conclusions arrived at by the author, since a deflection equal to the half width of the column is quite small in the sense introduced by the author.

F. K. G. ODQVIST, Royal Institute of Technology, Stockholm: Could you indicate how your solution differs from that of HOFF, and in particular where the curves of f as a function of τ deviate appreciably from one another?

N. J. HOFF, Stanford University: The author's very elegant treatment of the creep buckling of columns subject to elastic deformations and to creep according to a power law is related to earlier theories of creep buckling in the same manner as the elastica theory is to EULER's theory of the buckling of slender perfectly elastic columns. Both large deflection theories indicate that the infinitely large deflections predicted by the simpler theories cannot exist in reality. This does not rules out, however, the usefulness of the simpler approach to creep buckling in engineering design, just as the EULER load was not rendered useless by the development of the theory of the elastica. It should also not be forgotten that large deflections cause plastic deformations; these were found by the discussor to lead to infinitely large lateral velocities when inertia effects were disregarded. The most complete treatment of this elastoplastic small-deflection theory of creep buckling can be found in "A theoretical and experimental investigation of creep buckling", by J. C. CHAPMAN, BURTON ERICKSON and N. J. HOFF, International Journal of Mechanical Sciences, **1**, Nos. 2/3, April 1960.

Author's closing remarks: The differences between the curves τ versus f obtained from the small-deflection theory on the one hand and the finite-deflection theory on the other hand become significant as a rule when $f = \pi\,\delta/2\,l$ is greater than 0.5. However, under the action of the EULER force as well as in the supercritical range the finite-deflection theory alone can properly describe the phenomenon.

Sur un modèle apte à traduire le fluage sous charge constante des structures

Par

R. Mazet

Office National d'Etudes et de Recherches Aéronautiques,
Chatillon-sous-Bagneux, Seine, France

Lorsqu'on étudie les déformations d'une structure simple telle qu'un barreau sous des charges monoaxiales qui l'obligent à travailler au-delà de la limite élastique, on utilise habituellement la formule de NICHOLAS J. HOFF [1]

$$\dot{\varepsilon} = \frac{\dot{\sigma}}{E} + \left(\frac{\sigma}{\mu}\right)^m \left(\frac{\dot{\sigma}}{\mu}\right) + \left(\frac{\sigma}{\lambda}\right)^n, \tag{1}$$

ε étant la déformation, σ la contrainte, E le module de YOUNG, λ, μ, m, n des constantes caractéristiques du matériau, m étant généralement pris égal à 1 et n à 3 pour faciliter les calculs.

Cette formule serait rigoureusement valable pour un matériau obéissant à la description suivante: il flue *si petite que soit la contrainte*; toutefois, lorsque σ est petit, on a approximativement et instantanément:

$$\varepsilon = \frac{\sigma}{E} + \frac{1}{m+1}\left(\frac{\sigma}{\mu}\right)^{m+1}$$

et, lorsque σ est encore plus petit, $\varepsilon \simeq \frac{\sigma}{E}$ (fig. 1).

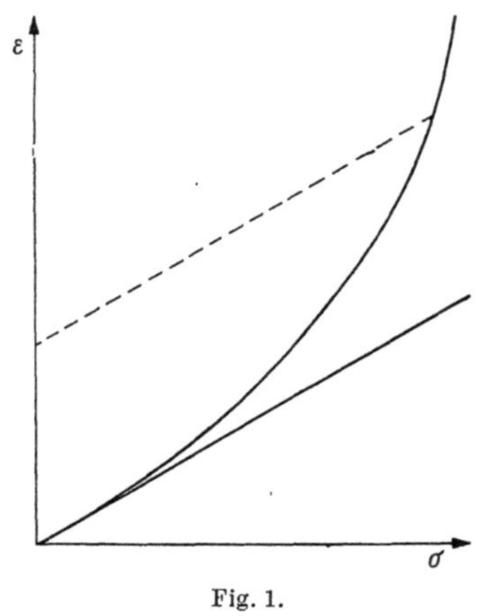

Fig. 1.

Il est en outre convenu que, si la pièce est déchargée, le second terme reste constant, ce qui traduit le phénomène de l'écrouissage (droite pointillée sur la fig. 1).

Ainsi le domaine élastique (équilibre possible sous contrainte constante sans déformation permanente) est pratiquement réduit à zéro et le domaine plastique (équilibre possible sous contrainte constante avec déformation permannte) disparaît lui-même en tant que domaine où un état d'équilibre sous charge, d'une durée illimitée, puisse avoir lieu.

On doit d'autre part observer que la formule (1) oblige à considérer la *contrainte* σ alors que la grandeur accessible aux mesures est la *charge* π. Ceci n'a pas d'importance tant que l'une est proportionnelle à l'autre, le facteur de proportionnalité (en l'espèce: la section du barreau) restant constant. Mais cette circonstance favorable ne se produit que dans le domaine élastique, et encore à condition de négliger la contraction de POISSON. Dès que la plasticité et le fluage interviennent, le matériau subit un écoulement notable qui détruit la proportionnalité et complique les calculs.

Une troisième observation touche à l'aspect intrinsèque d'un problème de mécanique régi par le principe de NEWTON: pour que l'évolution mécanique d'un système soit bien déterminée, il faut que les forces tant intérieures qu'extérieures responsables de cette évolution puissent être définies a priori en fonction de son état et de ses vitesses à l'instant considéré (et peut-être aux instants antérieurs). Si l'on considère un barreau dont l'état à l'instant t est défini par sa déformation ε, une loi naturelle de force intérieure doit de préférence exprimer la force (ici: la contrainte) en fonction de ε et de $\dot{\varepsilon}$ (et peut-être de leurs valeurs antérieures) plutôt que l'inverse, à savoir $\dot{\varepsilon}$ en fonction de σ et de $\dot{\sigma}$.

Pour toutes ces raisons, on a cherché à remplacer la formule (1) par une autre formule utilisant aussi étroitement que possible les données de l'expérience. Nous n'étudierons ici que les charges monoaxiales d'un barreau, excluant donc toute charge biaxiale de plaque ou triaxiale de pièce massive. Comme on le sait, le cas traité englobe non seulement les phénomènes de traction et compression simple d'une poutre droite, mais encore, d'une manière approchée, les phénomènes de flexion simple et de flexion avec compression (flambement) d'une poutre profilée — en remplacant la poutre par deux barreaux jumelés reliés par une âme d'aire nulle parfaitement rigide au cisaillement [1] — et, également, les cas de torsion d'une plaque où la torsion peut être remplacée par la flexion différentielle de deux longerons renforcés.

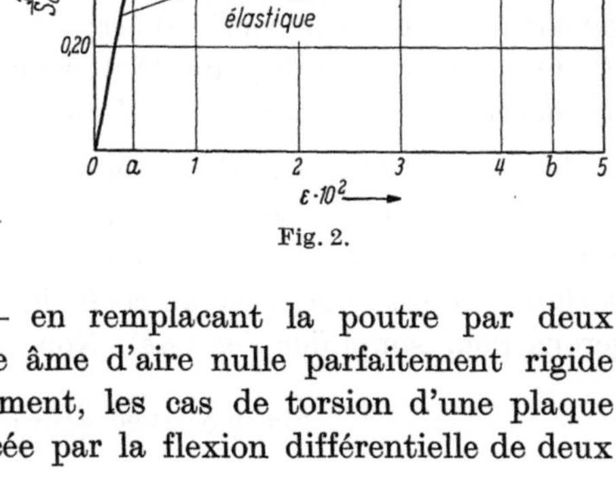

Fig. 2.

Les données expérimentales sont les suivantes [2]:

1. La courbe d'écrouissage (fig. 2) reliant, *dans tout état réalisable d'équi-*

libre, la charge π à la déformation $\varepsilon = \frac{l - l_0}{l_0}$ (l_0 = longueur du barreau non chargé et non écroui, l = longueur du barreau sous π):

$$\pi = E\, S_0\, C(\varepsilon)$$

(S_0: section du barreau non chargé et non écroui).

2. La courbe de fluage sous charge constante à la température considérée (fig. 3):

$$\varepsilon = D(\pi, t).$$

La courbe d'écrouissage n'est définie, en traction par exemple, que pour $\varepsilon \leq b$ tel que $C'(b) = 0$ et se compose de deux parties: $0 < \varepsilon \leq a$ déformation élastique, $a < \varepsilon \leq b$ déformation élasto-plastique. Les courbes de fluage, définies en supposant que la charge π est appliquée graduellement de 0 à $E\, S_0\, C(a)$, puis instantanément de cette valeur à π, se composent aussi de deux parties séparées par un point d'inflexion à allure très plate. Nous supposerons la pièce assez régulière et homogène pour que la striction n'apparaisse ni à gauche du point de la courbe d'écrouissage où la tangente est horizontale [$C'(b) = 0$], ni avant le point d'inflexion de la courbe de fluage. Les effets thermiques ne seront pas étudiés ici.

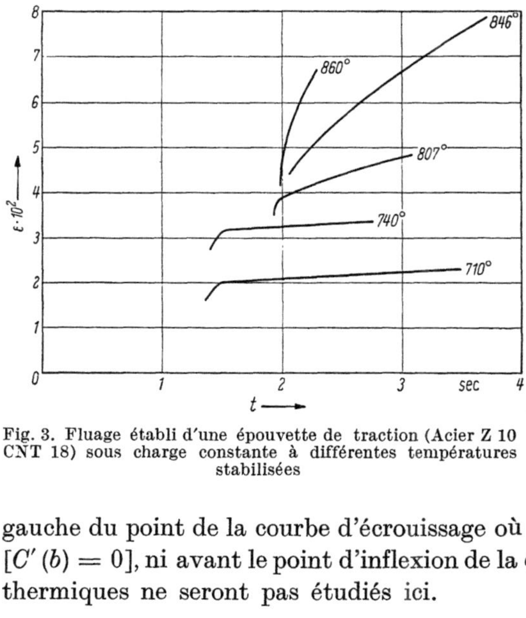

Fig. 3. Fluage établi d'une épouvette de traction (Acier Z 10 CNT 18) sous charge constante à différentes températures stabilisées

Le barreau sera considéré comme possédant, à l'instant t, un certain *taux* (irréversible) *de plasticité* η définissant la part s de la section totale S qui est *en fluage pur à volume constant*, l'autre part $S - s$ restant *parfaitement élastique* (avec un coefficient de contraction que nous négligerons pour simplifier l'exposé). Nous aurons ainsi:

$$s = \eta\, S \text{ avec } dS = -\frac{s}{1 + \varepsilon}\, d\varepsilon.$$

Moyennant quoi, considérant l'accroissement de s lorsque la déformation passe, *pour la première fois*, de u à $u + du$ (avec la vitesse $\dot{u}$) comme dû à l'entrée en fluage pur d'un «paquet de fibres» de section ds

dans lequel la contrainte obéit, au-dela du seuil $\sigma_0 = E\,u$, à la loi

$$\sigma = \sigma_0 + E(\tau\,\dot{\varepsilon})^{\frac{1}{n}}$$

[τ: constante caractéristique du matériau égale, par comparaison avec (1), à $\left(\frac{\lambda}{E}\right)^n$], on met la résistance globale du barreau sous la forme

$$\begin{aligned} R &= E\left\{(S - s)\,\varepsilon + \int_0^s \left[u + (\tau\,\dot{\varepsilon})^{\frac{1}{n}}\right] ds\right\} \\ &= E\left[S\,\varepsilon - \int_0^\varepsilon s(u, \dot{u})\,du + (\tau\,\dot{\varepsilon})^{\frac{1}{n}}\,s(\varepsilon, \dot{\varepsilon})\right] \end{aligned} \tag{2}$$

avec

$$S = S_0 - \int_0^\varepsilon \frac{s}{1+u}\,du \tag{3}$$

et les conventions suivantes exposées pour le cas $\varepsilon > 0$ (une transposition serait à faire pour le cas $\varepsilon < 0$):

1. Au premier passage de 0 à $\varepsilon \leq b$, $s(u, u)$ *ne dépend que de* u et l'on a:

$$\frac{R}{E} = S_0\,\varepsilon - \int_0^\varepsilon \left(\frac{\varepsilon}{1+u} + 1\right) s(u)\,du + (\tau\,\dot{\varepsilon})^{\frac{1}{n}}\,s(\varepsilon) \equiv S_0\,C(\varepsilon) + (\tau\,\dot{\varepsilon})^{\frac{1}{n}}\,s(\varepsilon). \tag{4}$$

Cette relation détermine la fonction $s(\varepsilon)$ dans l'intervalle $0 < \varepsilon \leq b$; on en déduit $S(\varepsilon)$ par la relation (3). En particulier, dans le domaine élastique $0 < \varepsilon \leq a$, on a simplement:

$$s(\varepsilon) \equiv 0, \quad S(\varepsilon) \equiv S_0.$$

2. Au premier passage de b à $\varepsilon > b$, *la résistance globale* R (et non la contrainte) *ne dépend que la vitesse de déformation* $\dot{\varepsilon}$ et non de ε, ni des états antérieurs depuis $u = b$. Cette hypothèse simplificatrice revient à «étaler» le point d'inflexion des courbes de fluage sous charge constante dans un intervalle *fini* $b < \varepsilon \leq c$ (dont la borne supérieure sera définie plus loin) (fig. 4); le domaine de déformation $b < \varepsilon \leq c$ (réalisable sous des charges assez élevées) a été appelé, dans un travail antérieur, *domaine du fluage établi.*

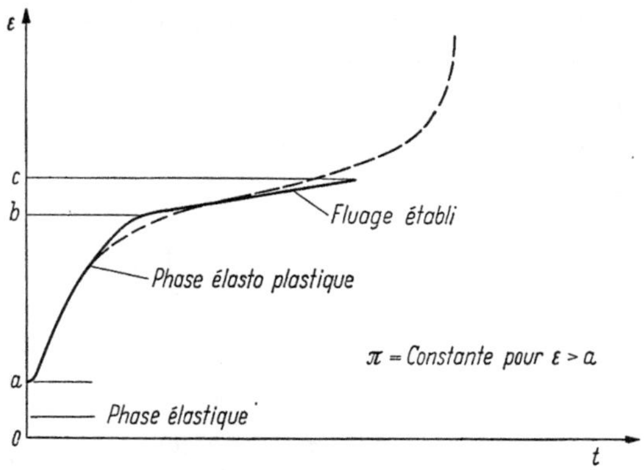

Fig. 4. Courbe de fluage

Lorsque ε passe pour la première fois de b à une valeur comprise entre b et c, $s(u, \dot{u})$ peut, dans la relation (2), être confondu avec $s(u, \dot{\varepsilon})$ et l'on peut écrire:

$$\frac{R}{E} = S\,\varepsilon - \int_0^{\varepsilon} s(u, \dot{\varepsilon})\, du + (\tau\,\dot{\varepsilon})^{\frac{1}{n}}\, s(\varepsilon, \dot{\varepsilon}) \equiv S_0\, C(b) + (\tau\dot{\varepsilon})^{\frac{1}{n}}\, s(b) \quad \text{(4 bis)}$$

ce qui détermine, dans l'intervalle $b < \varepsilon \leq c$, la fonction $s(\varepsilon, \dot{\varepsilon})$; on en déduit $S(\varepsilon, \dot{\varepsilon})$ par la relation (3).

3. *A tous les passages en ε quelconque autres que le premier:*

$$s(\varepsilon, \dot{\varepsilon}) \equiv 0\,.$$

[Faire attention, dans ce cas, au calcul de l'intégrale $\int_0^{\varepsilon} s(u, \dot{u})\, du$ *qui fait intervenir tout le passé du barreau* et doit être calculée par tronçons successifs dans tout le domaine d'évolution *réelle* de u de sa valeur initiale zéro à sa valeur actuelle ε].

4. S'il s'agit d'une structure à «courte vie», par exemple d'un engin astreint à remplir une mission unique en un temps donné relativement bref, la *déformation résistante ultime* c sera définie par la condition $s = S$.

C'est la déformation pour laquelle le taux de plasticité atteint la valeur unité. Lorsque la contrainte va constamment en croissant, le temps qui s'écoule entre l'instant où la charge π est intégralement appliquée (rappelons qu'on la suppose *graduellement* appliquée dans le domaine élastique, puis *instantanément* au-delà) et l'instant où $s = S$ est le *temps maximal* correspondant à cette charge.

L'application la plus directe concerne la traction simple d'un barreau sous charge constante. On peut résoudre l'équation (4) en valeurs discrètes de $s(\varepsilon)$ en recouvrant le plan d'un réseau de courbes d'écrouissage à section fluante constante $s = k\, S_0$ (fig. 5); on en déduit par lissage la courbe $s(\varepsilon)$, puis la courbe $S(\varepsilon)$ (fig. 6).

Si $\pi \leq E\, S_0 C(b)$, la déformation ne va pas plus loin que ε_l tel que $\pi = E\, S_0 C(\varepsilon_l)$ et cette déformation limite est atteinte au bout d'un temps infini.

Si $\pi > E\, S_0 C(b)$, le temps mis pour passer de a à ε quelconque $> a$ est donné par:

$$t_a^{\varepsilon} = \tau \int_a^{\varepsilon} \frac{du}{[\alpha(u, \pi)]^n}$$

avec:

$$\alpha(u, \pi) = \begin{cases} \dfrac{\pi - E\, S_0\, C(u)}{E\, s(u)} & \text{pour } a < u \leq b \\[2ex] \dfrac{\pi - E\, S_0\, C(b)}{E\, s(b)} = \alpha(b, \pi) & \text{pour } u \geq b\,. \end{cases}$$

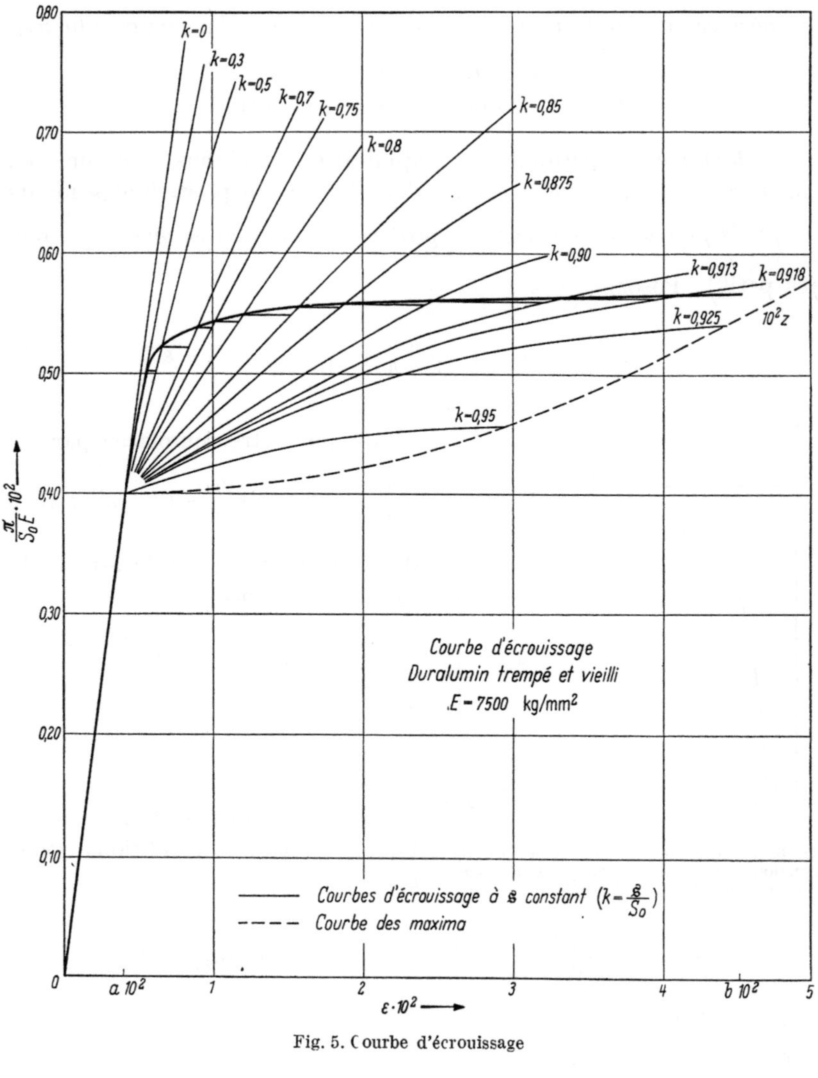

Fig. 5. Courbe d'écrouissage

Bornons-nous à calculer le temps mis pour passer de b à c:

$$t_b^c = \frac{\tau(c-b)}{[\alpha(b,\pi)]^n}.$$

La fonction $s(\varepsilon, \dot{\varepsilon})$ dans l'intervalle $b \leq \varepsilon \leq c$, remplacée, compte tenu de $\dot{\varepsilon} = \frac{1}{\tau}[\alpha(b,\pi)]^n$, par une fonction $s(\varepsilon, \pi)$, est donnée par l'équation

$$\frac{s\,\varepsilon}{1+\varepsilon} - \alpha(b,\pi)\frac{ds}{d\varepsilon} = S - s \tag{5}$$

qui, dérivée et compte tenu de (3), devient [α mis à la place de $\alpha(b, \pi)$]:

$$\alpha \frac{d^2 s}{d\varepsilon^2} - \left(1 + \frac{\varepsilon}{1+\varepsilon}\right) \frac{ds}{d\varepsilon} - \left[\frac{1}{1+\varepsilon} + \frac{1}{(1+\varepsilon)^2}\right] s = 0. \tag{6}$$

$\varepsilon - b$ et α étant petits dans les applications que nous avons en vue, nous regarderons $\varepsilon - b$ et α comme deux infiniment petits indépendants *tels que le premier ne soit pas très petit par rapport au second* $\left(\frac{\varepsilon - b}{\alpha}\right.$ non très petit$\left.\right)$. Posons:

$$1 + \frac{b}{1+b} = 1 + \varepsilon', \quad \frac{1}{1+b} + \frac{1}{(1+b)^2} = 2 - \varepsilon''.$$

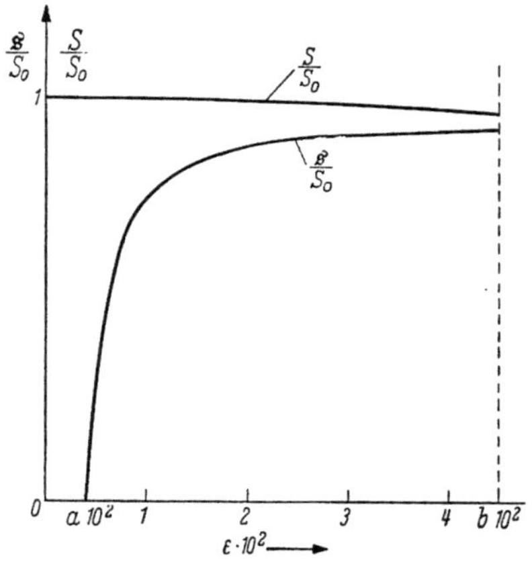

Fig. 6. Variation de la section fluante s et de la section totale S en fonction de la déformation ε dans la phase plastique $a \leqq \varepsilon \leqq b$

(6) peut être remplacée par:

$$\alpha \frac{d^2 s}{d\varepsilon^2} - (1 + \varepsilon') \frac{ds}{d\varepsilon} - (2 - \varepsilon'') s = 0$$

dont l'intégrale générale est de la forme (approchée):

$$s = A\, e^{-2\beta_1 \varepsilon} + B\, e^{\frac{1+\varepsilon'}{\alpha}\varepsilon}$$

avec

$$2\beta_1 = \frac{2 - \varepsilon''}{1 + \varepsilon'}.$$

Les constantes A et B se déterminent par les conditions pour $\varepsilon = b$:

$$s = s(b), \left(\frac{ds}{d\varepsilon}\right)_{\varepsilon = b} = \frac{1}{\alpha}\,[s(b)\,(1 + \varepsilon') - S(b)] = 0$$

d'après (3) et (4).

On en déduit:

$$s \simeq \frac{s(b)}{1 + \frac{2\beta_1 \alpha}{1+\varepsilon'}} \left[1 - 2\beta_1(\varepsilon - b) + \frac{2\beta_1 \alpha}{1+\varepsilon'}\, e^{\frac{(1+\varepsilon')(\varepsilon - b)}{\alpha}}\right]$$

et ensuite

$$S \simeq S(b) - \frac{s(b)}{1 + \frac{2\beta_1 \alpha}{1+\varepsilon'}}\, \frac{\varepsilon - b}{1 + b}$$

(les autres termes sont négligeables d'après les hypothèses faites).

La condition $s = S$, réalisée pour $\varepsilon = c$, s'écrit, en posant $\frac{(1+\varepsilon')(c-b)}{\alpha} = \varkappa$

$$\alpha = \frac{1+\varepsilon'}{2\beta_1} \, \frac{S(b) - s(b)}{s(b)\left(e^{\varkappa} - \dfrac{2\beta_1 - \dfrac{1}{1+b}}{2\beta_1}\,\varkappa\right) - S(b)}, \tag{7}$$

relation dont il est facile, par une inversion graphique, de tirer $\varkappa(\alpha)$. Il vient alors finalement

$$c - b = \frac{\alpha\,\varkappa(\alpha)}{1+\varepsilon'},$$

$$\left[\text{rappelons que } \alpha = \frac{\pi - E\,S_0\,C(b)}{E\,s(b)} > 0\right].$$

$$t_b^c = \frac{\tau(c-b)}{\alpha^n} = \frac{\tau\,\varkappa(\alpha)}{(1+\varepsilon')\,\alpha^{n-1}}.$$

Inversement, si l'on se fixe une *durée minimale de mission* t_μ, les relations précédentes fournissent la charge π_{maxi} à ne pas dépasser ou, plus exactement, à substituer à la «charge de rupture» classique comme référence pour la fixation de la «charge de sécurité».

Application numérique. Prenons un duralumin trempé et vieilli dont les caractéristiques sont les suivantes:

$$E = 7500 \text{ kg/mm}^2, \; a = 0{,}004, \; b = 0{,}045, \; n = 3, \; \tau = 8 \cdot 10^{-6}.$$

La fig. 2 reproduit la courbe d'écrouissage $\frac{\pi}{S_0 E} = C(\varepsilon)$, la fig. 6 fournit les valeurs de $\frac{s(\varepsilon)}{S_0}$ et de $\frac{S(\varepsilon)}{S_0}$, les fig. 7, 8 et 9 respectivement les valeurs de α, $c - b$ et t_b^c en fonction de $\varkappa$.

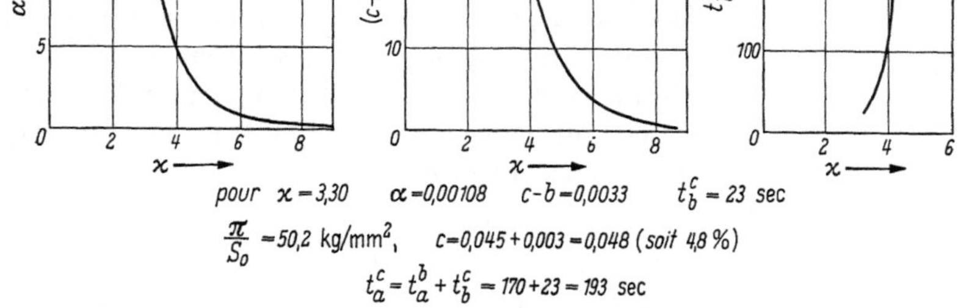

Fig. 7. Fig. 8. Fig. 9.

Fig. 7—9. Caractéristiques d'un duralumin trempé

Si l'on prend par exemple $\varkappa = 3{,}30$, il lui correspond

$$\frac{\pi}{S_0} = 50{,}2 \text{ kg/mm}^2 \text{ (fig. 10)}$$

soit un accroissement de 17% de la charge de rupture (donc aussi de la charge de sécurité). La déformation résistante ultime est

$$c = b + (c - b) = 0{,}045 + 0{,}003 = 0{,}048 \text{ (soit } 4{,}8\%).$$

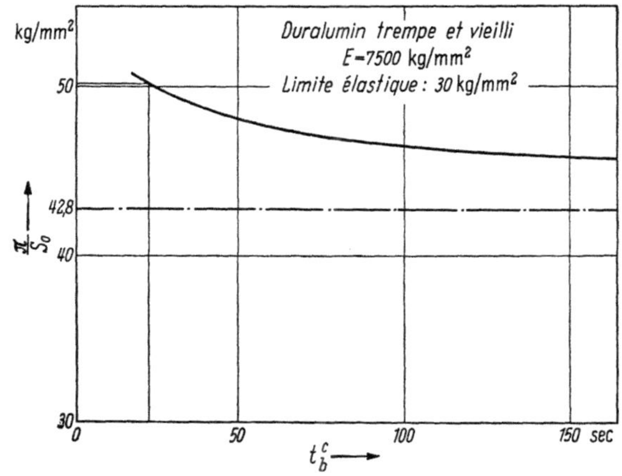

Fig. 10. Variation de la contrainte initiale maximale en fonction de la durée de fluage établi (cas de la traction simple)

Le temps total au bout duquel cette déformation serait atteinte est

$$t_a^c = t_a^b + t_b^c = 170 + 23 = 193 \text{ secondes.}$$

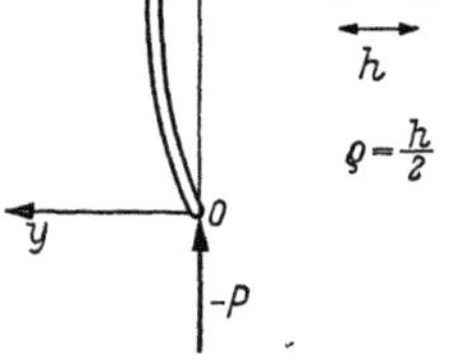

Fig. 11.
Flambement d'une poutre

Une autre application concerne le flambement d'une poutre droite de section constante $2\,S_0$ soumise à une charge P axiale en principe, mais légèrement excentrée d'une petite quantité $-\varrho\,\delta$ (ϱ: rayon de giration de la section transversale dans le plan de flexion $X\,0\,Y$). Nous admettrons, comme nous l'avons déjà dit, que la poutre se compose de deux barreaux identiques jumelés de longueur L_0 et de section S_0 reliés par une âme d'aire nulle et de largeur h parfaitement rigide au cisaillement (fig. 11). $Y = f(X)$ étant l'équation de la fibre moyenne rapportée à la ligne d'action de P, nous poserons:

$$\frac{X}{L_0} = x, \ \frac{Y}{\varrho} = w, \ \varrho = \frac{h}{2}.$$

La charge P étant inférieure à la charge critique d'EULER $2\,\pi^2\,E\,S_0 \frac{\varrho^2}{L_0^2}$, la poutre s'infléchit vers les Y positifs; nous distinguerons les caractérisstiques des deux barreaux par la lettre i pour le barreau *intérieur*, e pour le barreau *extérieur*. Le premier est toujours plus comprimé que le second qui peut même, au delà d'une certaine flèche, être tendu; en effet, les contraintes (négatives s'il y a compression) dans la section x sont respectivement:

$$\sigma_i = \frac{P}{S_i + S_e}(1 + w), \quad \sigma_e = \frac{P}{S_i + S_e}(1 - w) \quad \text{avec } P < 0.$$

Ecrivons la formule (2), compte tenu de (3), pour les deux barreaux:

$$\left.\begin{aligned} & P\frac{S(\varepsilon_i)}{S(\varepsilon_i) + S(\varepsilon_e)}(1 + w) = \\ & E\left[S_0\,\varepsilon_i - \int_0^{\varepsilon_i}\left(\frac{\varepsilon_i}{1+u} + 1\right)s(u, \dot{u}_i)\,du + (\tau\,\dot{\varepsilon}_i)^{\frac{1}{n}}\,s(\varepsilon_i, \dot{\varepsilon}_i)\right], \\ & P\frac{S(\varepsilon_e)}{S(\varepsilon_i) + S(\varepsilon_e)}(1 - w) = \\ & E\left[S_0\,\varepsilon_e - \int_0^{\varepsilon_e}\left(\frac{\varepsilon_e}{1+u} + 1\right)s(u, \dot{u}_e)\,du + (\tau\,\dot{\varepsilon}_e)^{\frac{1}{n}}\,s(\varepsilon_e, \dot{\varepsilon}_e)\right]. \end{aligned}\right\} \quad (8)$$

On sait d'autre part que

$$\frac{\partial^2 w}{\partial x^2} = \frac{2\,L_0^2}{h^2}(\varepsilon_i - \varepsilon_e). \quad (9)$$

Nous disposons donc de trois équations pour déterminer les trois fonctions inconnues w, ε_i, ε_e de x et de t. Les conditions aux limites sont:

$$w = \delta \text{ pour } x = 0 \text{ et pour } x = 1.$$

Quant aux conditions initiales, elles sont les suivantes: la charge π est appliquée graduellement jusqu'à la valeur pour laquelle le barreau intérieur atteint, en sa section médiane, le domaine plastique, puis instantanément jusqu'à sa valeur finale.

Appliquant la méthode de la compensation harmonique (solution exprimée par une série de FOURIER limitée à son premier terme), nous poserons, quel que soit l'état de la poutre,

$$w \cong A \sin k\,x + B \cos k\,x,$$

A et B étant déterminés par les conditions aux limites:

$$A = \delta\,\frac{1 - \cos k}{\sin k}, \quad B = \delta.$$

L'équation (9) peut être remplacée par les deux équations suivantes:

$$\left.\begin{aligned} -k^2\,w &= \frac{2\,L_0^2}{h^2}(\varepsilon_i - \varepsilon_e) \\ w &= \delta\left(\frac{1 - \cos k}{\sin k}\sin k\,x + \cos k\,x\right). \end{aligned}\right\} \quad (10)$$

Le système (8)—(10) détermine les quatre inconnues w, ε_i, ε_e et k (sous réserve que k soit indépendant de x, ce qui est pratiquement réalisé avec une approximation suffisante). Il suffit de résoudre pour le milieu de la poutre, $x = \frac{1}{2}$; affectons de l'indice M les valeurs correspondantes de w, ε_i, ε_e:

$$\left.\begin{aligned}
&P\,\frac{S(\varepsilon_{i_M})}{S(\varepsilon_{i_M})+S(\varepsilon_{e_M})}(1+w_M) = \\
&E\left[S_0\,\varepsilon_{i_M} - \int\limits_0^{\varepsilon_{i_M}}\left(\frac{\varepsilon_{i_M}}{1+u}+1\right)s(u,u_{i_M})\,du + (\tau\,\dot{\varepsilon}_{i_M})^{\frac{1}{n}}\,s(\varepsilon_{i_M},\dot{\varepsilon}_{i_M})\right] \\
&P\,\frac{S(\varepsilon_{e_M})}{S(\varepsilon_{i_M})+S(\varepsilon_{e_M})}(1-w_M) = \\
&E\left[S_0\,\varepsilon_{e_M} - \int\limits_0^{\varepsilon_{e_M}}\left(\frac{\varepsilon_{e_M}}{1+u}+1\right)s(u,u_{e_M})\,du + (\tau\,\dot{\varepsilon}_{e_M})^{\frac{1}{n}}\,s(\varepsilon_{e_M},\dot{\varepsilon}_{e_M})\right] \\
&-k^2\,w_M = \frac{2\,L_0^2}{h^2}(\varepsilon_{i_M}-\varepsilon_{e_M}), \\
&w_M = \delta\left(\frac{1-\cos k}{\sin k}\sin\frac{k}{2}+\cos\frac{k}{2}\right) = \delta\,\frac{2\sin\frac{k}{2}}{\sin k} = \frac{\delta}{\cos\frac{k}{2}}.
\end{aligned}\right\} \quad (11)$$

Plusieurs phases successives sont à distinguer dans l'évolution de la poutre:

I. Une phase préalable dans laquelle ε_{i_M} décroît graduellement de zéro à $-a$ tandis que ε_{e_M} décroît de zéro à $-a' > -a$ (les deux barreaux sont en déformation élastique).

Nous supposons P assez grand pour que la poutre dépasse cet état.

II. ε_{i_M} décroît de $-a$ à $-b$ en un temps t_{-a}^{-b}; pendant le même temps, ε_{e_M} croît de $-a'$ à $b' < b$ (le barreau intérieur est en déformation élasto-plastique, le barreau extérieur en déformation élastique ou élasto-plastique).

Il peut arriver, si P est relativement faible, que ε_{i_M} décroisse de $-a$ à $-b_l$ en un temps *infini* $t_{-a}^{-b_l}$. Lorsqu'il en est ainsi, l'équilibre est possible dans le domaine élasto-plastique et l'étude ne va pas plus loin.

III. ε_{i_M} décroît de $-b$ à $-c$ ne un temps fini t_{-b}^{-c}; cette phase termine l'étude si b'', valeur atteinte par ε_{e_M} au bout de ce temps, est inférieure à b (le barreau intérieur est en fluage établi, le barreau extérieur en déformation élastique ou élasto-plastique).

Dans le cas contraire, elle se subdivise en deux autres:

a) ε_{e_M} croit de b' à b en un temps fini $t_{b'}^{b}$; pendant le même temps, ε_{i_M} décroit de $-b$ à $-c' > -c$ (le barreau intérieur est en fluage établi, le barreau extérieur en déformation élasto-plastique).

b) ε_{i_M} décroît de $-c'$ à $-c$ en un temps fini $t_{-c'}^{-c}$; pendant le méme temps, ε_{e_M} croît de b à $c'' < c$ (les deux barreaux sont en fluage établi).

L'étude complète n'étant pas achevée, nous nous bornerons à donner quelques indications sur la phase II dans le cas où l'équilibre est possible. Pour déterminer les limites correspondantes de P, nous commencerons par étudier l'équilibre correspondant à une valeur donnée de k. Aux seconds membres des deux premières équations (11), les crochets sont à remplacer respectivement par $S_0\, C(\varepsilon_{i_M})$ et $S_0\, C(\varepsilon_{e_M})$, la fonction $C(\varepsilon)$ étant celle de la courbe d'écrouissage $\pi = E\, S_0\, C(\varepsilon)$ tracée également vers les ε et les π négatifs (cas d'une première compression du matériau non écroui)[1].

Le système (11) devient ainsi, en posant $S_0 \dfrac{C(\varepsilon)}{S(\varepsilon)} = F(\varepsilon)$:

$$\left.\begin{aligned}
&\frac{P}{E[S(\varepsilon_{i_M}) + S(\varepsilon_{e_M})]}(1 + w_M) = F(\varepsilon_{i_M}),\\
&\frac{P}{E[S(\varepsilon_{i_M}) + S(\varepsilon_{e_M})]}(1 - w_M) = F(\varepsilon_{e_M}),\\
&-k^2 w_M = \frac{2 L_0^2}{h^2}(\varepsilon_{i_M} - \varepsilon_{e_M}),\\
&w_M = \frac{\delta}{\cos\frac{k}{2}}.
\end{aligned}\right\} \qquad (12)$$

On déduit des trois premières équations:

$$(1 - w_M)\, F(\varepsilon_{i_M}) = (1 + w_M)\, F(\varepsilon_{e_M}),$$
$$\varepsilon_{i_M} - \varepsilon_{e_M} = -\frac{k^2 h^2}{2 L_0^2} w_M,$$

k étant donné, et par suite w_M, ε_{i_M} est fourni, dans le plan auxiliaire (ε_{i_M}, u), par l'intersection des deux courbes

$$u = F(\varepsilon_{i_M}),$$
$$u = \frac{1 + w_M}{1 - w_M} F\left(\varepsilon_{i_M} + \frac{k^2 h^2}{2 L_0^2} w_M\right).$$

On en déduit ε_{e_M}, puis P par

$$P = \frac{E[S(\varepsilon_{i_M}) + S(\varepsilon_{e_M})]}{2 w_M}[F(\varepsilon_{i_M}) - F(\varepsilon_{e_M})].$$

[1] On ne doit pas s'attendre á trouver exactement $C(-\varepsilon) = -C(\varepsilon)$, car la variation de section due á l'écoulement des fibres fluantes favorise cet écoulement en traction et le contrarie en compression.

En donnant à k des valeurs croissantes (depuis $k = 0$), on pourra tracer les courbes $\varepsilon_{i_M}(P)$ et $\varepsilon_{e_M}(P)$ et en déduire les valeurs de P pour lesquelles $\varepsilon_{i_M} = -a$ et $\varepsilon_{i_M} = -b$ (ou les valeurs qui jouent le rôle de a et b en compression) en même temps que les valeurs correspondantes $-a'$ et b' de ε_{e_M}.

Bibliographie

[1] Hoff, N. J.: Effets thermiques dans le calcul de la résistance des structures d'avions et d'engins. Rapport AGARD, No. 52, 1956.

[2] Jaoul, B.: L'écrouissage. Cahiers du Groupe français d'Etudes de Rhéologie, IV, No. 3 (1959).

[3] Comptes rendus des séances de l'Académie des Sciences, **249**, 942 (1959).

Some Simple Models for Torsional Creep Buckling

By

A. H. Chilver

Cambridge University, Cambridge, England

Summary

Up to the present time little attention has been given to problems of creep buckling involving torsional modes of instability, although the torsional creep buckling of open tubes has been studied recently by LIANIS (Ref. [*1*]), from whose work it is evident that torsional effects introduce many analytical problems.

To reduce the analytical difficulties, a number of simple structural models are introduced to study problems of torsional creep buckling.

The first problem discussed is that of uniform torsional buckling of a compressed "member". In the model the "member" consists of one, or more, longitudinal elemental tubes. It is shown that for large deformations the problem is complex, whereas for small deformations the behaviour is analogous to that of the flexural buckling of a column, and a critical time can be evaluated. This critical time can be compared with that derived for open tubes by LIANIS.

A second model is introduced to study torsional-flexural creep buckling. A column under constant axial load, and enforced to twist about a given axis, is discussed first. Secondly, the torsional-flexural creep buckling of a narrow beam under constant end moment is studied. When certain approximations are made, the analysis of these problems is similar to that for flexural creep buckling. The structural models again consist of connected elemental tubes.

All the models are studied from the point of view of secondary creep. The effects of primary creep are not taken into account. The secondary creep law is taken in the form

$$\dot{\varepsilon} = k\, J_2^m\, \sigma\,,$$

where ε is the strain corresponding to the component σ of the deviator stress tensor, J_2 is the second invariant of the deviator stress tensor, and k and m are constants. Throughout the analysis of the models, m is taken equal to unity.

1. Introduction

In recent years many studies have been made of flexural creep buckling problems; a general discussion of creep buckling by FRAEIJS DE VEUBEKE, (Ref. [2]), published in 1958, deals entirely with flexural creep buckling problems, and develops theories of flexural creep buckling for simple structural models.

Little attention has been given, up to the present, to torsional creep buckling problems. LIANIS (Ref. [1]) has recently attempted the solution of the problem of torsional creep buckling of open thin-walled tubes; LIANIS makes use of a variational theorem put forward by SANDERS, MCCOMB and SCHLECHTE.

In this paper simple structural models are introduced to show some of the problems in the torsional and torsional-flexural buckling of columns and beams due to secondary creep. The models are composed of elemental circular tubes, and the creep behaviour of these tubes is discussed first.

Notation

b, h, r, t, L, R	linear dimensions
ϱ	geometrical parameter
A	cross-sectional area
σ, τ, s	stresses
J_2	second invariant of deviator stress tensor
k, m	constants
E, G	elastic moduli
M	couple
ε, γ, e	strains
$\dot{\varepsilon}, \dot{\gamma}$	strain rates
θ, Φ, ξ, B	angles
θ'	angle per unit length
u, U	displacements
x, y, z	co-ordinate axes
μ, λ	time-dependent constants
T	time
f, g	time-dependent functions
	subscripts c, e refer to creep and elastic strains, respectively
	subscript E refers to elastic buckling stress
$\cdot$	denotes differentiation with respect to time
$'$	denotes differentiation with respect to distance.

2. Creep of a Thin Circular Tube under Combined Compression (or Tension) and Torsion

The structural models used to illustrate torsional creep buckling problems consist of one, or more, thin circular tubes. The creep behaviour of the model is governed, therefore, by the creep characteristics of these circular tubes.

The circular tube shown in Fig. 1 has a uniform wall-thickness t, and a mean radius r, $(r \gg t)$. The wall of the tube is subjected to a uniform compressive stress, σ, in a longitudinal direction, and a uniform shearing stress, τ, in a circumferential direction. In the absence of any other stresses, the components of the deviator stress tensor in the longitudinal and circumferential directions, respectively, are

$$\frac{2}{3}\,\sigma \text{ (compressive)},\ \tau.$$

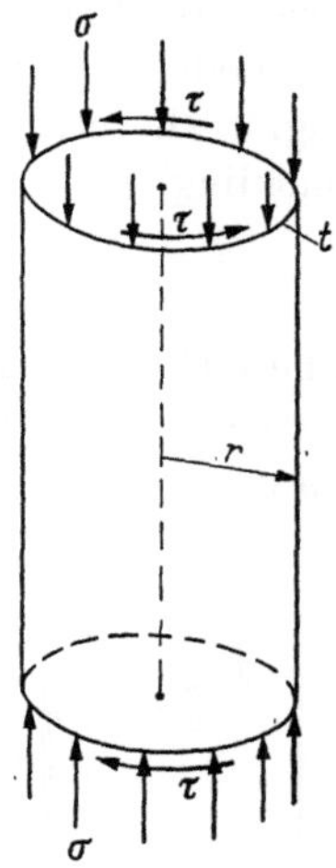

Fig. 1. Uniform compressive and shearing stresses in a thin-walled circular tube ($r \gg t$).

If ε and γ are the corresponding compressive and shearing strains, respectively, in the wall of the tube, then the strain rates due to secondary creep alone are

$$\dot{\varepsilon}_c = k\,J_2^m\left(\frac{2\,\sigma}{3}\right),\ \dot{\gamma}_c = k\,J_2^m(\tau), \tag{1}$$

where J_2 is the second invariant of the deviator stress tensor, and k and m are constants; the subscript c refers to creep strains. In the particular case of the circular tube under combined compression and torsion,

$$J_2 = \frac{1}{3}\,\sigma^2 + \tau^2. \tag{2}$$

If $\dot{\varepsilon}_0$ is the strain rate due to secondary creep under a uniaxial compressive stress σ_0, then

$$\dot{\varepsilon}_0 = 2\,k\,\sigma_0^{2m+1}/3^{m+1},\ \text{or}\ k = 3^{m+1}\,\dot{\varepsilon}_0/2\,\sigma_0^{2m+1}. \tag{3}$$

The analysis is restricted to the case when $m = 1$; then

$$\dot{\varepsilon}_c = k\,J_2\left(\frac{2\,\sigma}{3}\right),\ \dot{\gamma}_c = k\,J_2\,\tau, \tag{4}$$

and

$$k = 9\,\dot{\varepsilon}_0/2\,\sigma_0^3. \tag{5}$$

The total strains, ε and γ, are assumed to be made up of elastic strains, ε_e and γ_e, and creep strains, ε_c and γ_c. So that

$$\varepsilon = \varepsilon_c + \varepsilon_e,\quad \gamma = \gamma_c + \gamma_e. \tag{6}$$

The total strain rates are then given by

$$\dot{\varepsilon} = \dot{\varepsilon}_c + \dot{\varepsilon}_e,\quad \dot{\gamma} = \dot{\gamma}_c + \dot{\gamma}_e. \tag{7}$$

If

$$\varepsilon_e = \sigma/E,\quad \gamma_e = \tau/G, \tag{8}$$

where E and G are Young's modulus and the shearing modulus, respectively, then

$$\dot{\varepsilon} = \dot{\varepsilon}_c + \dot{\sigma}/E,\quad \dot{\gamma} = \dot{\gamma}_c + \dot{\tau}/G. \tag{9}$$

For the case of combined tension and torsion, Eqs. (4) and (9) again apply, but if σ is taken positive ε and ε_c are tensile strains.

3. Uniform Torsional Buckling of a Compressed Member about a Fixed Axis

The first problem studied is that of uniform torsional buckling of a compressed member, constrained to rotate about a fixed axis. In the elastic torsional buckling of a long thin strip, uniformly compressed on the ends, supported on one longitudinal edge, and free on the other, Fig. 2, the critical value, σ_E, of the end compressive stress for elastic buckling, is

$$\sigma_E = G\left(\frac{t}{b}\right)^2, \tag{10}$$

where G is the elastic shearing modulus. The mode of buckling involves a constant angle of twist per unit length; this is described as uniform torsional buckling.

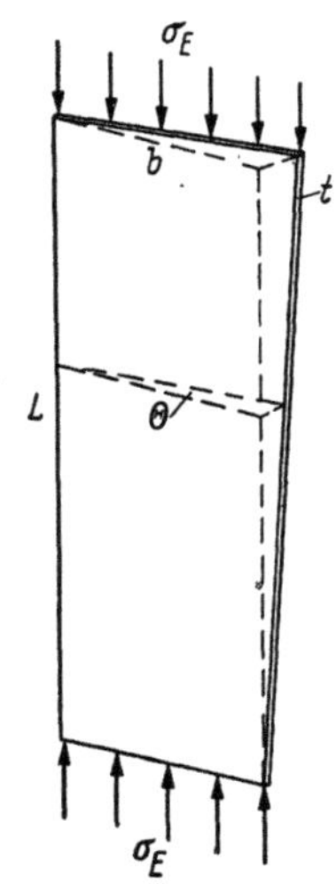

Fig. 2. Uniform torsional buckling of a long thin strip about a fixed axis of rotation ($L \gg b \gg t$).

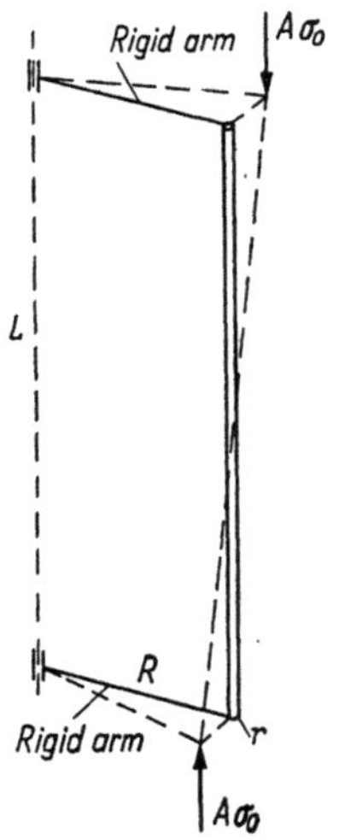

Fig. 3. Uniform torsional buckling of an elemental tube attached to a fixed axis by rigid arms ($L \gg R \gg r$).

Suppose the thin strip of Fig. 2 is replaced by a single thin circular tube of mean radius r, located a distance R from the axis of twist, Fig. 3; R is considerably greater than r. The tube is attached to the axis of twist by rigid arms at each end. The arms can rotate freely about the axis of twist, but in the buckled mode they apply torque to the elemental tube. A constant compressive load $A\,\sigma_0$ is applied to the tube, where A is the cross-sectional area of the tube; no longitudinal thrust is transmitted by the rigid arms to the axis of twist. Elastic torsional buckling of the model occurs when σ_0 has the critical value

$$\sigma_E = G\left(\frac{r}{R}\right)^2. \tag{11}$$

This elastic critical stress is analogous to that defined by Eq. (10) for a thin strip. In deriving Eq. (11) it is assumed that in the buckled mode there is a uniform angle of twist per unit length of the elemental tube.

In studying creep buckling of the model, it is assumed that before any compressive loads are applied there is an initial imperfection in the system in the form of a small uniform angle of twist, θ_0', per unit length of the member, Fig. 4a. Sometime after the application of the compressive load $A\,\sigma_0$, the tube has contracted and twisted; the angle of twist

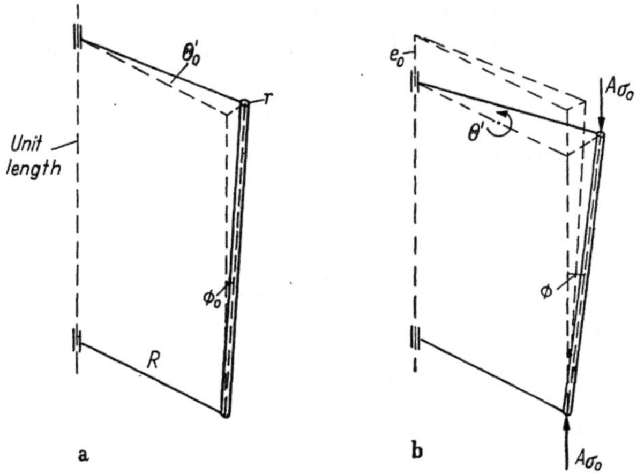

Fig. 4. (a) Initial form of the model in the unloaded condition; (b) form of the model under end compressive loads.

per unit length is then θ'. Immediately after the application of load, no secondary creep has taken place and θ' is given by

$$\theta_1' = \theta_0'/\left(1-\frac{\sigma_0}{\sigma_E}\right), \qquad (12)$$

where σ_E is given by Eq. (11). It is assumed that the compressive and shearing strains in the elemental tube are small, although distortions of the system may be large. If a unit length of the member shortens an amount e_0, measured along the axis of twist, Fig. 4b, then the linear compressive strain in the tube is

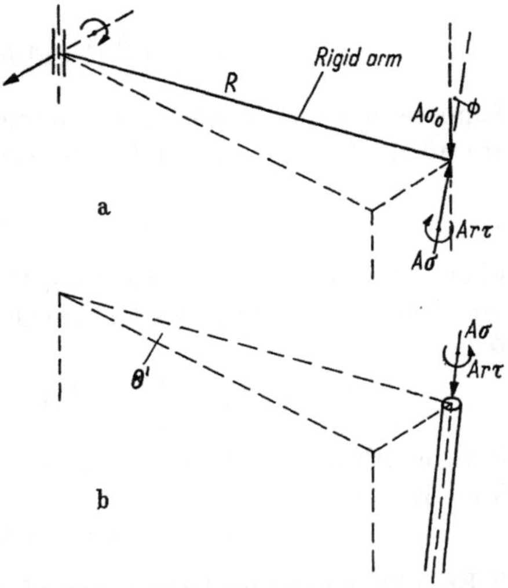

Fig. 5. (a) Loading actions on a rigid arm in the buckled mode; (b) loading actions on an elemental tube.

$$\varepsilon = 1 - [\{(1-e_0)^2 + R^2\,\theta'^2\}/(1 + R^2\,\theta_0'^2)]^{\frac{1}{2}}\,, \qquad (13)$$

and the shearing strain is taken in the form

$$\gamma = \frac{r}{2\,R}\,(\sin 2\,\Phi - \sin 2\,\Phi_0)\,, \qquad (14)$$

where Φ_0 and Φ are the helix angles before and after application of the compressive load, respectively.

The compressive stress, σ, and the shearing stress, τ, induced in the tube by the compressive load $A\,\sigma_0$ can be found by considering statical equilibrium of the rigid arms at each end of the tube, Fig. 5. If the arms are free to rotate about, and slide along, the axis of twist, then

$$\sigma = \sigma_0 \sec\Phi,\ \tau = \frac{R}{r}\,\sigma_0 \sec\Phi \tan\Phi. \tag{15}$$

The total shearing strain rate is given by

$$\dot{\gamma} = \dot{\gamma}_c + \dot{\tau}/G. \tag{9 bis}$$

But from Eq. (14),

$$\gamma = \frac{r}{R}\,\dot{\Phi}\cos 2\,\Phi, \tag{16}$$

and from Eq. (15)

$$\dot{\tau} = \frac{R\,\sigma_0}{r}\,\dot{\Phi}\sec^3\Phi\,(1+\sin^2\Phi). \tag{17}$$

The creep strain rate, $\dot{\gamma}_c$, is defined by Eqs. (2) and (4), so that Eq. (9) becomes finally

$$\left[\cos 2\,\Phi - \frac{\sigma_0}{\sigma_E}\sec^3\Phi\,(1+\sin^2\Phi)\right]\dot{\Phi} = \frac{k\,\sigma_0^3}{3}\left(\frac{R}{r}\right)^2 \sec^3\Phi\tan\Phi\left[1+\frac{3\,R^2}{r^2}\tan^2\Phi\right]. \tag{18}$$

Suppose first that simple approximations to the trigonometric functions are adequate; approximately, Eq. (18) becomes

$$\left(1-\frac{\sigma_0}{\sigma_E}\right)\dot{\Phi} = \frac{1}{2}\,\dot{\varepsilon}_0\,\varrho^2\,\Phi\,(1+\varrho^2\,\Phi^2), \tag{19}$$

where $\varrho^2 = 3\,R^2/r^2$. On separating the variables, and integrating, the time T taken to increase the helix angle from an initial value Φ_1 to a value Φ is

$$T = \frac{1}{\varrho^2\,\dot{\varepsilon}_0}\left(1-\frac{\sigma_0}{\sigma_E}\right)\log\left[(1+1/\varrho^2\,\Phi_1^2)/\left(1+\frac{1}{\varrho^2\Phi^2}\right)\right]. \tag{20}$$

If Φ_1 is taken immediately after the compressive load is applied, then from Eq. (12)

$$\Phi_1 \doteq R\,\theta_1' = R\,\theta_0'/(1-\sigma_0/\sigma_E). \tag{21}$$

If Eq. (20) is valid for large values of $\varrho^2\,\Phi^2$, then any such large value is approached in the critical time

$$T_{cr} = \frac{1}{\varrho^2\,\dot{\varepsilon}_0}\left(1-\frac{\sigma_0}{\sigma_E}\right)\log\left[1+\frac{1}{\varrho^2\,\Phi_1^2}\right]. \tag{22}$$

This value of critical time is similar in form, (but not precisely the same numerically), as the value of critical time for a cruciform column, as derived by Lianis (Ref. [1]); however, Lianis considers large strains in setting up the strain rates, and, in the case of a cruciform, creep is taking place simultaneously at all points of the cross-section.

A more accurate solution of Eq. (18) may be found; on separating the variables, Eq. (18) may be written

$$\frac{1}{2}\,\dot{\varepsilon}_0\,\varrho^2\,\delta T = \frac{\cos 2\Phi - \frac{\sigma_0}{\sigma_E}\sec^3\Phi\,(1+\sin^2\Phi)}{\sec^3\Phi\,\tan\Phi\,(1+\varrho^2\tan^2\Phi)}\,\delta\Phi, \tag{23}$$

where δT is an element of time. If $\varrho^2 \gg 1$, approximately,

$$\int \frac{\cos 2\Phi\, d\Phi}{\sec^3\Phi\,\tan\Phi\,(1+\varrho^2\tan^2\Phi)} = \log\tan\frac{\Phi}{2} - \frac{1}{2}\log\left[\frac{\varrho - \sqrt{\varrho^2-1}\,\cos\Phi}{\varrho + \sqrt{\varrho^2-1}\,\cos\Phi}\right], \tag{24}$$

and also, approximately,

$$\int \frac{(1+\sin^2\Phi)\,d\Phi}{\tan\Phi\,(1+\varrho^2\tan^2\Phi)} = \frac{1}{2}\log\frac{\sin^2\Phi}{(1+\varrho^2\sin^2\Phi)}. \tag{25}$$

Then the time, T, taken for the member to twist from Φ_1 to Φ is given by

$$\frac{1}{2}\,\dot{\varepsilon}_0\,\varrho^2\,T = \log\frac{\tan\frac{1}{2}\Phi}{\tan\frac{1}{2}\Phi_1} - \frac{1}{2}\log\left[\left(\frac{\varrho-\sqrt{\varrho^2-1}\,\cos\Phi}{\varrho+\sqrt{\varrho^2-1}\,\cos\Phi}\right)\left(\frac{\varrho+\sqrt{\varrho^2-1}\,\cos\Phi_1}{\varrho-\sqrt{\varrho^2-1}\,\cos\Phi_1}\right)\right]$$

$$-\frac{\sigma_0}{\sigma_E}\log\left[\frac{\sin\Phi}{\sin\Phi_1}\left(\frac{1+\varrho^2\sin^2\Phi_1}{1+\varrho^2\sin^2\Phi}\right)^{1/2}\right]. \tag{26}$$

If $\Phi_1 \ll 1$, and $\varrho^2 \gg 1$, this reduces to

$$\frac{1}{2}\,\dot{\varepsilon}_0\varrho^2\,T = \frac{1}{2}\log\left(1+\frac{1}{\varrho^2\Phi_1^2}\right) - \frac{\sigma_0}{\sigma_E}\log\left[\frac{\sin\Phi}{\Phi_1}\left(\frac{1+\varrho^2\Phi_1^2}{1+\varrho^2\sin^2\Phi}\right)^{1/2}\right]. \tag{27}$$

Then, if $\varrho^2\sin^2\Phi \gg 1$, the critical time is given by

$$\dot{\varepsilon}_0\,\varrho^2\,T_{cr} = \left(1-\frac{\sigma_0}{\sigma_E}\right)\log\left(1+\frac{1}{\varrho^2\Phi_1^2}\right). \tag{28}$$

This agrees with the value given by Eq. (22) for small deflections, and suggests that the concept of critical time as given by small distortion theory may be reasonably accurate for large distortions.

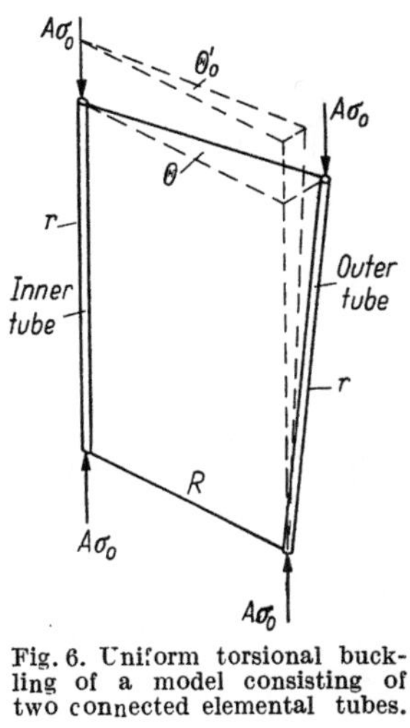

Fig. 6. Uniform torsional buckling of a model consisting of two connected elemental tubes.

The structual model of Fig. 3 is now modified by introducing a second elemental tube at the axis of twist of the member, Fig. 6. The two tubes have the same geometrical properties; they are connected by arms at each end, and undergo the same amounts of twisting; the tubes are also displaced by the same amounts longitudinally, as measured along the axis of twist. The model carries compressive loads $A\,\sigma_0$ at the end of each tube. Uniform torsional elastic buckling occurs initially when σ_0 has the value

$$\sigma_E = 2\,G\left(\frac{r}{R}\right)^2. \tag{29}$$

In the elastic post-buckling range σ_0 may be increased above the initial buckling stress, and for a large helix angle, Φ, approximately

$$\sigma_0 = 2\,G\left(\frac{r}{R}\right)^2\left[1 + \frac{E}{8\,G}\Phi^2\left(\frac{R}{r}\right)^2\right]. \tag{30}$$

Then $\sigma_0 = 2\,\sigma_E$ when

$$\Phi = \frac{r}{R}\sqrt{\frac{8G}{E}}. \tag{31}$$

This implies that the end compressive force can increase appreciably for small values of Φ.

The analysis for secondary creep buckling is restricted to small distortions. If ε_1 and ε_2 are the linear, compressive strains in the inner and outer tubes, respectively, with sufficient accuracy

$$\varepsilon_1 - \varepsilon_2 = \frac{1}{2}\,R^2(\theta'^2 - \theta_0'^2), \tag{32}$$

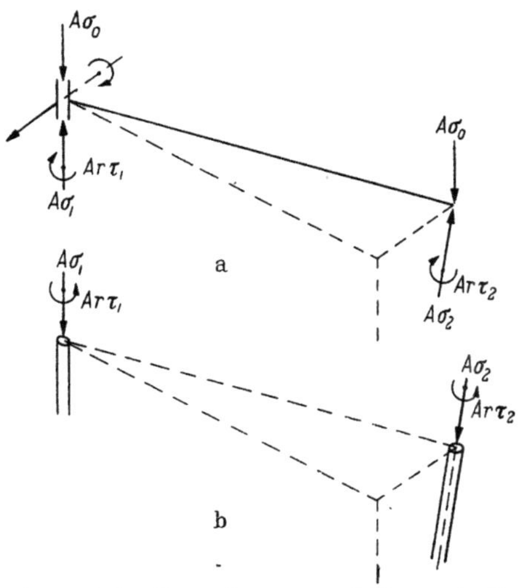

Fig. 7. (a) Loading actions on a rigid arm in the buckled mode; (b) loading actions on the elemental tubes.

where θ_0' is the initial angle of twist per unit length. The shearing strains in the two tubes are very nearly equal for small distortions, and are given with sufficient accuracy by

$$\gamma_1 = \gamma_2 = r(\theta' - \theta_0'). \tag{33}$$

The compressive stresses in the two tubes are not necessarily equal after the onset of creep buckling, but the total longitudinal force is assumed to remain constant. If distortions are small this implies that

$$\sigma_1 + \sigma_2 = \text{constant} = 2\,\sigma_0. \tag{34}$$

(Subscripts 1 and 2 refer to the inner and outer tubes, respectively.)

For statical equilibrium of the system, Fig. 7,

$$(\tau_1 + \tau_2)\,r = \sigma_2\,R^2\,\theta'. \tag{35}$$

The strain rates are then given by

$$\dot{\varepsilon}_1 = \frac{2\,k}{3}\,\sigma_1\left(\frac{\sigma_1^2}{3} + \tau_1^2\right) + \dot{\sigma}_1/E\,,$$

$$\dot{\varepsilon}_2 = \dot{\varepsilon}_1 - R^2\,\theta'\,\dot{\theta}' = \frac{2\,k}{3}\,\sigma_2\left(\frac{\sigma_2^2}{3} + \tau_2^2\right) + \dot{\sigma}_2/E\,, \tag{36}$$

$$\dot{\gamma}_1 = \dot{\gamma}_2 = r\,\dot{\theta}' = k\,\tau_1\left(\frac{\sigma_1^2}{3} + \tau_1^2\right) + \dot{\tau}_1/G = k\,\tau_2\left(\frac{\sigma_2^2}{3} + \tau_2^2\right) + \dot{\tau}_2/G\,.$$

Suppose the elastic strain rates are negligible, and that the total strain rates are given by the creep components only. Then Eqs. (36) may be written

$$\begin{aligned} R^2\,\theta'\,\dot{\theta}' &= \frac{2\,k}{3}\left[\sigma_1\left(\frac{\sigma_1^2}{3} + \tau_1^2\right) - \sigma_2\left(\frac{\sigma_2^2}{3} + \tau_2^2\right)\right], \\ r\,\dot{\theta}' &= k\,\tau_1\left(\frac{\sigma_1^2}{3} + \tau_1^2\right) = k\,\tau_2\left(\frac{\sigma_2^2}{3} + \tau_2^2\right). \end{aligned} \tag{37}$$

Suppose

$$\xi = R^2\,\theta'/r\,. \tag{38}$$

Then Eqs. (34), (35) and (37) may be written in the forms

$$\sigma_1 + \sigma_2 = 2\,\sigma_0,\ \ \tau_1 + \tau_2 = \xi\,\sigma_2\,, \tag{39a}$$

$$\frac{\sigma_1}{\tau} - \frac{\sigma_2}{\tau_2} = \frac{3}{2}\,\xi\,, \tag{39b}$$

$$\frac{r^2}{R^2}\,\dot{\xi} = k\,\tau_1\left(\frac{\sigma_1^2}{3} + \tau_1^2\right) = k\,\tau_2\left(\frac{\sigma_2^2}{3} + \tau_2^2\right). \tag{39c}$$

These five equations are necessary for the solution of the five unknowns ξ, σ_1, σ_2, τ_1, τ_2. A possible method of solution of Eqs. (39) is as follows: assume a value for σ_2; then Eqs. (39a) and (39c) give three equations from which σ_1, τ_1 and τ_2 can be found; the corresponding values of ξ and $\dot{\xi}$ can then be found from Eqs. (39b) and (39c). If ξ is small, then $\dot{\xi}$ can be found explicitly in terms of ξ with reasonable accuracy.

4. Torsional-Flexural Buckling of a Column about a Fixed Axis

A model is set up next to simulate a simple problem in torsional-flexural buckling. The model consists of two elemental thin circular tubes, each of mean radius r and length L, Fig. 8. The tubes are a distance 2b apart; the centroid of the two tubes is a distance h from the axis of twist, $O\,z$, ($h \gg b \gg r$). The two tubes are connected to each other, and to the axis of twist, by rigid arms; these arms do not resist compression. Before loading there is an initial imperfection of the form

$$u_0 = h\,\theta_0 = U_0 \sin\frac{\pi\,z}{L}\,, \tag{40}$$

where u_0 is the lateral displacement of the centroid of the two tubes, θ_0 is the angle of twist, and U_0 is a constant. Buckling takes place under the

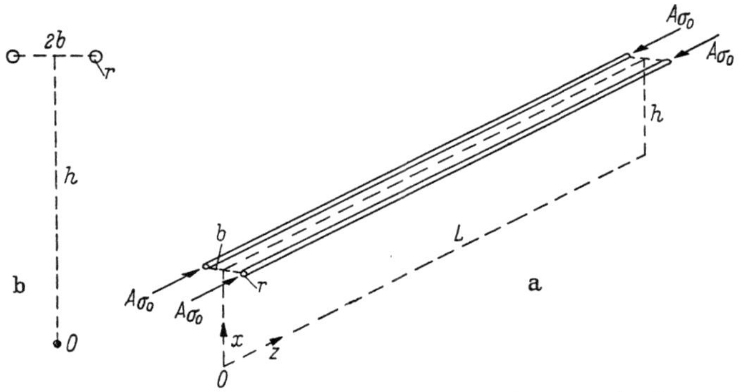

Fig. 8. (a) Column constrained to twist about a fixed axis; (b) cross-sectional geometry of the model.

action of equal compressive loads $A\,\sigma_0$ applied to the ends of the tubes, (A is the cross-sectional area of each tube). If σ_1, σ_2 are the compressive stresses in the tubes at any section in the buckled mode, Fig. 9, then

$$\sigma_1 + \sigma_2 = \text{constant} = 2\,\sigma_0, \tag{41}$$

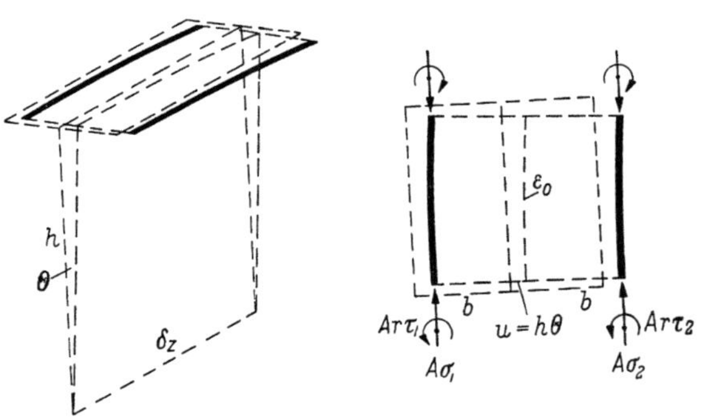

Fig. 9. Buckling of an elemental length of the model

and, if τ_1, τ_2 are the shearing stresses,

$$2\,\sigma_0\,u'' + b(\sigma_2'' - \sigma_1'') - \frac{r}{h}(\tau_1' + \tau_2') = 0, \tag{42}$$

where u is the lateral displacement of the centroid of the two tubes. (Primes denote differentiation with respect to z.) If ε_0 is the compressive strain of the centroid of the two tubes, the compressive strains in the two tubes are

$$\varepsilon_1 = \varepsilon_0 - b\,u'', \ \varepsilon_2 = \varepsilon_0 + b\,u''. \tag{43}$$

The shearing strain in both tubes is

$$\gamma_1 = \gamma_2 = r\,\theta' = \frac{r}{h}\,u'. \tag{44}$$

For elastic buckling of the model

$$\varepsilon_1 = \frac{\sigma_1}{E}\,,\ \varepsilon_2 = \frac{\sigma_2}{E}\,,\ \gamma_1 = \frac{\tau_1}{G}\,,\ \gamma_2 = \frac{\tau_2}{G}\,. \tag{45}$$

Then

$$\sigma_2'' - \sigma_1'' = 2\,b\,E\,u'''',\ \tau_1' + \tau_2' = 2\,G\left(\frac{r}{h}\right)u''. \tag{46}$$

In terms of u, Eq. (42) becomes

$$b^2\,E\,u'''' + \left(\dot{\sigma}_0 - G\frac{r^2}{h^2}\right)u'' = 0. \tag{47}$$

If $u = U \sin \pi\, z/L$, where U is constant, then the critical value of σ_0 is given by

$$\sigma_E = \pi^2\,E\left(\frac{b}{L}\right)^2 + G\left(\frac{r}{h}\right)^2. \tag{48}$$

Elastic buckling therefore involves both the flexural and torsional stiffnesses of the member.

In dealing with creep buckling we assume again that elastic strains are negligible. The compressive strain rates are then given by

$$\begin{aligned}\dot{\varepsilon}_1 &= \dot{\varepsilon}_0 - b\,\dot{u}'' = \frac{2\,k}{3}\left(\frac{\sigma_1^2}{3} + \tau_1^2\right)\sigma_1,\\ \dot{\varepsilon}_2 &= \dot{\varepsilon}_0 + b\,\dot{u}'' = \frac{2\,k}{3}\left(\frac{\sigma_2^2}{3} + \tau_2^2\right)\sigma_2,\end{aligned} \tag{49}$$

and on eliminating $\dot{\varepsilon}_0$,

$$b\,\dot{u}'' = \frac{k}{3}\left[\left(\frac{\sigma_2^2}{3} + \tau_2^2\right)\sigma_2 - \left(\frac{\sigma_1^2}{3} + \tau_1^2\right)\sigma_1\right]. \tag{50}$$

Furthermore,

$$\dot{\gamma}_1 = \dot{\gamma}_2 = \frac{r}{h}\,\dot{u}' = k\left(\frac{\sigma_1^2}{3} + \tau_1^2\right)\tau_1 = k\left(\frac{\sigma_2^2}{3} + \tau_2^2\right)\tau_2. \tag{51}$$

Suppose we take the approximate forms

$$\begin{gathered} u = U \sin \pi\, z/L,\\ \sigma_1 = \sigma_0(1 + \lambda \sin \pi\, z/L),\ \sigma_2 = \sigma_0(1 - \lambda \sin \pi\, z/L),\\ \tau_1 = s_1 \cos \pi\, z/L,\ \tau_2 = s_2 \cos \pi\, z/L,\end{gathered} \tag{52}$$

where U, λ, s_1 and s_2 are constants at any given instant. Eq. (50) then becomes

$$\begin{aligned}\dot{U} \sin\frac{\pi z}{L} = \frac{L^2}{\pi^2 b}\cdot\frac{k}{3}\Bigg[&\frac{2\,\sigma_0^3}{3}\lambda\left(3 \sin\frac{\pi z}{L} + \lambda^2 \sin^2\frac{\pi z}{L}\right)\\ &- \sigma_0(s_2^2 - s_1^2)\cos^2\frac{\pi z}{L} + \sigma_0\,\lambda(s_2^2 + s_1^2)\cos^2\frac{\pi z}{L}\sin\frac{\pi z}{L}\Bigg].\end{aligned} \tag{53}$$

If all the trigonometric functions on the right-hand side of Eq. (53) are represented by the first terms of their FOURIER expansions[1], we have

$$= \frac{L^2}{\pi^2 b}\cdot\frac{k}{3}\left[2\,\sigma_0^3\,\lambda\left(1 + \frac{\lambda^2}{4}\right) - \frac{4}{3\,h}\sigma_0(s_2^2 - s_1^2) + \frac{\lambda}{4}\sigma_0(s_2^2 + s_1^2)\right]. \tag{54}$$

[1] This method of analysis has been developed by HOFF, (Ref. [3]).

In a similar way, Eqs. (51) become

$$\begin{aligned}\dot{U} &= \frac{h\,L}{\pi\,r}\,k\,s_1\left[\frac{\sigma_0^2}{3}\left(1+\frac{8\,\lambda}{3\,\pi}+\frac{\lambda^2}{4}\right)+\frac{3}{4}\,s_1^2\right]\\ &= \frac{h\,L}{\pi\,r}\,k\,s_2\left[\frac{\sigma_0^2}{3}\left(1-\frac{8\,\lambda}{3\,\pi}+\frac{\lambda^2}{4}\right)+\frac{3}{4}\,s_2^2\right],\end{aligned} \tag{55}$$

and Eq. (42) becomes

$$U = \lambda\,b + \frac{L\,r}{\pi\,h}\cdot\left[\frac{s_1+s_2}{2\,\sigma_0}\right]. \tag{56}$$

Suppose

$$s_1 = s_0(1-\mu),\quad s_2 = s_0(1+\mu). \tag{57}$$

Then Eqs. (55) give

$$\left(\frac{s_0}{\sigma_0}\right)^2 = \left(\frac{32\,\lambda}{3\,\pi} - 4\,\mu - \lambda^2\,\mu\right)\Big/9\,\mu\,(3+\mu^2), \tag{58}$$

and Eq. (56) reduces to

$$U = \lambda\,b + \frac{L\,r}{\pi\,h}\left(\frac{s_0}{\sigma_0}\right). \tag{59}$$

Moreover, Eq. (54) becomes

$$\dot{U} = \frac{L^2}{\pi^2\,b}\cdot\frac{k\,\sigma_0^3}{3}\left[2\,\lambda\left(1+\frac{\lambda^2}{4}\right)+\left(\frac{s_0}{\sigma_0}\right)^2\left\{-\frac{16\,\mu}{3\,\pi}+\frac{\lambda}{2}\,(1+\mu^2)\right\}\right], \tag{60}$$

and the first of Eqs. (55) gives

$$\dot{U} = \frac{h\,L}{\pi\,r}\cdot\frac{k\,\sigma_0^3}{3}\,(1-\mu)\left[1+\frac{8\,\lambda}{3\,\pi}+\frac{\lambda^2}{4}+\frac{9}{4}\,(1-\mu)^2\left(\frac{s_0}{\sigma_0}\right)^2\right]\left(\frac{s_0}{\sigma_0}\right). \tag{61}$$

The four Eqs. (58), (59), (60), (61) are sufficient to solve for U, λ, μ and (s_0/σ_0). A solution may be found as follows: assume a value of (s_0/σ_0); then Eqs. (58) and Eqs. (60) and (61) combined give λ and μ; values of U and $\dot{U}$ are then found from Eq. (59) and either Eqs. (60) or (61).

5. Torsional-Flexural Buckling of a Narrow Beam

Finally, a simple model is set up to study the torsional-flexural creep buckling of a narrow beam. The model consists of four equal thin circular

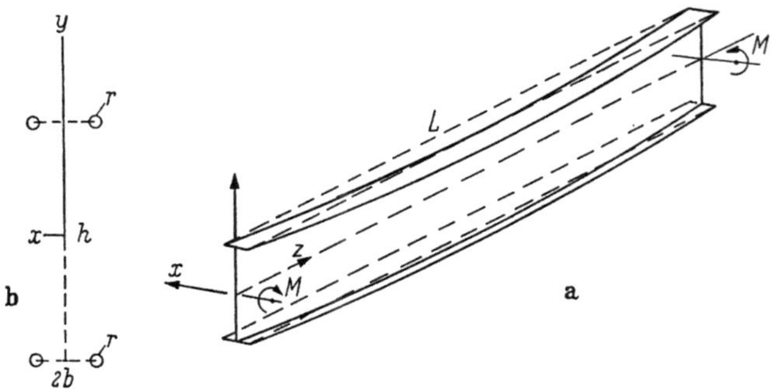

Fig. 10. (a) Lateral buckling of a narrow beam under constant couple; (b) cross-sectional geometry of the beam.

tubes, each of radius r, placed at the corners of a rectangle, Fig. 10. The tubes form a beam of breadth $2b$ and depth h; it is assumed that $h \gg b \gg r$. Constant couples M are applied about the x-axis at each end of the beam, so that the beam is under the action of a uniform bending moment. The beam is held so that no twisting can take place at the ends. The loading actions on an elemental length of the model in the buckled mode are shown in Fig. 11; the compressive stresses in the upper flange are σ_1, σ_2, and in the lower flange σ_3, σ_4. The shearing stresses in the tubes are τ_1, τ_2, τ_3, τ_4. The lateral displacement of the centroid of the beam is u, and all tubes are assumed to twist through the same angle θ at any cross-section. If A is the cross-sectional area of each tube, we have approximately

$$M = (\sigma_1 + \sigma_2)\, A\, h = (\sigma_3 + \sigma_4)\, A\, h = 2\, \sigma_0\, A\, h, \text{ (say)}, \qquad (62\text{a})$$

$$M\, \theta = (\sigma_1 - \sigma_2 + \sigma_4 - \sigma_3)\, A\, b, \qquad (62\text{b})$$

$$M\, u' = (\tau_1 + \tau_2 + \tau_3 + \tau_4)\, A\, r. \qquad (62\text{c})$$

The longitudinal strains in the upper and lower flanges are related by the equations

$$2\, b\, u'' = \varepsilon_2 - \varepsilon_1 = \varepsilon_3 - \varepsilon_4, \qquad (63)$$

and the shearing strain in each tube is

$$\gamma_1 = \gamma_2 = \gamma_3 = \gamma_4 = r\, \theta'. \qquad (64)$$

For elastic buckling of the model,

$$\varepsilon_1 = \sigma_1/E, \text{ etc.}, \gamma_1 = \tau_1/G, \text{ etc.}, \qquad (65)$$

Then

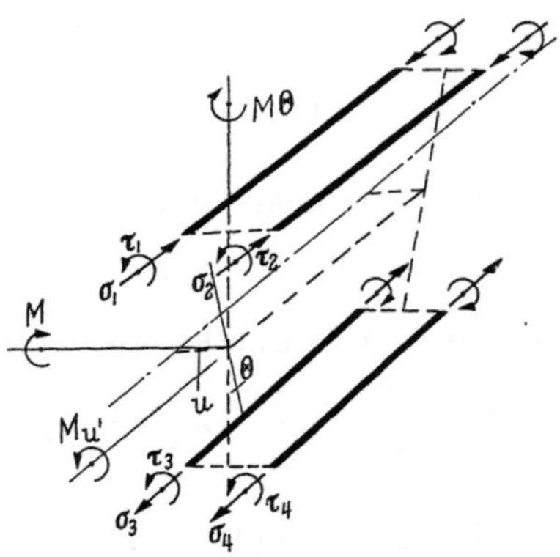

Fig. 11. Loading actions on an elemental length of the beam

$$\sigma_1 - \sigma_2 + \sigma_4 - \sigma_3 = -4\, b\, E\, u'', \quad \tau_1 + \tau_2 + \tau_3 + \tau_4 = 4\, G\, r\, \theta'. \qquad (66)$$

The last two of Eqs. (62) then give

$$\theta'' + \frac{M^2\, \theta}{(4\, E\, A\, b^2)\, (4\, G\, A\, r^2)} = 0. \qquad (67)$$

With the buckled form $\theta = B \sin \pi\, z/L$, the critical moment is

$$M_E = \frac{\pi}{L} \left[(4\, E\, A\, b^2)\, (4\, G\, A\, r^2)\right]^{1/2}. \qquad (68)$$

The equivalent flexural and torsional stiffnesses of the model are then $(4\, E\, A\, b^2)$ and $(4\, G\, A\, r^2)$, respectively.

In studying buckling due to creep, we neglect elastic strains. The creep strain rates are then given by

$$b\,\dot{u}'' = \frac{k}{3}\left[\left(\frac{\sigma_2^2}{3} + \tau_2^2\right)\sigma_2 - \left(\frac{\sigma_1^2}{3} + \tau_1^2\right)\sigma_1\right] = \frac{k}{3}\left[\left(\frac{\sigma_3^2}{3} + \tau_3^2\right)\sigma_3 - \left(\frac{\sigma_4^2}{3} + \tau_4^2\right)\sigma_4\right]$$

$$\dot{\gamma} = r\,\dot{\theta}' = k\left(\frac{\sigma_1^2}{3} + \tau_1^2\right)\tau_1 = \cdots. \tag{69}$$

Suppose we take the following forms:

$$\begin{aligned}
&\sigma_1 = \sigma_0(1 + \lambda \sin \pi\, z_4 L),\ \sigma_2 = \sigma_0(1 - \lambda \sin \pi\, z/L),\\
&\sigma_3 = \sigma_0(1 - \bar{\lambda} \sin \pi\, z_4 L),\ \sigma_4 = \sigma_0(1 + \bar{\lambda} \sin \pi\, z/L),\\
&\tau_1 = s_1 \cos\frac{\pi z}{L},\quad \tau_2 = s_2 \cos\frac{\pi z}{L},\ \ldots,\\
&u = U \sin\frac{\pi z}{L},\quad \theta = B \sin\frac{\pi z}{L},
\end{aligned} \tag{70}$$

in which λ, $\bar{\lambda}$, s_1, s_2, ..., U, B are constant at any given instant. Then Eqs. (62b) and (62c) give

$$\begin{aligned}
B &= \frac{b}{h}(\lambda + \bar{\lambda}),\\
U &= \frac{L\,r}{2\,\sigma_0\,\pi\,h}(s_1 + s_2 + s_3 + s_4).
\end{aligned} \tag{71}$$

For an approximate solution, suppose

$$\lambda = \bar{\lambda},\ s_1 = s_4,\ s_2 = s_3. \tag{72}$$

Then after a Fourier series treatment of Eqs. (69), we have

$$\begin{aligned}
\dot{U} &= \frac{L^2}{\pi^2 b}\cdot\frac{k\,\sigma_0^3}{3}\left[2\lambda\left(1 + \frac{\lambda^2}{4}\right) - \frac{4}{3\pi}\left(\frac{s_2^2}{\sigma_0^2} - \frac{s_1^2}{\sigma_0^2}\right) + \frac{\lambda}{4}\left(\frac{s_2^2}{\sigma_0^2} + \frac{s_1^2}{\sigma_0^2}\right)\right],\\
\dot{B} &= \frac{L}{\pi r}\cdot\frac{k\,\sigma_0^3}{3}\left[\left(1 + \frac{8\lambda}{3\pi} + \frac{\lambda^2}{4}\right) + \frac{9}{4}\frac{s_1^2}{\sigma_0^2}\right]\left(\frac{s_1}{\sigma_0}\right)\\
&= \frac{L}{\pi r}\cdot\frac{k\,\sigma_0^3}{3}\left[\left(1 - \frac{8\lambda}{3\pi} + \frac{\lambda^2}{4}\right) + \frac{9}{4}\frac{s_2^2}{\sigma_0^2}\right]\left(\frac{s_2}{\sigma_0}\right),
\end{aligned} \tag{73}$$

and Eqs. (71) become

$$B = \frac{2b}{h}\lambda,\ U = \frac{2\,L\,r}{\sigma_0\,\pi\,h}(s_1 + s_2). \tag{74}$$

The five equations given by Eqs. (73) and (74) are sufficient to solve for U, B, λ, s_1 and s_2. We put

$$s_1 = s_0(1 - \mu),\ s_2 = s_0(1 + \mu). \tag{75}$$

Then Eqs. (74) become

$$B = 2\,b\,\lambda/h,\ U = \frac{L\,r}{\pi\,h}\left(\frac{s_0}{\sigma_0}\right). \tag{76}$$

Eqs. (73) give

$$\dot{U} = \frac{L^2}{\pi^2 b} \cdot \frac{k \sigma_0^3}{3} \left[2 \lambda \left(1 + \frac{\lambda^2}{4}\right) + \left(\frac{s_0}{\sigma_0}\right)^2 \left\{ -\frac{16 \mu}{3 \pi} + \frac{\lambda}{2} (1 + \mu^2) \right\} \right],$$

$$\dot{B} = \frac{L}{\pi r} \cdot \frac{k \sigma_0^3}{3} \left[1 + \frac{8 \lambda}{3 \pi} + \frac{\lambda^2}{4} + \frac{9}{4} \left(\frac{s_0}{\sigma_0}\right)^2 (1 - \mu)^2 \right] \left(\frac{s}{\sigma_0}\right) (1 - \mu), \quad (77)$$

$$\left(\frac{s_0}{\sigma_0}\right)^2 = \left(\frac{32 \lambda}{3 \pi} - 4 \mu - \mu \lambda^2\right) / 9 \mu (3 + \mu^2).$$

Eqs. (76) and (77) imply that the velocities $\dot{U}$, $\dot{B}$ can be expressed in terms of U and B. Then

$$\dot{U} = f(U, B), \qquad \dot{B} = g(U, B). \quad (78)$$

Immediately after application of the couple M, suppose $U = U_0$, $B = B_0$. Then initially,

$$\dot{U}_0 = f(U_0, B_0), \quad \dot{B}_0 = g(U_0, B_0). \quad (79)$$

After a short time interval ΔT, the values of U and B are, approximately,

$$U = U_0 + \dot{U}_0 \Delta T, \quad B = B_0 + \dot{B}_0 \Delta T. \quad (80)$$

A process of numerical integration could thus be used to give further values of U and B.

6. Conclusions

In the case of uniform torsional buckling a simple structural model can be set up to demonstrate secondary creep buckling. If this model is very simple a critical time can be found which is analogous to that already found for flexural buckling. For more elaborate models it may be difficult to find explicit relations for critical time.

For torsional-flexural buckling about a fixed axis, in which lateral deflections and angles of twist are related, methods of analysis suggested by Hoff can be used profitably. Again, it appears difficult to find explicit expressions for critical time.

For torsional-flexural buckling of a narrow beam the problem requires solution of simultaneous differential equations in lateral displacement and angle of twist. Approximate distortions could be found by numerical integration.

References

[1] Lianis, G.: Torsional creep buckling of open tubes having arbitrary cross-section. Report No. S-59-1, School of Aeronautical Engineering, Purdue University, December, 1959.

[2] Fraeijs de Veubeke, B.: Creep buckling. High temperature effects in aircraft structures. Edited by N. J. Hoff and published for AGARD by Pergamon Press, 1958.

[3] Hoff, N. J.: Creep buckling. Aeronautical Quarterly, 7 (1956).

Discussion

F. K. G. Odqvist, Royal Institute of Technology, Stockholm: It is suggested that Eqs. (78) should be integrated numerically, dividing the two equations and thus obtaining an equation of first order for U as a function of B, to be integrated with the initial condition $U = U_0$, $B = B_0$. Subsequently either of the two Eqs. (78) could be used to determine U or B as functions of the time.

Author's closing remarks: The method suggested by Professor Odqvist for the solution of Eqs. (78) conveniently eliminates the time variable. Eqs. (78) imply that

$$\frac{dU}{dB} = F(U, B).$$

Beginning with the initial conditions $U = U_0$, $B = B_0$, the equation can be integrated numerically; this method of solution has the advantage that it shows directly the relationship between U and B.

Damping of the Vibrations of a Coiled Spring Due to Creep

By

Nicholas J. Hoff

Stanford University, Stanford, California

Summary

The vibrations of a concentrated mass suspended from a coiled spring are studied when the material of the spring is capable of deforming both elastically and in consequence of creep. When the creep law is linear, the attenuation of the vibrations is independent of the amplitude. When the creep law is non-linear in the sense that the creep rate is a power function of the force transmitted by the spring, with the exponent greater than unity, the attenuation increases with increasing amplitude.

Notation

g	acceleration of gravity
$i = (-1)^{1/2}$	
k_1	spring constant
k_2	dashpot constant
m	mass of particle
n	exponent in creep law
p	exponent
r	ratio defined in Eq. (81)
t	time
v	velocity of motion of particle
x	displacement of particle
A, B, C	constants of integration
C_0, C_1, C_2	coefficients of Fourier series
C_1^*	second approximation to value of C_1
D	kinetic energy dissipated by creep
D_1, D_2	non-dimensional coefficients of Fourier series defined in Eq. (115)
D_1^*	second approximation to value of D_1
F	force acting on spring
G	equivalent initial amplitude of displacement defined in Eq. (100)
I	integrand defined in Eq. (87)
J	definite integral defined in Eq. (86)
R	ratio of two consecutive maxima of velocity of oscillatory part of motion

[1] The work here presented was supported in whole by the United States Air Force under Contract No. AF 49(638)—223 monitored by the Air Force Office of Scientific Research of the Air Research and Development Command.

R^*	absolute value of ratio of amplitudes of velocity of oscillatory part of motion at $t = \pi/\omega$ and at $t = 0$
T	natural period of oscillation
V	increment in velocity at $t = 0$
W_{cr}	energy absorbed by creep deformations
α	variable
β	damping constant defined in Eq. (16)
δ	correction to value of D_1 defined in Eq. (128)
ε	increment in circular frequency
ξ	displacement of oscillatory part of motion defined in Eq. (7)
ψ	ratio of circular frequencies of damped and undamped systems
ω	circular frequency of damped system
ω_0	circular frequency of undamped system
Φ	force in spring in absence of gravity defined in Eq. (7)
Ω	parameter defined in Eq. (69)

Subscripts

cr	creep
$cr\ st$	steady creep
el	elastic
p	particular
st	static
T	at end of period when $t = T$

1. Introduction

Since creep in metals is essentially an irrecoverable process, it absorbs energy and thus it must attenuate mechanical vibrations. The study of the damping due to creep is, however, complicated by the fact that the stress dependence of the creep velocity is highly non-linear.

In the present paper the problem is analyzed by means of the example of a coiled spring from which a concentrated mass is suspended. The material of the spring is assumed to deform elastically in accordance with HOOKE's linear law and in creep according to a power law. First the case of linear creep, or viscosity, is studied to develop a feeling for the order of magnitude of the quantities involved. Next the attenuation is calculated from energy considerations both for linear and nonlinear creep. Finally the differential equations of the motion are derived for the general non-linear case and are solved through the use of FOURIER series.

2. The Linear Problem

(a) Statement of the Problem

The oscillating system to be analyzed is shown in Fig. 1. Since the spring from which the mass m is suspended deforms both elastically and in consequence of creep, it can be represented symbolically by the mechanical model of a linear spring and a linear dashpot in series as illustrated in Fig. 2. In the calculations one must start from NEWTON's equation of motion:

$$m \ddot{x} = m g - F, \tag{1}$$

where m is the mass, g the acceleration of gravity, F the force transmitted by the spring, x the displacement of the mass and dots indicate differentiation with respect to time t. Since the elastic elongation of the spring is governed by HOOKE's law

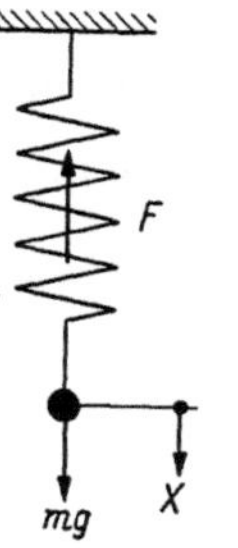

Fig. 1. Oscillating system

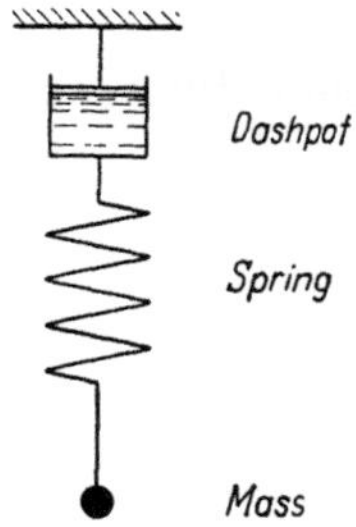

Fig. 2. Model of spring with viscous damping

$$F = k_1 x_{el}, \tag{2}$$

where x_{el} is the elastic part of the displacement and k_1 is the spring constant, and the creep velocity will be assumed to follow the linear law of viscosity

$$\dot{x}_{cr} = F/k_2, \tag{3}$$

where x_{cr} is the part of the displacement due to creep and k_2 is a constant characterizing the dashpot, the deformation law of the system can be written as

$$\dot{x} = \dot{x}_{el} + \dot{x}_{cr} = (\dot{F}/k_1) + (F/k_2). \tag{4}$$

The third-order system consisting of Eqs. (1) and (4) will be integrated in the presence of the following three initial conditions:

$$x_{el} = m g/k_1, \; x_{cr} = 0, \; \dot{x}_{el} = V, \; t = 0. \tag{5}$$

The physical meaning of these conditions is the following: The mass is descending at a uniform rate corresponding to a creep velocity $\dot{x}_{cr} = m g/k_2$ when at time $t = 0$ suddenly a vertical impact is imparted to the mass with the result that its vertical velocity is instantaneously increased by the amount V. The creep displacements are counted from $t = 0$; at that time the elastic elongation of the spring is the static extension $m g/k_1$.

(b) Solution

Because of the linearity of the problem it is convenient to proceed to the solution in two steps. First a simple particular solution is given in the form

$$x_p = (m g/k_1) + (m g/k_2) t. \tag{6}$$

This represents the steady-state motion in the absence of vibrations. In the solution of the remaining homogeneous differential equations one may introduce two new dependent variables according to the definitions

$$\xi = x - x_p, \; \Phi = F - m g. \tag{7}$$

Of these, ξ is the displacement of the vibratory motion superimposed on the steady creep motion, and Φ is the force that would act on the spring in the absence of the gravitational field of the earth ($g = 0$). Since

$$\xi_{el} = \Phi/k_1, \; \dot{\xi}_{cr} = \Phi/k_2 \tag{8}$$

substitutions and elimination lead to the following initial value problem:

$$\ddot{\Phi} + (k_1/k_2)\dot{\Phi} + \omega_0^2 \Phi = 0, \tag{9}$$

$$\dot{\xi} = (\dot{\Phi}/k_1) + (\Phi/k_2), \tag{10}$$

$$\xi_{el} = 0, \; \xi_{cr} = 0, \; \xi = V, \; t = 0, \tag{11}$$

where

$$\omega_0^2 = k_1/m \tag{12}$$

is the circular frequency of the undamped system.

The assumption

$$\Phi = e^{pt} \tag{13}$$

reduces Eq. (9) to

$$p^2 + (k_1/k_2)\, p + \omega_0^2 = 0 \tag{14}$$

from which

$$p = -\beta \pm i\, \omega, \tag{15}$$

where

$$\beta = (1/2)\,(k_1/k_2) \tag{16}$$

and

$$\omega = \omega_0\, \psi, \qquad \text{with} \qquad \psi = [1 - (\beta/\omega_0)^2]^{1/2}. \tag{17}$$

Because of well-known properties of linear differential equations with constant coefficients, the solution can be written as

$$\Phi = e^{-\beta t}\,(A \cos \omega\, t + B \sin \omega\, t). \tag{18}$$

Because of the first of the initial conditions in Eqs. (11) and because of the elasticity condition represented by the first of Eqs. (8) one has

$$\Phi = 0, \; t = 0. \tag{19}$$

Thus Eq. (18) simplifies to

$$\Phi = B\, e^{-\beta t} \sin \omega\, t. \tag{20}$$

If it is observed that in consequence of Eq. (16)

$$(1/k_2) - (\beta/k_1) = (\beta/k_1) \tag{21}$$

substitution of Φ from Eq. (20) into Eq. (10) yields

$$\dot{\xi} = (B/k_1)\, e^{-\beta t}\,(\omega \cos \omega\, t + \beta \sin \omega\, t). \tag{22}$$

The third of the initial conditions in Eqs. (11) requires that

$$B = (k_1/\omega)\, V. \tag{23}$$

Integration of Eq. (22) gives

$$\xi = -(V/\omega)\,\omega_0^2\, e^{-\beta t}\,[2\,\beta\,\omega\cos\omega\,t + (\beta^2-\omega^2)\sin\omega\,t] + C. \tag{24}$$

Since ξ must vanish when $t = 0$ it follows that

$$C = (2\,\beta/\omega_0^2)\,V = (m/k_2)\,V. \tag{25}$$

Upon substitution of C, Eq. (24) becomes

$$\xi = (V/\omega)\,(\omega^2+\beta^2)^{-1}\,\{2\,\beta\,\omega\,[1-e^{-\beta t}\cos\omega\,t]-(\beta^2-\omega^2)\,e^{-\beta t}\sin\omega\,t\}. \tag{26}$$

Because of Eqs. (6) and (7) the total displacement and the total velocity can be written as

$$x = (V/\omega)\,(\omega^2+\beta^2)^{-1}\,[2\,\omega\,\beta\,(1-e^{-\beta t}\cos\omega\,t) + (\omega^2-\beta^2)\,e^{-\beta t}\sin\omega\,t] + (m\,g/k_2)\,t + (m\,g/k_1), \tag{27}$$

$$\dot{x} = (V/\omega)\,e^{-\beta t}\,(\omega\cos\omega\,t + \beta\sin\omega\,t) + (m\,g/k_2). \tag{28}$$

The parts of the displacement due to elasticity and due to creep can be easily calculated. Differentiation of Eq. (28) gives

$$\ddot{x} = -(V/\omega)\,(\omega^2+\beta^2)\,e^{-\beta t}\sin\omega\,t. \tag{29}$$

From Eq. (1)

$$F = V\,(m/\omega)\,(\omega^2+\beta^2)\,e^{-\beta t}\sin\omega\,t + m\,g. \tag{30}$$

From Eqs. (3) and (30)

$$\begin{aligned} x_{cr} &= (V/k_2)\,(m/\omega)\,(\omega^2+\beta^2)\int_0^t e^{-\beta t}\sin\omega\,t\,dt + (m\,g/k_2)\,t \\ &= (V/k_2)\,(m/\omega)\,[\omega\,(1-e^{-\beta t}\cos\omega\,t) - \beta\,e^{-\beta t}\sin\omega\,t] + (m\,g/k_2)\,t. \end{aligned} \tag{31}$$

Similarly, from Eq. (2)

$$x_{el} = (V/k_1)\,(m/\omega)\,(\omega^2+\beta^2)\,e^{-\beta t}\sin\omega\,t + (m\,g/k_1). \tag{32}$$

The following identities are consequences of the earlier definitions:

$$\begin{aligned} &\omega^2+\beta^2 = \omega_0^2, && (m/k_1)\,(\omega^2+\beta^2) = 1, \\ &(m\,\omega/k_2) = 2\,\beta\,\omega/(\omega^2+\beta^2), && (m\,\beta/k_2) = 2\,\beta^2/(\omega^2+\beta^2). \end{aligned} \tag{33}$$

Thus Eqs. (31) and (32) can also be written in the alternative forms

$$x_{cr} = (V/\omega)\,(\omega^2+\beta^2)^{-1}\,[2\,\omega\,\beta\,(1-e^{-\beta t}\cos\omega\,t) - 2\,\beta^2\,e^{-\beta t}\sin\omega\,t] + (m\,g/k_2)\,t, \tag{34}$$

$$x_{el} = (V/\omega)\,e^{-\beta t}\sin\omega\,t + (m\,g/k_1). \tag{35}$$

The sum of x_{cr} and x_{el} is obviously x as given in Eq. (27).

(c) Magnitudes of Physical Quantities

It follows from Eq. (17) that the motion is oscillatory provided that

$$(\beta/\omega_0)^2 < 1. \tag{36}$$

The damping is critical when

$$\beta = \omega_0 \tag{37}$$

that is when

$$k_2 = (1/2)\,(k_1\,m)^{1/2} = (1/2)\,(k_1/\omega_0) = (1/2)\,m\,\omega_0. \tag{38}$$

One measure of the attenuation is R, the ratio of two consecutive maxima of the velocity of the oscillatory part of the motion. From Eq. (28) this ratio is

$$R = e^{-\beta T} = e^{-2\pi\beta/\omega}, \tag{39}$$

where T is the natural period of the oscillations

$$T = 2\,\pi/\omega = 2\,\pi/\omega_0\,\psi. \tag{40}$$

In most practical applications R is not smaller than 0.5. From tables of the exponential function the exponent corresponding to $R = 0.5$ is -0.69. It follows then from Eq. (17) that

$$\begin{aligned}\psi = [1-(\beta/\omega_0)^2]^{1/2} = [1-(0.69/2\,\pi)^2]^{1/2} &\simeq 1-(1/2)\,(0.69/2\,\pi)^2 \\ &= 1-0.006\end{aligned} \tag{41}$$

if under the root sign β/ω is replaced by β/ω_0. But this substitution is justified because the difference between ω and ω_0 is less than 1 percent according to Eq. (41). It can be concluded therefore that in most practical applications the damping has no noticeable effect on the frequency of the oscillations, and thus

$$\omega \simeq \omega_0. \tag{42}$$

A few physical quantities will next be defined that are suitable for the characterization of the motion. One of them is the static displacement

$$x_{st} = m\,g/k_1 = g/\omega_0^2. \tag{43}$$

The natural period defined in Eq. (40) can also be given in the following alternative forms:

$$T = 2\,\pi/\omega \simeq 2\,\pi/\omega_0 = 2\,\pi/(k_1/m)^{1/2} = 2\,\pi\,(x_{st}/g)^{1/2} \simeq 0.32\,(x_{st})^{1/2} \tag{44}$$

if T is measured in seconds and x_{st} in inches and if the value of g is taken as

$$g = 386 \text{ in. sec}^{-2}. \tag{45}$$

Another useful symbol is

$$\dot{x}_{cr\,st} = v_{cr\,st} = m\,g/k_2 \tag{46}$$

which represents the uniform (steady) creep velocity caused by the weight $m\,g$.

For a numerical example the following system will be studied:

$$m\,g = 50\text{ lb},\; k_1 = 10\text{ lb/in.},\; k_2 = 50\text{ lb sec/in.} \tag{47}$$

It follows then that

$$x_{st} = 5\text{ in.},\; \dot{x}_{cr\,st} = 1\text{ in./sec},\; T = 0.716\text{ sec},$$
$$\omega = \omega_0 = 8.78\text{ sec}^{-1}. \tag{48}$$

The ratio of two consecutive maxima of the velocity is

$$R = 0.931. \tag{49}$$

(d) Energy Considerations

When the creep law is non-linear, exact solutions of the problem of the oscillations of the mass cannot be found as a rule. In such cases recourse can be had to strain energy considerations. To clarify the principles involved in the strain energy approach, the attenuation will now be calculated from the changes that take place in the energy of the linear system during the oscillations.

First the work W_{cr} done by the force F in consequence of the creep deformations, that is the strain energy absorbed by the creep process, will be calculated:

$$W_{cr} = \int_0^{x_{cr\,T}} F\,dx_{cr} = \int_0^T F\,\dot{x}_{cr}\,dt = (1/k_2)\int_0^T F^2\,dt. \tag{50}$$

From Eq. (30) the force F in the spring is

$$F = V(m/\omega)\,(\omega^2 + \beta^2)\,e^{-\beta t}\sin\omega\,t + m\,g. \tag{30}$$

In the evaluation of the right-hand member of Eq. (50) the following ndefinite integrals are needed:

$$\int e^{-\beta t}\sin\omega\,t\,dt = -(\omega^2+\beta^2)^{-1}\,e^{-\beta t}\,(\omega\cos\omega\,t + \beta\sin\omega\,t), \tag{51}$$

$$\int e^{-\beta t}\cos\omega\,t\,dt = -(\omega^2+\beta^2)^{-1}\,e^{-\beta t}\,(\beta\cos\omega\,t - \omega\sin\omega\,t), \tag{52}$$

$$\int e^{-2\beta t}\sin^2\omega\,t\,dt = -(1/4\,\beta)\,e^{-2\beta t} + (1/4)\,(\omega^2+\beta^2)^{-1}\,e^{-2\beta t} \times (\beta\cos 2\,\omega\,t - \omega\sin 2\,\omega\,t). \tag{53}$$

The corresponding definite integrals between the limits 0 and T are:

$$\int_0^T e^{-\beta t}\sin\omega\,t\,dt = [\omega/(\omega^2+\beta^2)]\,(1 - e^{-\beta T}), \tag{54}$$

$$\int_0^T e^{-\beta t}\cos\omega\,t\,dt = [\beta/(\omega^2+\beta^2)]\,(1 - e^{-\beta T}), \tag{55}$$

$$\int_0^T e^{-2\beta t}\sin^2\omega\,t\,dt = (1/4\,\beta)\,[\omega^2/(\omega^2+\beta^2)]\,(1 - e^{-2\beta T}). \tag{56}$$

Substitution in Eq. (50) gives the energy dissipated by creep in one cycle:

$$W_{cr} = (m^2/k_2)\,[V^2 (1/4\,\beta)\,(\omega^2 + \beta^2)\,(1 - e^{-2\beta T}) + 2\,V\,g\,(1 - e^{-\beta T}) + g^2\,T]. \tag{57}$$

Part of this energy is furnished by the decrease in the potential of the force $m\,g$:

$$m\,g\,x_{cr\,T} = m\,g \int_0^T \dot{x}_{cr}\,dt = 2\,V\,m\,g\,\beta(\omega^2 + \beta^2)^{-1}\,(1 - e^{-\beta T}) + (m\,g)^2\;(T/k_2). \tag{58}$$

Since

$$2\,\beta/(\omega^2 + \beta^2) = m/k_2 \tag{59}$$

subtraction of the right-hand member of Eq. (58) from that of Eq. (57) gives

$$D = W_{cr} - m\,g\,x_{cr\,T} = (1/2)\;V^2\,m\,(1 - e^{-2\beta T}) + V\,m\,(g\,m/k_2)\,(1 - e^{-\beta T}). \tag{60}$$

This amount of dissipated energy must be furnished by the decrease in the kinetic energy of the system. Indeed, at $t = 0$ the velocity is $V + (m\,g/k_2)$ while at $t = T$ it is $V\,e^{-\beta T} + (m\,g/k_2)$; the difference of the squares of the two velocities, multiplied by $m/2$, is the decrease in the kinetic energy of the mass and is equal to D, as a simple calculation will show.

When in a non-linear system the attenuation of the motion cannot be calculated rigorously but the kinetic energy D dissipated in one cycle, as defined in Eq. (60), can be estimated, the ratio R of two consecutive maxima of the velocity follows from the equation

$$(2/m)\,D = [V + (m\,g/k_2)]^2 - [V\,R + (m\,g/k_2)]^2. \tag{61}$$

The solution of this quadratic in R is

$$R = \{[1 + (v_{cr\,st}/V)]^2 - [D/(m\;V^2/2)]\}^{1/2} - (v_{cr\,st}/V). \tag{62}$$

If both ratios under the square root sign are small, this expression simplifies to

$$R \simeq 1 - (1/2)\,\frac{D}{m\;V^2/2}. \tag{63}$$

3. The Non-Linear Problem

(a) Derivation of the Differential Equations

The system is again represented by Figs. 1 and 2 and NEWTON's equation [Eq. (1)] and HOOKE's law [Eq. (2)] remain unchanged. However, the dashpot is now a non-linear one corresponding to the non-linear creep law

$$\dot{x}_{cr} = v_{cr} = (F/k_2)^n \tag{64}$$

which replaces Eq. (3). It is assumed that n is an odd integer. When this is not a good representation of the experimental data available for the material, F must be replaed by the absolute value of the force $|F|$ and it must be required in addition that the sign of $\dot{x}_{cr}$ should be the same as that of F.

The deformation law can be written as

$$\dot{x} = \dot{x}_{el} + \dot{x}_{cr} = v = (\dot{F}/k_1) + (F/k_2)^n. \tag{65}$$

In one approach to the derivation of the differential equations one may again introduce the force

$$\Phi = F - m\,g. \tag{66}$$

The deformation law becomes

$$v = \dot{x} = \dot{x}_{el} + \dot{x}_{cr} = (\dot{\Phi}/k_1) + [(\Phi + m\,g)/k_2]^n. \tag{67}$$

Substitution in NEWTON's law and rearrangement of the terms yield

$$\Phi + n\,\omega_0\,\Omega\,[1 + (\Phi/m\,g)]^{n-1}\,\dot{\Phi} + \omega_0^2\Phi = 0, \tag{68}$$

where

$$\omega_0^2 = k_1/m, \quad \Omega = \omega_0\,v_{cr\,st}/g. \tag{69}$$

The initial conditions can be written as

$$\Phi = 0,\; x = x_{st},\; \dot{x} = V + v_{cr\,st},\; t = 0. \tag{70}$$

Eqs. (67), (68) and (70) define the initial value problem.

In a second approach Eq. (1) is solved for F:

$$F = m\,(g - \dot{v}). \tag{71}$$

Hence

$$\dot{F} = -m\,\ddot{v}. \tag{72}$$

Substitution in the deformation law Eq. (65) leads to the differential equation

$$\ddot{v} - \omega_0^2\,v_{cr\,st}\,[1 - (\dot{v}/g)]^n + \omega_0^2\,v = 0. \tag{73}$$

Together with the equation

$$\dot{x} = v \tag{74}$$

and the initial conditions

$$v = V + v_{cr\,st},\; F = m\,g,\; x = x_{st},\; t = 0, \tag{75}$$

where F is defined in Eq. (71), Eq. (73) represents another formulation of the same initial value problem.

(b) Estimate of the Attenuation of the Oscillations

On the basis of the order of magnitude estimate made earlier, it will be assumed that the natural frequency of the oscillations will be little influenced by the non-linear damping. In addition, the character of the motion will be assumed to be the same, in a first approximation, as that of the undamped oscillations. Hence in a rough approximation, Eq. (68) will be replaced by

$$\ddot{\Phi} + \omega_0^2 \Phi = 0. \tag{76}$$

If the first of the initial conditions contained in Eq. (70) is taken into account, the solution of this equation can be written as

$$\Phi = A \sin \omega_0 t. \tag{77}$$

It follows then from Eq. (66) that

$$F = A \sin \omega_0 t + m g. \tag{78}$$

From this approximate solution of the problem the creep velocity can be calculated with the aid of Eq. (64). It is

$$v_{cr} = (1/k_2)^n (A \sin \omega_0 t + m g)^n. \tag{79}$$

The net displacement of the mass during the first period in consequence of creep is

$$x_{cr\,T} = v_{cr\,st} \int_0^T (1 + r \sin \omega_0 t)^n \, dt, \tag{80}$$

where

$$r = A/m g. \tag{81}$$

The decrease in the potential of the force $m g$ during the same period in consequence of the creep deformations is

$$m g x_{cr\,T} = m g v_{cr\,st} \int_0^T (1 + r \sin \omega_0 t)^n \, dt. \tag{82}$$

The total energy dissipated by creep is

$$W_{cr} = \int_0^T F \dot{x}_{cr} \, dt = m g v_{cr\,st} \int_0^T (1 + r \sin \omega_0 t)^{n+1} \, dt. \tag{83}$$

The portion D of the initial kinetic energy that is dissipated in the first cycle is therefore

$$D = m g v_{cr\,st} \int_0^T [(1 + r \sin \omega_0 t)^{n+1} - (1 + r \sin \omega_0 t)^n] \, dt \tag{84}$$

which can also be written in the form

$$D = \pi r^2 m g (v_{cr\,st}/\omega_0) J, \tag{85}$$

where

$$J = (1/\pi r^2) \int_0^{2\pi} I(\tau)\, d\tau = (1/\pi r^2) \int_0^{2\pi} [(1 + r \sin \tau)^{n+1} - (1 + r \sin \tau)^n]\, d\tau . \tag{86}$$

For $n = 1, 3, 5$ the expression under the integral sign becomes

$$\begin{aligned} I_1 &= r \sin \tau + r^2 \sin^2 \tau , \\ I_3 &= r \sin \tau + 3 r^2 \sin^2 \tau + 3 r^3 \sin^3 \tau + r^4 \sin^4 \tau , \\ I_5 &= r \sin \tau + 5 r^2 \sin^2 \tau + 10 r^3 \sin^3 \tau + 10 r^4 \sin^4 \tau \\ &\quad + 5 r^5 \sin^5 \tau + r^6 \sin^6 \tau . \end{aligned} \tag{87}$$

Since

$$\begin{gathered} \int_0^{2\pi} \sin \tau\, d\tau = \int_0^{2\pi} \sin^3 \tau\, d\tau = \int_0^{2\pi} \sin^5 \tau\, d\tau = 0 , \\ \int_0^{2\pi} \sin^2 \tau\, d\tau = \pi, \quad \int_0^{2\pi} \sin^4 \tau\, d\tau = (3/4)\, \pi, \quad \int_0^{2\pi} \sin^6 \tau\, d\tau = (5/8)\, \pi \end{gathered} \tag{88}$$

the definite integral of Eq. (86) becomes

$$\begin{aligned} J_1 &= 1 , \\ J_3 &= 3 + (3/4)\, (V \omega_0/g)^2 , \\ J_5 &= 5 + (15/2)\, (V \omega_0/g)^2 + (5/8)\, (V \omega_0/g)^4 . \end{aligned} \tag{89}$$

In the present approximation the amplitude of the oscillations is A/k_1 if the amplitude of the force Φ is A. Consequently the initial amplitude of the velocity of the oscillatory motion is

$$V = A\, \omega_0/k_1 . \tag{90}$$

It follows then from Eq. (81) that

$$r = V k_1/\omega_0\, m\, g = V \omega_0/g = (1/2\pi) \frac{V}{x_{st}/T} . \tag{91}$$

Thus Eq. (85) becomes

$$D = 2\pi \frac{\omega_0\, v_{cr\,st}}{g} \frac{m\, V^2}{2} J = \frac{v_{cr\,st}\, T}{x_{st}} \frac{m\, V^2}{2} J . \tag{92}$$

Since in the linear case, $n = 1$, one has

$$\beta = (1/2)\, (k_1/k_2), \quad v_{cr\,st} = m\, g/k_2, \quad J = 1 \tag{93}$$

the expression for D reduces to

$$D = 2\pi (\beta/\omega_0)\, m\, V^2 = \beta\, T\, m\, V^2, \quad n = 1 . \tag{94}$$

The ratio R of two consecutive maxima of the oscillatory velocity given in Eq. (62) can now be written in the form

$$R = \left[\left(1 + \frac{v_{cr\,st}}{V}\right)^2 - \frac{v_{cr\,st}\,T\,J}{x_{st}}\right]^{1/2} - \frac{v_{cr\,st}}{V}$$
$$= \left[\left(1 + \frac{v_{cr\,st}}{V}\right)^2 - 2\,\pi\,\Omega\,J\right]^{1/2} - \frac{v_{cr\,st}}{V}. \qquad (95)$$

When all the ratios of the physical quantities appearing in Eq. (95) are small compared to unity, the equation reduces to

$$R = 1 - \frac{v_{cr\,st}\,T}{2\,x_{st}}\,J = 1 - \pi\,\Omega\,J. \qquad (96)$$

In this form R is independent of the initial velocity when $n = 1$ while it is a function of V when $n \neq 1$ as can be seen from the values given for J in Eqs. (89). In the linear case Eq. (96) becomes

$$R = 1 - \pi\,\Omega = 1 - \beta\,T, \quad n = 1. \qquad (97)$$

It is to be noted that the terms $1 - \beta\,T$ are the first two terms of the TAYLOR series for the exponential function $e^{-\beta T}$, which was the value of R in the exact solution of the linear problem. Hence the energy solution agrees with the exact solution in the linear case when the damping is small. When $\beta\,T$ is not small compared to unity, the large damping significantly changes the motion and the energy solution is inaccurate.

If the numerical example of the linear problem is continued, the nonlinear creep law can be so chosen that again

$$\dot{x}_{crst} = v_{crst} = 1 \text{ in./sec.} \qquad (98)$$

It follows then from Eqs. (48) that

$$x_{st}/T = 6.985 \text{ in./sec}, \quad v_{crst}/(x_{st}/T) = 0.143,$$
$$\Omega = 0.02275. \qquad (99)$$

The values of R were calculated from the equations just derived for four different values of the initial velocity increment V, and the results are collected in Table 1. There are also included the values of the equivalent initial amplitude G of the displacement of the oscillatory motion as computed from the formula

$$G = V\,T/2\,\pi = V/\omega_0 = A/k_1. \qquad (100)$$

Table 1. *Attenuation of Oscillatory Motion. Energy Solution*

Initial Velocity	V in./sec	1	5	10	25
Approx. Equivalent Displ. Ampl.	G in.	0.114	0.57	1.14	2.85
$V/v_{cr\,st}$		1	5	10	25
R_1 $(n = 1)$		0.963	0.939	0.933	0.929
R_3 $(n = 3)$		0.892	0.805	0.781	0.747
R_5 $(n = 5)$		0.815	0.644	0.563	0.063

It can be seen from Table 1 that for a constant value of n the attenuation increases when the initial velocity increment V increases, and that this effect is more pronounced for high values of n than for low ones. However, a slight amplitude effect is noticeable in the table even for $n = 1$ where the attenuation should be independent of the amplitude in accordance with the rigorous solution. In the linear case, of course, the variation of R_1 with V is a consequence of the inaccuracy of the energy analysis. The energy absorption was calculated from the forces of the undamped motion; hence the value obtained must be higher than the true value, and the deviation between the two must increase as the difference between damped and undamped motion increases. At the same time the approximate calculation also yields an exaggerated value for the decrease in the potential of the force $m\,g$ in consequence of the creep deformations. Consequently the values of R cannot be reliable when they are small.

It may be noted that the simplified solution of Eq. (97) yields $R = 0.9285$. If this value, as well as those listed in the table for $n = 1$, is compared with the value 0.931 obtained from the rigorous solution in Eq. (49), it can be concluded that in the linear case the energy approach yields reasonable values for the attenuation.

(c) Integration by Fourier Series

Since moderate amounts of damping do not seem to change significantly the character of the undamped motion, it appears profitable to solve the initial value problem stated in Eqs. (73) to (75) by setting

$$v = C_0 + C_1 \cos \omega\, t + C_2 \cos 2\, \omega\, t + \cdots \quad 0 \leq t \leq \pi/\omega . \qquad (101)$$

When the damping is absent, the first two terms above represent the rigorous solution. The effect of the damping is to decrease the velocity with time, and that is accomplished by the third term which makes the velocity smaller in the range $\pi/2\,\omega < t \leq \pi/\omega$ than in the range $0 \leq t \leq \pi/2\,\omega$. It can be expected, therefore, on these physical grounds that C_2/C_1 will be small for moderate amounts of damping, and that the other terms of the Fourier series not explicitly given in Eq. (101) will be so small that they cannot be of any interest in an engineering solution of the problem.

Because of the experience gained in the earlier parts of this paper, the circular frequency of the oscillatory part of the motion will be taken as

$$\omega = \omega_0 + \varepsilon, \ \varepsilon/\omega \ll 1 . \qquad (102)$$

It follows then that

$$\omega_0 = \omega - \varepsilon, \ \omega_0^2 \simeq \omega^2 - 2\,\omega\,\varepsilon . \qquad (103)$$

The derivatives of the velocity are

$$\dot{v} = -\,\omega\ [C_1 \sin \omega\, t + 2\, C_2 \sin 2\, \omega\, t + \cdots], \tag{104}$$

$$\ddot{v} = -\,\omega^2 [C_1 \cos \omega\, t + 4\, C_2 \cos 2\, \omega\, t + \cdots]. \tag{105}$$

When $n = 3$, part of the damping term is

$$\begin{aligned}[1 - (\dot{v}/g)]^3 &= [1 + (\omega/g)\,(C_1 \sin \omega\, t + 2\, C_2 \sin 2\, \omega\, t + \cdots)]^3 \\ &= 1 + 3(\omega/g)\,(C_1 \sin \omega\, t + 2\, C_2 \sin 2\, \omega\, t + \cdots) \\ &+ 3\,(\omega/g)^2\,(C_1^2 \sin^2 \omega\, t + 4\, C_1\, C_2 \sin \omega\, t \sin 2\, \omega\, t + 4\, C_2^2 \sin^2 2\, \omega\, t + \cdots) \\ &+ (\omega/g)^3\,(C_1^3 \sin^3 \omega\, t + 6\, C_1^2\, C_2 \sin^2 \omega\, t \sin 2\, \omega\, t \\ &+ 12\, C_1\, C_2^2 \sin \omega\, t \sin^2 2\, \omega\, t + 8\, C_2^3 \sin^3 2\, \omega\, t + \cdots).\end{aligned} \tag{106}$$

Since C_2/C_1 is expected to be much smaller than unity, terms multiplied by C_2^2 or C_2^3 will be disregarded compared to terms multiplied by C_1^2 or C_1^3. Use will be made of the following trigonometric identities:

$$\begin{aligned}&\sin^2 \omega\, t = (1/2) - (1/2) \cos 2\, \omega\, t, \\ &\sin \omega\, t \sin 2\, \omega\, t = (1/2) \cos \omega\, t - (1/2) \cos 3\, \omega\, t, \\ &\sin^3 \omega\, t = -\,(1/4) \sin 3\, \omega\, t + (3/4) \sin \omega\, t, \\ &\sin^2 \omega\, t \sin 2\, \omega\, t = (1/2) \sin 2\, \omega\, t - (1/4) \sin 4\, \omega\, t.\end{aligned} \tag{107}$$

The sine terms will be expaned in cosine series:

$$\begin{aligned}\sin \alpha &= (2/\pi) - (4/3\,\pi) \cos 2\,\alpha + \cdots, \\ \sin 2\,\alpha &= (8/3\,\pi) \cos \alpha + \cdots, \\ \sin 3\,\alpha &= (2/3\,\pi) + (12/5\,\pi) \cos 2\,\alpha + \cdots, \\ \sin 4\,\alpha &= (16/15\,\pi) \cos \alpha + \cdots,\end{aligned} \tag{108}$$

provided

$$0 \leq \alpha \leq \pi. \tag{108a}$$

Substitutions yield

$$\begin{aligned}\sin^3 \omega\, t &= (4/3\,\pi) - (8/5\,\pi) \cos 2\, \omega\, t + \cdots, \\ \sin^2 \omega\, t \sin 2\, \omega\, t &= (16/15\,\pi) \cos \omega\, t + \cdots,\end{aligned} \tag{109}$$

provided

$$0 \leq t \leq \pi/\omega. \tag{109a}$$

With the aid of these expressions Eq. (106) becomes:

$$\begin{aligned}[1 - (\dot{v}/g)]^3 - 1 &+ 3(\omega/g)\,[(2/\pi) - (4/3\,\pi) \cos 2\, \omega\, t]\, C_1 \\ &+ (16/3\,\pi) \cos \omega\, t\, C_2 + 3(\omega/g)^2\,[(1/2)\,(1 - \cos 2\, \omega\, t)\, C_1^2 \\ &+ 2 \cos \omega\, t\, C_1\, C_2] + (\omega/g)^3\,[(1/15\,\pi)\,(20 - 24 \cos 2\, \omega\, t)\, C_1^3 \\ &+ (32/5\,\pi) \cos \omega\, t\, C_1^2\, C_2].\end{aligned} \tag{110}$$

From this equation all terms containing trigonometric functions of an argument greater than $2\,\omega\,t$ and all terms multiplied by C_2^2 or C_2^3 have been omitted. If the multipliers of like trigonometric functions are collected, the result is

$$[1-(\dot{v}/g)]^3 \simeq 1+(6/\pi)\,(\omega/g)\,C_1+(3/2)\,(\omega/g)^2\,C_1^2+(4/3\pi)\,(\omega/g)^3\,C_1^3$$
$$+\,[(16/\pi)\,(\omega/g)\,C_2+6\,(\omega/g)^2\,C_1\,C_2+(32/5\,\pi)\,(\omega/g)^3\,C_1^2\,C_2]\cos\omega\,t$$
$$-\,[(4/\pi)\,(\omega/g)\,C_1+(3/2)\,(\omega/g)^2\,C_1^2+(8/5\,\pi)^3\,C_1^3]\cos 2\,\omega\,t. \tag{111}$$

This expression, as well as those given in Eqs. (101) and (105), can now be substituted in the differential equation of the problem, Eq. (73). Since the equation must be satisfied identically, the multipliers of like trigonometric terms must be set equal to zero. The result is the following three simultaneous equations:

$$C_0-v_{crst}[1+(6/\pi)\,D_1+(3/2)\,D_1^2+(4/3\,\pi)\,D_1^3]=0, \tag{112}$$

$$\varepsilon\,C_1+\omega_0\,v_{crst}[(8/\pi)\,D_2+3\,D_1\,D_2+(16/5\,\pi)\,D_1^2\,D_2]=0, \tag{113}$$

$$3\,C_2-v_{crst}[(4/\pi)\,D_1+(3/2)\,D_1^2+(8/5\,\pi)\,D_1^3]=0, \tag{114}$$

where

$$D_1=(\omega/g)\,C_1,\quad D_2=(\omega/g)\,C_2. \tag{115}$$

Thus C_0 and C_2 can be given explicitly in terms of D_1:

$$C_0=v_{crst}[1+(6/\pi)\,D_1+(3/2)\,D_1^2+(4/3\,\pi)\,D_1^3], \tag{116}$$

$$C_2=v_{crst}[(4/3\,\pi)\,D_1+(1/2)\,D_1^2+(8/15\,\pi)\,D_1^3]. \tag{117}$$

The change ε in the circular frequency is a function of D_1 and D_2:

$$\varepsilon=-(\omega_0\,v_{crst}/C_1)\,[(8/\pi)\,D_2+3\,D_1\,D_2+(16/5\,\pi)\,D_1^2\,D_2]. \tag{118}$$

It is to be noted that the assumed solution, Eq. (101), automatically satisfies one of the initial conditions listed in Eqs. (75). It follows from the requirement $F=m\,g$ at $t=0$ and from Eq. (71) that

$$\dot{v}=0,\ t=0. \tag{119}$$

This is obviously true from Eq. (104). A second initial condition, $x=x_{st}$ when $t=0$ can easily be satisfied after integration of Eq. (74) with the expression for v substituted from Eq. (101). However, at the present time we are not interested in finding an expression for the displacement.

There remains the third initial condition to satisfy, namely

$$v=V+v_{crst},\ t=0. \tag{120}$$

Because of Eq. (101) this becomes

$$V=C_0+C_1+C_2-v_{crst}. \tag{121}$$

Substitutions from Eqs. (116) and (117) yield the cubic

$$V = v_{cr\,st}\{[(1/\Omega) + (22/3\,\pi)]\,D_1 + 2\,D_1^2 + (28/15\,\pi)\,D_1^3\} \qquad (122)$$

with Ω defined in Eq. (69).

When D_1 is small, in a first approximation the solution of Eq. (122) is

$$D_1 = \frac{V/v_{cr\,st}}{(1/\Omega) + (22/3\,\pi)}\,. \qquad (123)$$

The corresponding value of C_1 follows from the first of Eqs. (115):

$$C_1 = \frac{V}{1 + (22/3\,\pi)\,\Omega}\,. \qquad (124)$$

In a similar first approximation one has

$$C_0 = v_{cr\,st} + (6/\pi)\,\Omega\,C_1\,, \qquad (125)$$

$$C_2/C_1 = (4/3\,\pi)\,\Omega\,, \qquad (126)$$

$$\varepsilon/\omega_0 = -\,(32/3\,\pi)\,\Omega^2\,. \qquad (127)$$

Since in most experiments Ω is a small number, C_2 is indeed much smaller than C_1, and ε is very much smaller than ω_0. Hence the order of magnitude assumptions made at the beginning of this section were justified.

A second approximation for D_1, denoted D_1^*, can be had if one sets

$$D_1^* = D_1(1 + \delta) \qquad (128)$$

with

$$\delta \ll 1\,. \qquad (128\text{a})$$

Substitution and the omission of small terms yield

$$\frac{V}{v_{cr\,st}} = \left(\frac{1}{\Omega} + \frac{22}{3\pi}\right)D_1 + \left(\frac{1}{\Omega} + \frac{22}{3\pi}\right)D_1\,\delta + 2\,D_1^2 \qquad (129)$$

from which δ follows as

$$\delta = \frac{-\,2\,D_1}{(1/\Omega) + (22/3\,\pi)} = \frac{-\,2\,\Omega^2\,(C_1/v_{cr\,st})}{1 + (22/3\,\pi)\,\Omega} = \frac{-\,2\,\Omega^2\,(V/v_{cr\,st})}{[1 + (22/3\,\pi)\,\Omega]^2}\,. \qquad (130)$$

Hence the second approximation for C_1, denoted C_1^* becomes

$$C_1^* = \frac{V}{1 + (22/3\,\pi)\,\Omega}\left\{1 - \frac{2\,\Omega^2(V/v_{cr\,st})}{[1 + (22/3\,\pi)\,\Omega]^2}\right\}. \qquad (131)$$

Of course, there is no difficulty involved in the numerical solution of the cubic of Eq. (122). With D_1 known, the values of C_0, C_1, C_2 and ε follow from the equations presented.

At the end of the half period, $t = \pi/\omega$, the velocity is from Eq. (101):

$$v = C_0 - C_1 + C_2,\ \ t = \pi/\omega\,. \qquad (132)$$

If R^* designates the absolute value of the ratio of the velocities of the oscillatory part of the motion, and if Eq. (120) is taken into account, one can write

$$-R^* = \frac{v_{T/2} - v_{cr\,st}}{v_0 - v_{cr\,st}} = (1/V)\,(C_0 - C_1 + C_2 - v_{crst}). \tag{133}$$

Substitutions yield

$$V\,R^* = -v_{crst}\,\{[-(1/\Omega) + (22/3\,\pi)]\,D_1 + 2\,D_1^2 + (28/15\,\pi)\,D_1^3\}. \tag{134}$$

It follows then that

$$R^* = (C_1/V)\,[1 - (22/3\,\pi)\,\Omega - 2\,\Omega\,D_1 - (28/15\,\pi)\,\Omega\,D_1^2]. \tag{135}$$

The numerical example can now be continued. It follows from the values of the physical quantities given in Eqs. (47) and (48) that

$$\Omega = \omega_0\,v_{crst}/g = 0.02275. \tag{136}$$

Numerical solution of the cubic Eq. (122) and substitution in the formulas defining the other quantities of interest results in the values listed in Table 2.

Table 2. *Characteristics of Motion when $n = 3$. Fourier Series Solution*

V in./sec	1	5	10	25
D_1	0.02162	0.108	0.2147	0.5275
C_1 in./sec	0.95	4.74	9.425	23.2
C_1/V	0.95	0.948	0.9425	0.928
C_2/C_1	0.0099	0.01094	0.0123	0.0167
$-10^3\,\varepsilon/\omega_0$	0.59	0.72	0.907	1.68
R^*	0.90	0.894	0.883	0.853
R	0.81	0.797	0.78	0.727

In Table 2 the value of R, the ratio of the oscillatory parts of the velocity at $t = 2\,\pi/\omega$ and at $t = 0$, that is at the end and at the beginning of the full period, was computed as the square of the value of R^*. This is a good estimate provided the attenuation is not too rapid.

It can be concluded from Table 2 that the time plot of the velocity is very well represented in all the cases studied by the simple cosin function whose coefficient is C_1; the ratio C_2/C_1 does not exceed 0.0167 in the table. The importance of the cosin function of the double angle increases slightly with increasing V (or increasing amplitude of the oscillations). Similarly, the attenuation increases slowly with increasing V. The agreement between the values obtained for R from the FOURIER analysis and from the energy calculations is satisfactory.

(d) Quintic Creep Law

When $n = 5$, the integration of the differential equation with the aid of the FOURIER expansion can be carried out in a similar manner.

One obtains first

$$[1-(\dot{v}/g)]^5 = 1 + 5\,[D_1 \sin \omega\, t + 2\, D_2 \sin 2\, \omega\, t]$$
$$+ 10\,[D_1^2 \sin^2 \omega\, t + 4\, D_1\, D_2 \sin \omega\, t \sin 2\, \omega\, t] + 10\,[D_1^3 \sin^3 \omega\, t$$
$$+ 6\, D_1^2\, D_2 \sin^2 \omega\, t \sin 2\, \omega\, t] + 5\,[D_1^4 \sin^4 \omega\, t \qquad (137)$$
$$+ 8\, D_1^3\, D_2 \sin^3 \omega\, t \sin 2\, \omega\, t]$$
$$+ [D_1^5 \sin^5 \omega\, t + 10\, D_1^4\, D_2 \sin^4 \omega\, t \sin 2\, \omega\, t].$$

In addition to those listed in Eqs. (107) and (108), use will be made of the following trigonometric identities and series expansions:

$$\begin{aligned} \sin^4 \omega\, t &= (3/8) - (1/2) \cos 2\, \omega\, t + \cdots, \\ \sin^3 \omega\, t \sin 2\, \omega\, t &= (32/\pi^2)\,(19/225) \cos \omega\, t, \\ \sin^5 \omega\, t &= (16/15\, \pi) - (32/21\, \pi) \cos 2\, \omega\, t, \\ \sin^4 \omega\, t \sin 2\, \omega\, t &= (64/105) \cos \omega\, t. \end{aligned} \qquad (138)$$

Substitution in Eq. (137) and setting equal to zero the coefficients of like trigonometric terms lead to the following three simultaneous equations

$$- 4\, C_2\, \omega^2 + \omega_0^2\, v_{crst}\,[(20/3\, \pi)\, D_1 + 5\, D_1^2 + (16/\pi)\, D_1^3 + (5/2)\, D_2^4$$
$$+ (32/21\, \pi)\, D_1^5] + \omega_0^2\, C_2 = 0, \qquad (139)$$

$$- \omega^2\, C_1 - \omega_0^2\, v_{crst}\,[(80/3\, \pi)\, D_2 + 20\, D_1\, D_2 + (64/\pi)\, D_1^2\, D_2 \qquad (140)$$
$$+ 40 (32/\pi^2)\,(19/225)\, D_1^3\, D_2 + (128/21\, \pi)\, D_1^4\, D_2] + \omega_0^2\, C_1 = 0,$$

$$- \omega_0^2\, v_{crst}\,[1 + (10/\pi)\, D_1 + 5\, D_1^2 + (40/3\, \pi)\, D_1^3 + (15/8)\, D_1^4$$
$$+ (16/15\, \pi)\, D_1^5\,] + \omega_0^2\, C_0 = 0. \qquad (141)$$

From these equations C_2 and C_0 can be expressed in terms of D_1 as in the case when $n = 3$:

$$C_2 = (v_{crst}/3)\,[(20/3\, \pi)\, D_1 + 5\, D_1^2 + (16/\pi)\, D_1^3 + (5/2)\, D_1^4$$
$$+ (32/21\, \pi)\, D_1^5], \qquad 142)$$

$$C_0 = v_{crst}\,[1 + (10/\pi)\, D_1 + 5\, D_1^2 + (40/3\, \pi)\, D_1^3 + (15/8)\, D_1^4$$
$$+ (16/15\, \pi)\, D_1^5]. \qquad (143)$$

C_1 and D_1 are determined from the initial condition Eq. (121):

$$V/v_{crst} = [(110/9\, \pi) + (1/\Omega)]\, D_1 + (20/3)\, D_1^2 + (56/3\, \pi)\, D_1^3$$
$$+ (65/24)\, D_1^4 + [(16/15\, \pi) + (32/63\, \pi)]\, D_1^5. \qquad (144)$$

If the cube of D_1 and its higher powers are ignored and D_1^2 is considered small, the following results are obtained:

$$C_1 = \frac{V}{1+(110/9\,\pi)\,\Omega}\left[1-\frac{(20/3)\,\Omega^2}{[1+(110/9\,\pi)\,\Omega]^2}\,\frac{V}{v_{cr\,st}}\right], \tag{145}$$

$$C_2 = (20/9\,\pi)\,\Omega\, C_1\,[1+(3\,\pi/4)\,\Omega\,(C_1/v_{cr\,st})], \tag{146}$$

$$C_0 = v_{cr\,st} + (10/\pi)\,\Omega\, C_1 + 5\,\Omega^2\,(C_1^2/v_{cr\,st}). \tag{147}$$

The attenuation can be calculated from Eq. (133):

$$R^* = (C_1/V)\,[1-(110/9\,\pi)\,\Omega-(20/3)\,\Omega\, D_1-(56/3\,\pi)\,\Omega\, D_1^2 - (65/24)\,\Omega\, D_1^3-\left(\frac{16}{15\,\pi}+\frac{32}{63\,\pi}\right)D_1^4\Big]. \tag{148}$$

The ratio R of the amplitudes of the velocities of the oscillatory part of the motion at the beginning and at the end of the period can be written for not too large attenuation in the form

$$R = R^{*2}. \tag{149}$$

Numerical values for the system treated before have been computed; the results are collected in Table 3.

Table 3. *Characteristics of Motion when $n = 5$. Fourier Series Solution*

V in./sec	1	5	10	25
D_1	0.0204	0.1031	0.2026	0.4754
C_1 in./sec	0.916	4.522	8.884	20.86
C_1/V	0.916	0.904	0.888	0.834
C_2/C_1	0.0169	0.0203	0.0256	0.0449
$-10^3\,\varepsilon/\omega_0$	1.72	2.10	3.00	7.68
R^*	0.833	0.809	0.777	0.634
R	0.694	0.654	0.604	0.400

All the values computed for R by the strain energy method and by the FOURIER analysis are shown in Fig. 3.

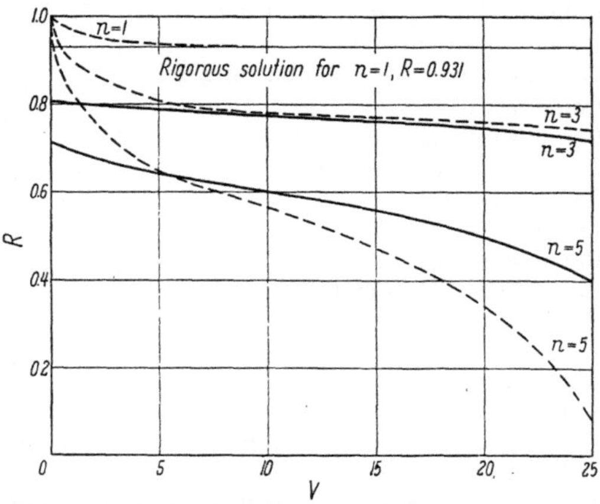

Fig. 3. Ratio R of successive velocity amplitudes of oscillatory part of motion of numerical example. (Full line represents solution by FOURIER series and dotted line that by energy analysis.)

Acknowledgement

The author acknowledges his indebtedness to Hanagud V. SATHYANARAYANA who checked all the calculations and carried out the FOURIER series analysis for the case $n = 5$.

Subject Index

MIX
Papier aus verantwortungsvollen Quellen
Paper from responsible sources
FSC® C105338

If you have any concerns about our products,
you can contact us on
ProductSafety@springernature.com

In case Publisher is established outside the EU,
the EU authorized representative is:
Springer Nature Customer Service Center GmbH
Europaplatz 3, 69115 Heidelberg, Germany

Printed by Libri Plureos GmbH
in Hamburg, Germany